THE EXPLANATION OF LOW ENERGY NUCLEAR REACTION

An Examination of the Relationship Between Observation and Explanation

Edmund Storms

ISBN 978-1-892925-10-7

Published and Distributed by Infinite Energy Press
Concord, NH, USA
www.infinite-energy.com

Cover design by Ruby Carat
www.coldfusionnow.org

Book available through http://lenrexplained.com/

ACKNOWLEDGEMENTS

A book of this kind is written by many people as a result of endless hours of trying to persuade nature to give up a few secrets. Often they were rewarded by confusing results or by premature termination of the work by management. In spite of these distractions, a picture has gradually taken shape. Describing this picture is not easy because many rules used in science do not seem to apply to LENR. Absence of such guidance has encouraged many people to propose explanations having very little relationship to what science understands and accepts. Nevertheless, these suggestions helped eliminate many false leads and focus on the true path, for which I'm grateful. I'm also grateful to the group of true skeptics, to which I consider myself a charter member, who forced some reality upon the unrestrained speculation and prevented the big laboratories from having all the fun. Nevertheless, their excessive skepticism has delayed an important gift to mankind.

I'm especially grateful to Ruby Carat who read every word and added a few of her own to make the description clearer. I expect her representation of the Hydroton on the book cover will become the modern icon of the new energy age. Thomas Grimshaw also provided many insightful suggestions and helped make the text more readable.

Christy Frazier did the final editing and corrected the mistakes hardest to find while turning the manuscript into a book.

I'm grateful to Mahadeva Srinivasan, Jean-Paul Biberian, and my wife Carol Storms for reading an early version and making important suggestions for improvement.

FOREWORD

It is rare to be able to identify an individual as having more knowledge than any other alive on a topic, or even a major sub-topic. Expansion of knowledge occurs at such a rate that most aspirants to that title, having passed on, leave no one person capable of wearing the mantle. In his lifetime John Bockris could have claimed that title in the field of Physical Electrochemistry, although some might have argued for Martin Fleischmann. Edward Teller might have deserved that title in some aspects of Nuclear Physics, and Julian Schwinger in others. To have such individuals exist and to have known them in person is a distinct privilege. That this came about as a cause and effect of my study of the field of cold fusion is remarkable and not coincidental. It took people of that stature to bring us to this point.

Ed Storms knows more about cold fusion or Low Energy Nuclear Reactions (LENR as his book title prefers) than any person alive. He has studied longer, harder, and deeper than any other. One should buy and read this book for that aspect alone but there are several other compelling reasons. One is Dr. Storms' willingness and ability to systematize the literature of the field. Probably the most effective attack made against us has been the effective prohibition of publication in major journals. The reason is as simple as it is cynical. No young person with academic inclinations wants to enter a field where publication is challenged or discouraged. As a result of this attack the literature of our common endeavor is in a state of some disarray although the heroic efforts of Jean-Paul Biberian and the *Journal of Condensed Matter Nuclear Science* (*JCMNS*) are starting to move this vector in a positive direction.

In this book, *The Explanation of Low Energy Nuclear Reaction*, Dr. Edmund Storms evaluates 904 references, not by way of review but in the manner of his subtitle, to examine the "*relationship between observation and explanation*." There is no better synthesis of knowledge and understanding presently available to us and I know of no other person capable of making an evaluation at this level. If for no other reason, read the book to help make sense of the literature. As his analysis of the state of theory indicates (Chapter 4) LENR is still very much an experimental science. One role of theory, perhaps the one scientific role, is to provide a thought structure from which to advance experiment. Dr. Storms is an active experimentalist and his (sometimes ruthless) evaluation of the huge diversity of models so far proposed to account for LENR is particularly well adapted to the needs of experimental science. For me this is a blessing and offers a considerable relief of my time and effort.

The opportunity to learn directly from *the* most knowledgeable person in arguably *the most important* emerging field, and to share his concise and well considered condensation of a difficult and scattered literature are not the only or primary reasons to comprehend *The Explanation of Low Energy Nuclear Reaction*. Laid out clearly and gently in Chapter 5, "Description of an Explanation," is the first physical science based description of a potential explanation for cold fusion, LENR, condensed matter nuclear reaction (CMNR). If correct then Dr. Storms will be credited with two advances, each of inestimable value: (i) the identification of a new state of matter leading to a new means of nuclear interaction in the solid state; (ii) the first working hypothesis capable of providing theory support for the practical development of a new, nuclear, primary energy source capable of satisfying mankind's insatiable and increasing demands while preserving our fossil fuel heritage for more rational purposes.

The central and crucial role played by a linear (or planar) extended coherent structure that is invoked by Dr. Storms has considerable historical support. For those of us actively engaged in researching the field from day one, it quickly became clear that: (a) the Fleischmann Pons Heat Effect (FPHE) exists at a level consistent with nuclear but not chemical processes, and (b) that this was not a property of the perfectly ordered palladium-deuterium system. The long initiation times for the effect, the active role of alloying — particularly surface alloying, and surface modification of various types — leads directly to the consideration of defect or void structures of various characters particularly at or near the metallic surface. Invoking Teller again, the last words I heard from him on the topic of cold fusion were that "*he could very well explain our results within the framework of nuclear physics as he understood it*" as the result of "*nuclear catalysis at an interface*." I did not understand that last phrase, or pursue it, but the distinct possibility exists that Ed Storms is describing in this book the mechanism and location of "nuclear catalysis at an interface."

The potential role of linear hydrogen arrays also has historical support from another of the great people in our field, also departed, my good friend Andrei Lipson. Andrei became convinced that the linear structure at the core of dislocations in the metallic lattice that was formed by repeated hydrogen loading and unloading of palladium (and other metals): (a) were capable of accreting significant densities of hydrogen atoms, (b) that the structure formed was in some manner similar to hydrogen metal, (c) was itself a high temperature superconductor, and (d) was a suitable environment in which to perform CMNR. Andrei developed considerable experimental support for "c" (with George Miley), which we were able to confirm in our own laboratory (with Paolo Tripodi). Experimental evidence for "d" was obtained by using carbon nano-tubes loaded with deuterium in a palladium composite matrix to create the linear deuterium array. Considerable excess energy and power gains were seen in such structures first by Energetics in Israel and later at SRI. Are these, too, early examples of effective, extended, nuclear active environments of the type proposed by Dr. Storms and described in this book?

More must still be done to understand the process of LENR and its cause. The complete book on this topic will not be written until industrial processes are achieved and/or this generation of the nuclear physics community departs the scene and Nobel prizes are proffered. Until then *The Explanation of Low Energy Nuclear Reaction* by Dr. Edmund Storms is the best description available by far, and I thoroughly recommend that everybody read it. Our critics are not the only ones with reactionary impulses and many experts in the LENR field will not be moved to pick this book up and read it, believing that "*we know better*." Quoting Dr. Storms in a private comment "*these ideas will be debated for years. In fact, most of the ideas are not obvious without study and debate*." To have such a debate we need a common language and understanding. This book provides just such a framework. Let the discussion of the next phase of our common undertaking begin here.

Dr. Michael McKubre
Director, Energy Research Center
SRI International, Menlo Park, CA, USA
June 2014

PREFACE

The phenomenon called cold fusion by the popular press continues to be rejected by many people. Many careers in science have been ruined as a result of this attitude. This response would only be an embarrassing note in history if the need for this ideal energy were not so great. Instead the stakes are high because mankind has now reached the edge of the cliff where a decision must be made to either explore this energy source in earnest or live with the dire consequences of pollution caused by conventional sources.

First, we need to decide whether this phenomenon is real or not. This book provides many answers to the rational questions posed by skeptics about the reality, shows how an explanation is best structured, and describes some basic features commercial application must take into account. Cold fusion is not a mystery because it can actually be understood using the concepts applied to normal science.

Although this book is about science, it is also about a break-through discovery that will affect your life as well as the lives of future generations. To make understanding easier, you will be pleased to find very few mathematical equations. For the present, the complex puzzle is best solved by finding the missing pieces and fitting them together in a logical and self-consistent way without using mathematics. In the process, unexpected relationships between chemistry and nuclear physics are revealed – relationships that were never before considered possible.

The goal of this book is to educate students, interested investors, and anyone else with an open mind. Hopefully, future studies and new explanations can be brought closer to what has been observed so that common false paths to understanding can be avoided and useless speculation can be reduced. To simplify reading, the conceptual discussions use normal English words with their normal meaning. The goal is to provide an easy to read general understanding of how low energy nuclear reactions

(LENR) function (*aka*, cold fusion)[1]. This approach does not provide a rigorous proof though such evidence does exist in the face of mainstream thought to the contrary. Indeed, so much evidence has been assembled that no additional proof would seem needed. Instead, the focus here is rather to achieve understanding.

To those with serious doubts about whether LENR is a real phenomenon, my previous book about the subject(*1*) assembles the evidence obtained before 2007. The paper by Storms and Grimshaw(*2*) puts this information into context. Regardless of how hard this claim is to understand or place in the context of present knowledge, 25 years of persistent study has resulted in overwhelming proof of an important phenomenon of nature. An explanation is proposed in Chapter 5, although this is not required to accept what is observed. Nevertheless, such understanding is required to apply the behavior. I hope this book can accelerate eventual application of what promises to be a source of ideal energy.

The field of study now called LENR started on March 23, 1989 when the announcement by Profs. Martin Fleischmann and Stanley Pons (F-P)(*3*) created a controversy about whether nuclear reactions could be initiated in what appeared to be an ordinary solid material. In their case, nuclear fusion between two deuterons was claimed to occur in a conventional electrolytic cell using palladium as the cathode and heavy-water (D_2O + LiOD) as the electrolyte. Up to that time, such a fusion reaction could only be caused in very expensive and complex machines, shown in Figs. 1 and 2, by applying very high energy to plasma, a very inefficient, expensive, and complex process.

Nonscientists might have never learned of the controversy had the implications not been so important. Instead, the entire world was awakened, thanks to the popular media, to the possibility of solving the environmental problems caused by present energy sources. Mankind had finally discovered an

[1] The term LENR is used in the text when the entire phenomenon is identified while the term "cold fusion" is used when only the fusion part of the phenomenon is discussed.

unlimited energy source without any environmental impact and with extraction of the fuel and conversion of the fuel to energy having no harmful byproducts. Then this hope for a better future was dashed. A false myth was created about how the claims were not real. As a result, small minds and economic self-interest took precedent over the future of mankind and anyone interested in the subject was rejected from conventional science.

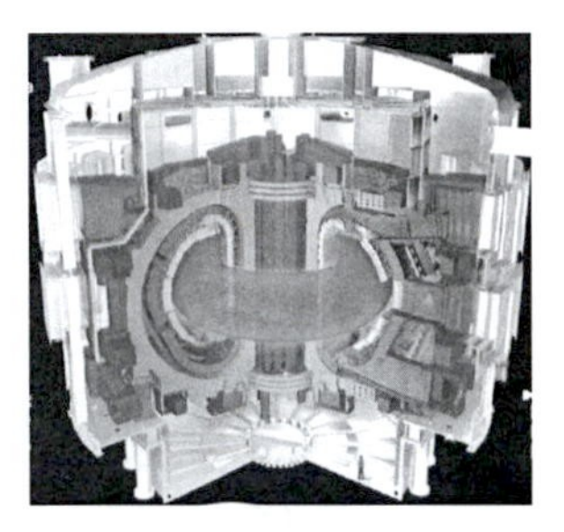

Figure 1. Picture of the ITER reactor under construction in France at a present cost of over 15 Billion Euro. Fusion occurs between D^+ and T^+ plasma within a magnetic bottle to which energy is applied until a useful fusion rate is achieved. Even after over 70 years of study, useful energy has not been made.[2]

Figure 2. The National Ignition Facility located at the Lawrence Livermore National Laboratory uses lasers to heat tritium to a temperature able to produce fusion. Recent success in making more energy than used during one shot was announced but a practical energy source is far in the future.[3]

---

[2]http://www.iter.org/media/www/downloads/poster_2009_new_scientist_iter.pdf
[3]https://lasers.llnl.gov/about/nif/about.php,http://en.wikipedia.org/wiki/National_Ignition_Facility

Even the discoverers of this phenomenon were forced to leave the US. Stanley Pons immigrated permanently to France with his family, gave up his US citizenship, and continued work in France where skepticism is not as harsh, while Fleischmann went back to his home in the UK. They continued their work in France with the help of Technova Inc., a Japanese company, and Johnson-Matthey, a British company. The sorry history of this discovery will not be repeated here. The heroes and villains have been clearly identified in other books and papers.(*4-6*)

We can forgive skepticism and rejection when a new discovery is first announced. This is the way of modern physics — reject until proven true. Nevertheless, publication of false information about the claims, refusal by peer reviewers to allow publication of information describing well documented behavior(*7, 8*), and personal attack have no business being used to stifle research. Science is not embarrassed or diminished by incorrect claims, but it is damaged by arrogant attack. Rather than providing protection from what is considered by some to be bad science, these attempts to keep science "pure" will now be remembered as the true examples of bad science. Treatment of cold fusion has become a diagnostic tool for revealing how science is actually practiced by some people in contrast to how they are expected to behave.

This rejection was based on several assumptions now known to be wrong. Conventional scientists assumed only one kind of fusion is possible, *i.e.* the one using high energy to overcome the Coulomb barrier — so-called hot fusion. Rejection followed from this assumption because the observations did not match behavior expected from this kind of fusion. The behavior conflicted with conventional theory[4] — end of discussion. Besides, if what F-P claimed were true, the hot fusion program would be in

---

[4] The word "theory" is used with its most general meaning without a distinction being made between hypothesis, explanation, or proposed suggestion. The word only describes an effort to show a logical relationship between various behaviors that can be used to predict behavior not yet observed.

financial trouble, with many professors being forced to relearn what they think they already know. This problem still exists.

In spite of this general rejection, a small group of scientists undertook to investigate the claim by risking their jobs, reputations, and money to discover what was real and how it might be made useful, in keeping with the ideals of scientific discovery. In the process, the initial claims have been replicated many times, with the same general behaviors being observed. Although replication still requires skill, this basic requirement demanded by science has been met many times. Efforts published before 2007 are summarized in my first book(*1*), including my studies of the effect. This present book was written to emphasize the important part of this information and describe an explanation that can guide future studies and application.

I discovered after hundreds of attempts to replicate the effect that more than occasional success required a better understanding than provided by the popular theories. After studying all published explanations, I found they all had basic conflicts with how LENR is observed to behave as well as with accepted understanding of conventional chemistry and physics. This left only one option — create a better explanation.

This book is not an objective and complete review, but instead is a selection of important observations that must be explained by any useful theory. The selection is chosen to reveal patterns of behavior that show a logically consistent mechanism. When judgment expressed here conflicts with what other people conclude, the reader alone must decide what to believe. I make no claim for knowing the truth or being correct in every opinion. Nevertheless, Chechin et al.(*9*) provide an excellent review of theories proposed before 1993 that I find in basic agreement with my own analysis. Yet, in spite of this critique, many theories continue to be proposed containing previously identified flaws. I hope this new analysis will guide studies in more fruitful directions so that the effect can be effectively controlled and then applied for beneficial use.

The approach used here avoids use of mathematical justification. Debating details of a mathematical treatment is useless if the basic idea on which the equations are based cannot be accepted. In this case, trying to explain LENR can be viewed as prospecting for gold, with the gold being the final correct explanation. The landscape can be explored at random, which has been the current approach, or location of discovered nuggets can be used to limit the search to certain regions, which is the approach used here. Of course, a map must take into account features in the landscape known not to contain gold. As with prospecting, many maps can be expected to stake claims to regions of the landscape and show where digging should start. In the case of LENR, this analogy has been considered; the location of found nuggets is identified in Chapter 2, the unproductive locations have been eliminated in Chapter 3, and all available maps have been examined in Chapter 4. All of these features have a logical connection based on a few simple assumptions. Identifying the assumptions and their connection is expected to result in a new and perhaps better map. This book is offered as the better map described in Chapter 5. My opinion of what comes next is reserved for Chapter 6. A book by Jed Rothwell goes into the future of cold fusion in more detail.(*10*)

For readers who are not interested in scientific detail and want to quickly learn what all this fuss means to an ordinary person, I suggest skipping the experimental descriptions and reading the conclusions at the end of each section. These conclusions are assembled into a brief description of LENR in Chapter 1, with an overview provided in Chapter 6. I hope that learning how and why a new and revolutionary discovery functions can be as much fun for you as writing this book has been for me.

Edmund Storms
Santa Fe, NM, USA
June 2014

TABLE OF CONTENTS

FOREWORD IV

PREFACE VIII

LIST OF FIGURES XXI

LIST OF TABLES XXVII

CHAPTER 1 1

OVERVIEW OF LENR AND REQUIREMENTS OF AN EXPLANATION 1

CHAPTER 2 9

OBSERVED BEHAVIOR REQUIRING EXPLANATION 9

2.0.0 INTRODUCTION 9

2.1.0 TRITIUM 10
- 2.1.1 What is tritium? 10
- 2.1.2 How is tritium measured? 11
  - Ion Counting 11
  - Liquid Scintillation 11
  - Mass Spectrometer 12
  - Photographic Film 13
- 2.1.3 How is tritium produced using LENR and at what rate? 13
- 2.1.4 How can tritium be produced by LENR? 20
- 2.1.5 CONCLUSIONS ABOUT TRITIUM 22

2.2.0 HELIUM 23

2.2.1 How is helium measured?..24
2.2.2 Methods found to produce helium in LENR....................24
2.2.4 CONCLUSIONS ABOUT HELIUM.....................................41

2.3.0 TRANSMUTATION 43
2.3.1 How is transmutation measured? 44
EDX...44
SIMS...45
Neutron Activation 45
XRF 45
ICP-MS...45
AES 46
XPS or ESCA 46
2.3.2 What types of transmutation occur during LENR? 46
Transmutation produced by fusion-fission 47
Transmutation produced by fusion without fission.....54
2.3.3 What methods are found to produce transmutation? ...54
Transmutation produced by gas loading 54
Transmutation produced by the glow discharge method
....62
Transmutation produced by electrolytic method........64
Transmutation produced by plasma electrolysis and arc-plasma
in water 64
2.3.4 Transmutation using applied laser radiation 67
2.3.5 CONCLUSIONS ABOUT TRANSMUTATION......................72

2.4.0 RADIATION.. 74
2.4.1 How is radiation detected? 74
CR-39 74
Geiger-Müller Detector 75
Proportional Detector...75
Silicon Barrier Detector (SBD) and Germanium Detector
(Ge) 75
Scintillation Detectors 75
Neutron Detection..75
2.4.2 Source of radiation 76
2.4.3 Detected radiation using CR-39 76

2.4.4 SUMMARY OF CR-39 STUDIES 81
2.4.5 Real-time detection of radiation 82
2.4.6 "Strange" radiation 91
2.4.7 CONCLUSIONS ABOUT EMITTED RADIATION 92

2.5.0 APPLIED ENERGY 94
2.5.1 Applied electron flux 95
2.5.2 Applied laser radiation to produce fusion 96
2.5.3 CONCLUSIONS ABOUT APPLIED RADIATION 105

2.6.0 HEAT ENERGY 105
2.6.1 How is energy measured? 106
Isoperibolic calorimeter 107
Adiabatic calorimeter 108
Seebeck calorimeter 108
Flow calorimeter 108
Phase change calorimeter 109
IR radiation calorimeter 109
2.6.2 What methods are found to produce heat? 109
Electrolysis 109
Plasma electrolysis 110
Electroplating 110
Gas (glow) discharge 110
Gas loading 110
Sonic 111
Electromigration and diffusion 111
2.6.3 Production of energy 111
2.6.4 Energy produced by electrolysis using D_2O 112
2.6.5 Effect of bulk metal treatment 118
2.6.6 Effect of temperature 119
2.6.7 Effect on power production of adding light hydrogen to D_2O 120
2.6.8 Energy produced using H_2O and H_2 120
2.6.9 SUMMARY OF ENERGY PRODUCTION 121

2.7.0 PROPERTIES OF PALLADIUM HYDRIDE 121
2.7.1 Phase relationship 122
2.7.2 Lattice parameter 129

2.7.3 Diffusion constant 130
2.7.4 Thermal conductivity 132
2.7.5 Calculations describing the atom arrangement in PdH 133
2.7.6 Laws of Thermodynamics as applied to chemical structures 134
2.7.7 CONCLUSIONS ABOUT THE PdD SYSTEM 136

2.8.0 EFFECT OF ION BOMBARDMENT & FRACTOFUSION......137

2.9.0 SUMMARY & CONCLUSIONS ABOUT OBSERVED BEHAVIOR 141

CHAPTER 3 143

REQUIREMENTS OF AN EXPLANATION....................... 143

3.0.0 GENERAL REQUIREMENTS.............................. 143

3.1.0 PARTICULAR REQUIREMENTS IMPOSED ON LENR 147
3.1.1 Limits to energy concentration 147
3.1.2 Limits to cluster formation...149
3.1.3 Limits on dissipation of mass-energy 151
3.1.4 Limits on production of radioactive products 152
3.1.5 Limits on production of radiation 152
3.1.6 Limits on the number & kind of nuclear mechanisms required 154

3.2.0 COMMON FEATURES USED IN THEORY 154
3.2.1 Resonance of deuterons in PdD lattice to achieve nuclear interaction 155
3.2.2 Tunneling or electron screening.................................157
3.2.3 Formation of clusters of certain sizes and located at certain sites
in the lattice 157
3.2.4 Formation of neutrons or interaction with stabilized neutrons already present in the lattice 158

3.2.5 Formation of special sites, such as super abundant vacancies, where fusion takes place....160
3.2.6 Transmutation of the metal atoms in the lattice....162
3.2.7 Shift of energy to a non-conventional state by formation of a special structure....163
3.2.8 Role of novel heavy particles....164
3.2.9 Dissipation of energy....165

3.3.0 ONLY EXOTHERMIC NUCLEAR REACTIONS ARE POSSIBLE....166

3.4.0 SUMMARY AND CONCLUSIONS....167

CHAPTER 4....169

EVALUATION OF PROPOSED EXPLANATIONS....169

4.0.0 INTRODUCTION....169

4.1.0 THEORIES WITH FOCUS ON CLUSTER FORMATION....170
P. Hagelstein....170
M. Swartz....172
G. Miley and H. Hora....174
A. Takahashi....178
Y. Kim....183

4.2.0 ROLE OF RESONANCE....184
G. Preparata....185
J. Schwinger....187
S. Chubb and T. Chubb....189
R. Bush and R. Eagleton....190
R. Bass....192
V. Violante....194

4.3.0 ROLE OF NEUTRONS....196
A. Widom and L. Larsen....196
R. Godes....197

J. Fisher 198
H. Kozima..199

4.4.0 ROLE OF SPECIAL ELECTRON STRUCTURES 199
A. Meulenberg and K. P. Sinha 200
J. Dufour 202
R. Mills..203

4.5.0 ROLE OF TRANSMUTATION.......................... 204
F. Piantelli 204
A. Rossi..205

4.6.0 ROLE OF TUNNELING 205
X. Z. Li...206

4.7.0 ROLE OF CRACKS AND SPECIAL STRUCTURES .. 207
F. Frisone 208
E. Storms 208

4.8.0 SUMMARY OF PROPOSED THEORIES 208

CHAPTER 5 210

DESCRIPTION OF A NEW EXPLANATION 210

5.0.0 BASIC ASSUMPTIONS...210
5.0.1 General description 211
5.0.2 Item #1: Variables affecting the process outside the NAE..212
5.0.3 SUMMARY OF ITEM #1 217
5.0.4 Item #2: Production of nuclear active sites................218
5.0.5 SUMMARY OF ITEM #2 220
5.0.6 Item #3: A process for converting mass-energy into heat-energy without radiation 221
5.0.7 SUMMARY OF ITEM #3 226

5.1.0 CAUSE OF TRANSMUTATION........................ 228
5.1.1 Transmutation type #1 230

5.1.2 Transmutation type #2 232
5.1.3 SUMMARY OF TRANSMUTATION REACTIONS 238

5.2.0 DETAILED EXPLORATION OF THE HYDROTON STATE 240
5.2.1 Metallic hydrogen 240
5.2.2 Resonance of negative and positive charges 241
5.2.3 Role of the surface 242
5.2.4 Role of the laser and magnetic fields 243

5.3.0 HOW CAN LENR BE MADE REPRODUCIBLE? 245
5.3.1 SUMMARY OF NEW THEORY 249

CHAPTER 6 251

FUTURE OF LENR 251

6.0.0 SCIENTIFIC 252

6.1.0 COMMERCIAL 254

6.2.0 POLITICAL 255

REFERENCES 256

INDEX 314

LIST OF FIGURES

Figure 1. Picture of the ITER reactor under construction in France at a present cost of over 15 Billion Euro..x
Figure 2. The National Ignition Facility located at the Lawrence Livermore National Laboratory uses lasers to heat tritium to a temperature able to produce fusion..x
Figure 3. Comparison between shape of beta energy spectrum for tritium and ^{14}C.. 12
Figure 4. A foil of Pd-Ag alloy (top) was heated in vacuum to 600°C, exposed to D_2 gas for 88 hrs, and placed on photographic film.. 14
Figure 5. Tritium production in an open electrolytic cell containing D_2O+LiOD and a palladium cathode 16
Figure 6. Tritium production in a closed electrolytic cell containing D_2O-LiOD and a palladium cathode 17
Figure 7. Cell similar to those used at LANL exposed to an environment containing tritium. 18
Figure 8. Effect of adding tritium to a closed electrolytic cell..... 19
Figure 9. Effect of electrolyzing a cathode of palladium containing tritium...20
Figure 10. Comparison between the amounts of D in the electrolyte and in a palladium cathode..21
Figure 11. Cross-section for the reaction of ^{7}Li with a neutron having the energy shown...22
Figure 12. Comparison between excess power and helium concentration in the evolving gas 36
Figure 13. Measured helium compared to the amount of helium expected from the measured energy if the D+D reaction were the source...36
Figure 14. The amount of energy based on two methods for its determination as a function of accumulated helium from Pd deposited on carbon...38
Figure 15. Relationship between measured excess power and helium production as a function of time as reported by DeNinno et al...40

Figure 16. Three studies during which laser light was applied during electrolyses are compared to the helium and energy detected.......41
Figure 17. Comparison between current produced by the helium peak in a mass spectrometer and the energy produced.......41
Figure 18. Example of structures on palladium after electrolysis in D_2O where localized concentration of assumed transmutation products are claimed to form........48
Figure 19. Number of reported detections of unexpected elements on the surface of a Pd cathode after electrolysis........50
Figure 20. Miley summarized many studies of transmutation to show
four regions of atomic number where large numbers of claimed transmuted elements are found.......51
Figure 21. Element distribution obtained using H_2O in the electrolyte........51
Figure 22. Effect of adding d, p, or n to isotopes of palladium CaO and Pd deposited on Pd.......53
Figure 23. Cross-section of layers of CaO and Pd deposited on Pd.......56
Figure 24. The isotopic ratio for iron is shown, with results obtained when D_2O is used as the electrolyte, shown on the left, and when H_2O is used shown on the right.......56
Figure 25. Number of atoms of the initial Cs and the final Pr on the surface of the Pd complex as a function of time while exposed to D_2.......58
Figure 26. Depth profile of Cs and Pr in the Pd-CaO complex surface. B.G. is the initial Cs profile and F.G. shows the profile after exposure to D_2........59
Figure 27. Number of atoms of initial Sr and final Mo on the surface of the Pd-CaO complex as a function of time while being exposed to D_2........60
Figure 28. Isotopic distribution of Ba in samples applied to the Pd complex compared to the resulting isotopic distribution of Sm after exposure
to D_2.......60
Figure 29. Summary of nuclear reactions observed by Iwamura et al. to occur on the surface of the Pd-CaO complex.......61

Figure 30. A common appearance of a Pd surface after glow discharge 63
Figure 31. Autoradiograph of a Pd cathode after gas discharge using X-ray film with a 2 mm lead (Pb) absorber 63
Figure 32. Transmutation products produced during plasma electrolysis using a tungsten cathode, Pt anode, and K_2CO_3 in H_2O 66
Figure 33. SEM view of the surface where transmutation products are detected in Pd thin films 68
Figure 34. Transmutation produced without laser using 125 nm layer of PdD on Si. Atomic number of Pd is noted for reference. 69
Figure 35. Typical aspects of the surfaces of Pd film deposited on polished silicon wafers, after the completion of the gas-loading phase 70
Figure 36. EDX spectrum after exposing Pd in H_2 to laser radiation 71
Figure 37. Typical result from a CR-39 sample showing distribution of track (pit) diameters and the calculated energy based on calibration using particles of known energy 77
Figure 38. Spectrum of deuterons emitted from a metal target during DC glow discharge in D_2 83
Figure 39. Examples of radiation reported by Piantelli et al. using specially treated Ni exposed to H_2 84
Figure 40. Count rate in excess of background for active Cu/Pd cathode electrolyzed in H_2O-Li 85
Figure 41. Count rate in excess of background for an active Ni cathode electrolyzed in H_2O-Li 86
Figure 42. Photon radiation detected by Bush and Eagleton from a Ni cathode in H_2O-Li electrolyte while making excess power 86
Figure 43. Ni coated by Cu and exposed to H_2 gas. Reaction caused cracks and radiation. 87
Figure 44. ln flux vs time for the decay in Fig. 43 between 1250 and 1500 minutes. 88
Figure 45. Surface appearance of Ni-Cu after reacting with H_2.... 89
Figure 46. Decay plot of a nickel cathode after electrolyzing using H_2O-Rb_2CO_3 89
Figure 47. Gamma events as a function of time while Pd is electrolyzed in D_2O-Li 90

Figure 48. Alpha spectrum resulting from PdD using CR-39 to which error limits are applied..94
Figure 49. Effect of laser applied to PdD having the D/Pd ratio shown 97
Figure 50. Proposed effect of laser photons on electrons holding d nuclei together in a metallic-like bond near the surface of a material...99
Figure 51. An image of a Pd surface coated by gold and used by Letts 100
Figure 52. Effect of applying 680-686 nm laser light (35 mW) to an activated Pt cathode during electrolysis in D_2O+Li 101
Figure 53. Effect of applying 661.5 nm laser light (30 mW) to Pd coated with gold while electrolyzing in D_2O+Li containing a rare earth additive 102
Figure 54. Surface analysis of an active cathode to which certain elements were added to the electrolyte 102
Figure 55. Excess power produced by two lasers superimposed on the same spot on a Pd-Au cathode and with the same polarization. 103
Figure 56. Relationship between frequency and wavelength in the electromagnetic spectrum..104
Figure 57. IR picture of the Pd surface during co-deposition of Pd from an electrolyte containing D_2O + $PdCl_2$ + LiCl......................112
Figure 58. Relationship during electrolysis of Pd between excess power and deuterium composition of the average bulk cathode wire based on resistance change...113
Figure 59. One of many examples of how applied current changes excess heat production 115
Figure 60. Example of the effect of applied current on excess power produced by a thin film of active material deposited on an inert substrate...115
Figure 61. Excess power and D/Pd ratio of a wire being electrolyzed in D_2O compared over time..................................116
Figure 62. Effect of cell temperature on the log of excess power production 119
Figure 63. Phase diagram of the Pd-H system showing applied pressure
for various temperatures..122

Figure 64. Example of a face-centered cubic structure (fcc) consisting of two identical interpenetrating lattices with H (small sphere) being in the octahedral sites relative to the Pd 123
Figure 65. Partial phase diagram of Pd-H and Pd-D systems showing the relationship between applied pressure and the resulting atom ratio
at 300K 123
Figure 66. Volume of sample vs H/Pd content 125
Figure 67. Excess volume in $PdD_{0.85}$ for a variety of Pd samples treated initially in different ways 126
Figure 68. Resistance ratio, R/R_o, as a function of H/Pd atom ratio
.. 127
Figure 69. Resistance ratio and temperature coefficient as a function of D/Pd and H/Pd atom ratio. Lines added to clarify relationship to phase diagram.. 128
Figure 70. Lattice parameter of fcc phase for various conditions and hydrogen contents... 129
Figure 71. Lattice parameter at 77K of beta PdH and PdD as a function of atom ratio. 130
Figure 72. Log Fick's diffusion constant for deuterium in PdD as a function of atom ratio at various temperatures 131
Figure 73. Effect of temperature on the log Fick's diffusion constant for β-PdD and β-PdH 132
Figure 74. Relationship between thermal diffusivity ratio and electrical resistance ratio for PdH... 133
Figure 75. Rate of hot fusion between isotopes of hydrogen as a function of applied energy in plasma.. 138
Figure 76. Increase in fusion rate relative to that in plasma as a function of energy of D+ used to bombard various targets....... 139
Figure 77. Increased rate of hot fusion as a result of D+ bombardment of Ti compared to the rate expected to result in plasma 140
Figure 78. Comparison between hot and cold fusion 145
Figure 79. Pores in PdH claimed to have resulted from decomposition of $Pd_3(\square H_x)$ after annealing in hydrogen 162
Figure 80. Excess power/applied power vs estimated applied power using data shown in Fig. 59 174
Figure 81. Proposed elemental distribution in the universe 176

Figure 82. Cartoon provided by Miley et al. of a lattice site in PdD where a super nucleus is proposed form *178*
Figure 83. Sequence of the proposed reactions that make heat and helium *180*
Figure 84. A proposed assembly of hydrogen that initiates fusion in a PdD lattice *194*
Figure 85. Effect of temperature on the amount of power *216*
Figure 86. A cartoon of the Hydroton *223*
Figure 87. Examples of Hydrotons in which target atoms are captured and of one example of a Hydroton without attached target *229*
Figure 88. Fragment distribution when a single p-e-p enters Pd followed by fission *235*
Figure 89. Fragment distribution when two (p-e-p) enter Pd followed by fission *235*
Figure 90. Fragment distribution when 2(p-e-p) are added to Pt followed by fission *236*
Figure 91. Fragment distribution after adding 2(p-e-p) to Ni followed by fission *236*
Figure 92. Distribution of energy resulting from fragmentation of Ni stable element combinations *237*

LIST OF TABLES

Table 1. Nuclear reactions that can make tritium (^3H) and the energy/event (Q) 21
Table 2. Possible nuclear reactions that produce helium 25
Table 3. Pd electrolyzed in D_2O and H_2O 30
Table 4. Additional samples from Miles et al 32
Table 5. List of samples that produced neither energy nor helium 34
Table 6. List of samples that produced both energy and helium. 35
Table 7. Excess heat and helium/energy for samples measured by Bush and Lagowski 37
Table 8. List of samples examined for heat and helium production by Isobe et al 39
Table 9. Summary of values for He/watt-sec 42
Table 10. List of elements detected using EDX in holes in a thin layer of PdD or PdH containing boron on a smooth Si substrate . 71
Table 11. Proposed fusion reactions and resulting energy. The open circles represent neutrons and the closed circles represent protons 225
Table 12. Result of adding ^4H to Pd isotopes followed by beta emission.
The half-life and decay mode are shown for the isotopes of Pd 231
Table 13. Information used to calculate relative amount of each fragment combination for $^{46}Pd + 2(p+e+p) = {}^{40}Zr + {}^8O$ 234
Table 14. The number of ways the isotopes of Ni can fragment and the resulting average energy for each event is listed for each isotope. 237

CHAPTER 1

OVERVIEW OF LENR AND REQUIREMENTS OF AN EXPLANATION

The book is arranged in sequence. First, a general overview of the subject is provided here in Chapter 1. Next, Chapter 2 describes observed behavior needing explanation and the relevant natural laws. Based on this information, Chapter 3 summarizes requirements all theories must acknowledge and shows how some concepts conflict with what is known. These general requirements are useful in evaluating theories[5] without having to discuss every detail.

Hundreds of explanations have been published, with only the major ideas described and evaluated next in Chapter 4 using the general requirements identified in Chapter 3. Chapter 5 describes in more detail a new theory and how it can be used to explain the engineering and scientific behavior of LENR. Methods to make replication reliable are described using the theory. Finally, the scientific, political, and commercial implications of this discovery are summarized in Chapter 6. Each major section is briefly summarized at its end so that a reader can quickly move through the book without being distracted by too much detail. Repetition is avoided to some extent by citing relevant sections throughout the text. Nevertheless, replication of major concepts is done to make sure all their aspects are understood.

The book reveals four basic errors that have seriously distorted how theory is developed. These are: (1) emphasis on bulk material being the location of the nuclear process, (2) emphasis on the properties of the pure hydride being important to the process, (3) emphasis on small particle size being an important requirement, and (4) a confusion between hot fusion and cold fusion.

---

[5] Many additional theories can be found at http://www.journal-of-nuclear-physics.com/?cat=3

In fact, the LENR process takes place in the surface region having a depth of a few microns where the properties significantly differ from pure hydride. When a powder is used, the particle size may be important but not just because of its large surface area. Furthermore, hot fusion and cold fusion have no relationship to each other.

The relationship between hot and cold fusion is particularly effective in causing confusion. These two processes produce different nuclear products and clearly involve different mechanisms. Hot fusion produces energetic fragments, while cold fusion produces helium that does not fragment and very little energetic radiation is detected. Consequently, an explanation has to treat these two forms of fusion separately by not mixing the mechanisms or using observations obtained from hot fusion to explain or reject cold fusion. Although both processes are called fusion, the cold fusion process is unique and requires its own explanation.

The process occurs only in a unique and rare part of the overall sample. For the sake of discussion, this region is called the Nuclear Active Environment (NAE).(*11-13*) This concept is general and does not identify the nature of the environment or where in the material it forms. The term simply allows location of the active conditions to be distinguished from inert conditions that normally exist in most of the material. Nevertheless, the NAE is generally found in the surface region where the basic material is highly modified.

The LENR mechanism can be considered to have three separate events, with each operating in collaboration and in sequence. First, the NAE has to be created in the material, with the creation process controlled by the laws of ordinary chemistry and physics. The LENR process can only occur in these rare sites. A single kind of NAE along with a single general mechanism is proposed to cause LENR regardless of the hydrogen isotope[6] present. Only the resulting nuclear products are different.

Second, a structure must form in the NAE with the ability to overcome the Coulomb barrier and dissipate the resulting nuclear energy in small units over time. This nuclear process is the focus of most proposed theories even though it does not have to be understood to make the effect work as a source of energy. Instead, understanding at the engineering level is sufficient to show how the process can be initiated and controlled.

Finally, the fuel, consisting of hydrogen isotopes, must find the NAE sites. This process involves normal diffusion. These three events combine to determine whether the rate of the nuclear process is large enough to be detected. Too few NAE, too little fuel, or too slow diffusion of fuel in the material would result in no LENR being detected. An increase in any one of these three variables will cause the amount of power to increase. Because three independent variables are involved in determining the rate of LENR, the observed rate cannot be assigned to a single variable. In addition, the various methods used to initiate LENR each affect these events in different ways. To make sense of the complex behavior, each variable needs to be examined to see which stage of the process it influences. Apparently, some behaviors have no relationship to how the NAE is created or how the nuclear process functions in the NAE, yet influence the observed reaction rate. Unfortunately, the complex inner-relationship between these separate events has confused many explanations — a problem this book will try to repair.

The process occurring in the NAE alone contains the mystery and follows the unknown rules of nuclear interaction made apparent by LENR. Every event leading up to this process must follow the rules of normal chemical behavior. Many observed behaviors make sense only after this insight is accepted. Emphasis on explaining the final nuclear process in many theories is

---

[6] The word "hydrogen" is used to describe all isotopes of hydrogen. The word "deuterium" is used only when the behavior of this single isotope is being described.

misplaced because this insight is ignored and, as a result, the models do not show how the effect can be replicated and how power can be increased.

When people complain about how seldom LENR is replicated, they are acknowledging mainly the difficulty in creating the NAE. Nevertheless, all factors making LENR difficult to initiate are important and must be understood. To further elaborate on this requirement, absence of any one of these three events would stop the LENR process no matter how large the other two might be. For example, the electrolytic process requires a high concentration of deuterium (D) because the concentration of NAE is small and the temperature is low. Increased temperature is found to improve power production because the resulting increased diffusion rate allows deuterium to reach the NAE more rapidly. Likewise, an increase in amount of NAE would reduce the concentration of fuel required for power to be detected. Regardless of how the nuclear process functions in the NAE, these conditions will determine how much power is produced and whether or not LENR can be detected.

Because each of these conditions is related to the properties of material in the NAE, the basic properties of this material need to be considered. These properties are not those of pure, ideal material. Most theories focus on the properties of pure PdD, which are discussed in Section 2.7.0 as a stand-in for the complex material actually involved in the LENR process. It is essential to realize that pure PdD is not present where the NAE forms and its properties are only indirectly related to the nuclear process.

The cold fusion process clearly requires an unusual and heretofore unidentified condition to function in the NAE. This unique and rare condition must operate outside of the normal chemical structure to avoid limitations imposed by rules operating in a chemical environment. Justifying this conclusion is one goal of the book.

The challenge is to discover the conditions causing the NAE to form. As an example of a failed approach, many explanations propose the NAE will form and the nuclear process will be initiated when the D/Pd ratio gets close to unity in the ordinary chemical structure. However, such high compositions have been reached without a nuclear reaction being detected and nuclear reactions are produced at much lower compositions on occasion. Apparently, increased deuterium alone cannot initiate the nuclear reaction, although it can clearly increase the rate of a reaction already underway.

To further limit and complicate understanding, all aspects of the fusion reaction must function in harmony. The process used to overcome the Coulomb barrier must, at the same time and place, also help dissipate energy to avoid production of the hot fusion-type reaction, as would happen should this collaboration fail. While many proposed processes are able to lower the barrier, they are not able to dissipate energy without additional features being required, which are frequently supported in theory only by unjustified assumptions. This halfway approach is rather like designing an airplane without the wings — proposing instead some undefined force to keep it aloft. Obviously, this unidentified force would need to be identified and controlled before anyone would invest in such a project. This book attempts to design the wings.

Many explanations propose to bring two deuterons close together by using a quantum mechanical process. This suggestion is not consistent with what is observed because once two atoms simply get close, they produce a hot fusion-type reaction, not cold fusion. An example of simply reducing the distance between deuterons can be seen in the effect of muons. In this case, the electron holding two deuterons together in the D_2 molecule is replaced by a muon, which has the same charge but 200 times the mass of an electron. This extra mass requires the two deuterons to move close enough for fusion to occur on a few occasions. The excess mass-energy is released by fragmentation of the resulting nuclear product, *i.e.* by hot fusion. As this example shows, simply reducing the distance alone cannot explain cold fusion. Understanding and accepting this conclusion is essential to effectively explaining the nuclear process.

Another essential requirement involves where in a chemical structure a mechanism can function to lower the barrier. In the absence of muons, deuterons do not get close enough to fuse at a detectable rate. This has been verified by calculations (Section 2.7.5) and by the failure of deuterium to fuse when placed in a wide variety of chemical structures and exposed to extreme pressure and/or temperature. Fusion is only possible when significant energy is applied to the deuterons by using ion bombardment, as discussed in Section 2.8.0. However, in this case fusion results from tunneling or screening and produces only hot fusion.

Penetration of the barrier by what is called tunneling requires energy be applied to the process (Sections 2.8.0 and 3.2.2). The amount of required energy is greatly in excess of the amount available in a chemical structure. A process proposed to concentrate this energy would conflict with the Second Law of Thermodynamics (Section 2.7.6). This limitation also applies to proposed creation of neutrons in the structure. Such conflict with a basic law needs to be acknowledged and eliminated before the proposed explanation can be accepted.

Even though LENR seems to produce only feeble radiation, a few energetic products are sometimes detected. Some of these energetic radiations might result from conventional hot fusion taking place at the same time cold fusion is underway and some can result from LENR itself, thereby confusing interpretation. In addition, "strange" radiation has been reported with the ability to initiate nuclear reactions in material, but in a manner different from how neutrons act. On occasion, radioactive products are produced, with their radiation being confused with that resulting directly from the LENR reaction. As a result, much more study is required to reveal how radiation is related to the cold fusion process. Concluding that the amount of generated radiation is not sufficient to account for heat production is premature.

Transmutation has been largely ignored in current theories because a clear relationship to the fusion process and a method to overcome the large Coulomb barrier have not been found. The theory in this book explains transmutation as a natural consequence of the fusion process and uniquely explains why two different types of transmutation products are produced containing a large collection of nuclear products, both stable and radioactive. Predicted products based on this model are compared to observation and used to explain some mysteries. Although transmutation has been largely ignored, it is common and provides a valuable path to understanding the LENR process. Nevertheless, transmutation cannot occur at a rate sufficient to make detectable energy.

Before discussing the proposed theories in detail, the behaviors in need of explanation must be identified. A wide variety of behaviors is reported, causing some people to believe many different kinds of novel nuclear reactions can occur under different conditions, with each requiring a different explanation. When single measurements are viewed in isolation, this interpretation appears plausible. However, when all observations are viewed as part of a pattern, with many measurements showing the same kind of behavior, all with a logical relationship, a common mechanism can be identified. This process is simplified when the volunteered explanations frequently provided by the authors are ignored, with focus being placed only on the observations themselves. Just how many different mechanisms might be operating has not yet been determined. Nevertheless, this book will focus on a single mechanism to explain fusion and transmutation involving all isotopes of hydrogen, following the common goal of seeking the simplest explanation first.

The important behaviors are examined in Chapter 2 with a goal of extracting patterns from the many observations. Observations fall into three categories: production of nuclear products including transmutation products, production of radiation, and production of energy in excess of a conventional source. The first two behaviors clearly demonstrate occurrence of nuclear reactions where none would be expected. Transmutation has a very small reaction rate typically at few events/sec, which is too small to be detected by measuring energy. In contrast, when energy is detected, it demonstrates a very high rate, amounting to nearly 10^{11} events per sec when 1 watt is made by fusion of deuterium. Less energetic reactions would require a greater rate to produce this amount of power. Indeed, the process has been found to have very wide range of reaction rates involving several different nuclear reactions, all of which must be explained. A theory designed to account for only a few events/sec is not useful to identify a source of practical power and simply focusing on power ignores the rich assortment of possible nuclear reactions.

To avoid these problems and limitations, the NAE is proposed to be located in nanogaps of very small and limited gap width. These cracks form as a result of stress relief in the surface region. Once formed, a linear structure of hydrogen isotopes forms by covalent bonding. This structure loses mass-energy by emitting photons generated as the structure resonates. This process is also instrumental in producing the two types of transmutation products. The details are described in Chapter 5.

This theory is used to explain how the process can be made reproducible (Section 5.6.0), which has become the essential requirement for acceptance of claims as well as explanations. Other problems needing to be mastered before commercial application is possible are discussed in Section 6.0.0.

The LENR process has revealed a new and important way for nuclei to interact. This phenomenon can be expected to keep scientists busy for years as details about the process are discovered and lead to other unanticipated gifts of nature.

CHAPTER 2

OBSERVED BEHAVIOR REQUIRING EXPLANATION

2.0.0 INTRODUCTION

Strong evidence has now demonstrated the reality of LENR, only details about the mechanism and some of the resulting nuclear reactions remain to be determined and explained. Four different nuclear products have been identified. These are helium-4 (^4He), tritium (^3H), various transmutation products, and low-level radiation, all of which require a nuclear source. In addition, more energy has been measured on many occasions than could result from other than a nuclear reaction.

Three products of LENR can be identified and used to define how an explanation of this elusive reaction has to be structured. First, the process produces various nuclear products. These are summarized in Sections 2.1.5, 2.2.4, and 2.3.5. Second, the mass-energy released by these reactions appears mostly as photon radiation, not as energetic particles typical of hot fusion, as summarized in Section 2.4.7. Third, heat energy is produced at rates too great to be caused by chemical sources, as summarized in Section 2.6.9. These summaries can be read to hasten passage to the discussion of theory in Chapters 3, 4 and 5. Relevant sections in Chapter 2 are cited when detailed understanding is required in other chapters.

All of the nuclear products are produced without significant energetic radiation being detected outside the apparatus. Because radiation can be contained in the generator, a safe source of energy is possible. Heat-energy is produced sometimes at significant and sustained rates, which promises inexpensive energy having a wide range of application. The transmutation process has the potential to remediate radioactive contamination resulting from accidental release from fission reactors, and tritium can be made with much less cost than by present methods. These features should encourage serious consideration of this phenomenon even if proof is not considered adequate. The potential benefit is too overwhelming to ignore for any reason. Unfortunately, each of these behaviors is still difficult to produce with reliability. This problem is discussed in Section 5.6.0.

Of the nuclear products, tritium provides the clearest proof for an unanticipated nuclear reaction being initiated in ordinary materials and gives insight about the mechanism. For this reason, its detection and behavior are discussed first.

2.1.0 TRITIUM

Of the nuclear products produced by LENR, tritium is the easiest to detect, is not seriously compromised by contamination, and is an isotope that can only be present above the very low background as a result of a nuclear reaction. An inquisitive person might be encouraged to ask, "If tritium can be made this way, would not other nuclear reactions be possible as well and be caused by the same mechanism?" An answer becomes easier when we discover that tritium is made in the same general location on the surface where the other nuclear products, such as helium and transmutation products, are found after similar treatment. This fact makes tritium behavior an especially important guide to understanding the entire LENR phenomenon.

2.1.1 What is tritium?

Tritium is an isotope of hydrogen consisting of one proton and two neutrons. The isotope is radioactive with a half-life of 12.346 years(*14*) by emission of an electron (18.59 keV maximum beta energy) and a neutrino. Because electron emission (beta) has too little energy to pass through even a piece of paper or travel more than 6 mm in air, detection requires special methods. Tritium concentration in the environment is too low to be important, although it does occur in some heavy-water (D_2O) where its concentration can be increased slightly by open-cell electrolysis. This selective concentration process(*15-22*) is well known and corrections for the increased amount of tritium are easy to make.

2.1.2 How is tritium measured?

Tritium, when present in a gas, can be measured by detecting the emitted beta several different ways or by using a mass spectrometer if the concentration is large. Detection of radioactive emission is possible using ion counting or liquid scintillation. In both cases, the tritium must be removed from the material in which it is located. Removal can be done by dissolving the material in acid or by heating in vacuum. The released gas is collected and can be converted to water for later analysis. The beta energy and decay rate combined with accumulating ^{3}He can be used to demonstrate unambiguously the presence of tritium.

All methods have errors and suffer from a small background signal that must be subtracted. Claims for tritium are based mostly on using the liquid scintillation method, which has a detection limit of about 10^{7} atoms in whatever volume is placed in the detector, and the patience of the operator.

Ion Counting

Tritium in air or in any gas can be detected as a voltage pulse produced when ions are created in the gas by the emitted beta. The small pulse produced by the weak beta from tritium is easy to separate from larger pulses caused by other radioactive elements that might be present as unwelcome contaminants, such as radon.

Liquid Scintillation

Tritium in water is mixed with an organic fluid that emits light pulses from where energetic particles pass. The intensity of the light pulse is related to the beta energy and the number/sec is related to the concentration of tritium. Beta emitters give a spectrum of energies thanks to the emitted neutrino taking a variable part of available energy. The shape and energy of the spectrum allows the source to be identified, as shown in Fig. 3 where a comparison is made between beta radiation from tritium and ^{14}C.

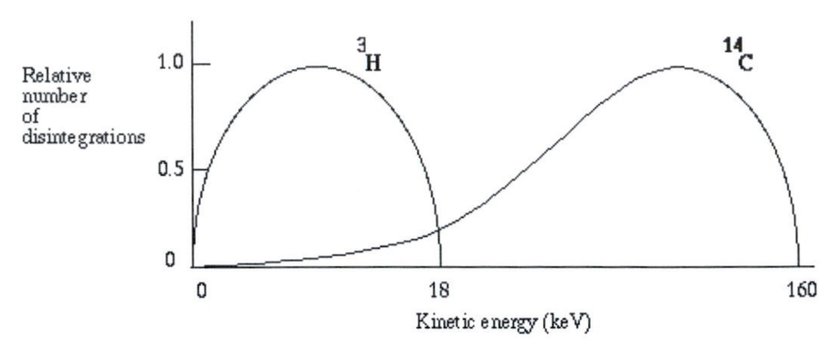

Energy spectra for two commonly used beta emitters. Note that the horizontal scale is not linear.

Figure 3. Comparison between shape of beta energy spectrum for tritium and ^{14}C. [7]

When a sample is initially mixed with detection fluid, a chemical reaction can on occasion produce light pulses. This effect can be eliminated by waiting about an hour until the chemical reaction has subsided or by distilling the water to remove active chemicals. Some of this brief activity might even result from decay of transmutation products collected in the sample, a possibility that has been ignored.

This method is the gold standard for tritium detection with high reliability and accuracy.[8] Because the half-life is long, the sample is available for measurement as often as necessary to demonstrate the presence of tritium.

Mass Spectrometer

Low concentrations of T (^3H) are expected to be present as the DT and/or HT molecule, depending on which additional hydrogen isotope is present in the gas. This molecule will be fragmented by the bombarding electrons to give some T^+ at m/e = 3. Other mass/charge (m/e) ratios at 3, 4, and 5 might also occur. Unfortunately, m/e = 4 conflicts with the D_2^+ ion and m/e = 3 conflicts with DH^+ and H_3^+, which makes detection of tritium in H_2 difficult unless the concentration is high. This leaves m/e = 5 as the

---

[7]http://www.ruf.rice.edu/~bioslabs/methods/radioisotopes/rad1.html
[8]http://en.wikipedia.org/wiki/Liquid_scintillation_counting, http://www.epa.gov/safewater/radionuclides/training/transcripts/tutorial_4.2.pdf

best place to look for the tritium, but only when D_2 is the other hydrogen isotope. However, this m/e can have a small background from DDH^+. Adjustments of the mass spectrometer, too complex to describe here, can be made to improve the sensitivity. Nevertheless, use of the mass spectrometer is limited to samples containing significant tritium.(*23*)

Photographic Film

If a sample containing tritium is laid on photographic film, location of tritium can be identified using what is called an autoradiograph, shown in Fig. 4. Exposure is found not to result from the D_2 gas being desorbed and reacting with the film. Because the beta range is so short, only tritium present within a few microns of the surface can be detected.

2.1.3 How is tritium produced using LENR and at what rate?

Tritium is not often sought because the ability to make such measurements is limited generally to major laboratories. Storms(*1*) in 2007 listed 67 studies reporting meaningful amounts of tritium using electrolysis, gas discharge, and gas loading when deuterium was used. Occasionally, tritium is made even when normal water is used in the electrolyte.

Only a few researchers have studied tritium production since 2007. Most notable is the work of Claytor(*24*), who has continued to study pulsed gas discharge of D_2 using various special alloys of palladium as the cathode in search of the most effective way to make tritium. This work has reached a high level of reproducibility and is expected to be published soon. Afonichev and Galkin(*25*) also used the gas discharge method to make tritium. They cleaned palladium foil in vacuum at 700°C and then subjected it to glow discharge (350 V) in low pressure D_2. External detection of neutrons and gamma did not exceed background. Tritium was detected at $>10^9$ atoms using the scintillation detection method.

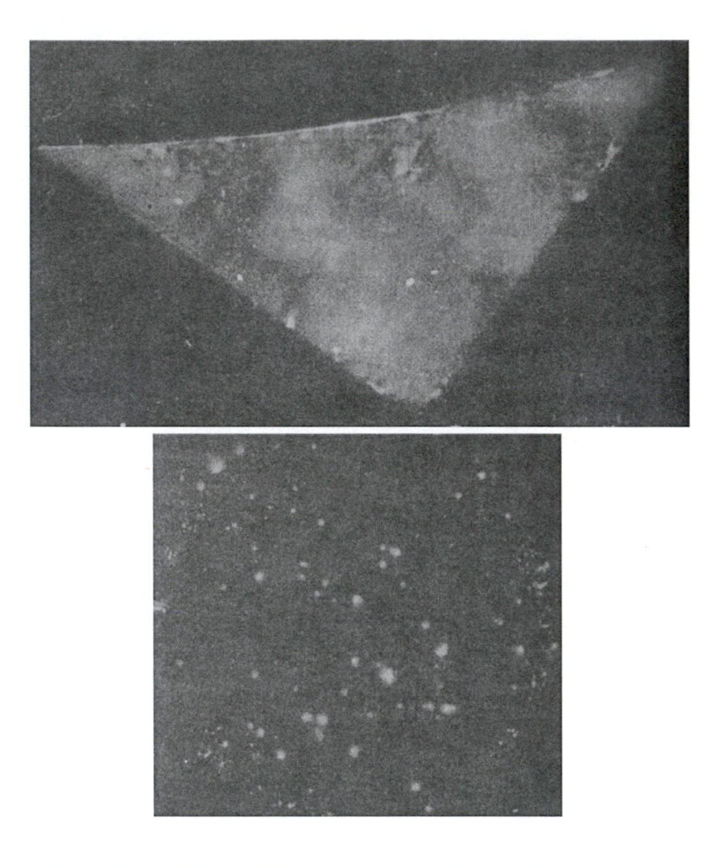

Figure 4. A foil of Pd-Ag alloy (top) was heated in vacuum to 600°C, exposed to D_2 gas for 88 hrs, and placed on photographic film.(*26*) The tritium (light areas) is spread over the surface but distribution is not uniform. A total of $1.5x10^{12}$ tritium atoms are estimated to be in the sample based on calibration using a sample made to contain a known amount of tritium. The longest dimension is about 4 cm. Titanium (bottom) shows the same ability to make tritium but the sites were very local and fewer in number.

Studies by Romodanov et al.(*27-29*) show that tritium production is sensitive to the D/H ratio in the cell. Ironically, people who worked hard to keep their D_2O pure were not rewarded because both D and H must be present to produce tritium. Even when normal water is used, enough deuterium (156 D/10^6 H) may be present to make detectable tritium. As explained in Chapter 5, tritium might also be produced in normal water if p-e-p fusion makes d, which then fuses with p and an electron to make tritium.

Iwamura et al.(*23*) reacted Pd with D_2 to produce D/Pd = 0.66, then vapor coated the metal with gold, and finally extracted D_2 by heating in vacuum. Tritium was found in the gas and a few neutrons were emitted. Coating the Pd with copper showed the same behavior.(*30*) The authors propose the nuclear reactions occurred in the palladium. Once again, the problem is to correctly indentify where the NAE is located. Perhaps the nuclear reaction occurred as deuterium, and perhaps some H, passed through the layer instead of through the Pd as the authors suggest. The amount of tritium, neutrons, and X-rays (21 keV) was related to the initial D/Pd and the degassing rate. This behavior has features perhaps related to fractofusion as the source of detected neutrons and some tritium.

Clarke et al.(*31-33*) studied samples supplied by McKubre et al.(*34*) after their replication of the Arata method.(*35*) This method uses palladium-black contained in a sealed palladium tube that is exposed to high-pressure D_2 gas generated by electrolysis. Accumulated ^4He and tritium were detected using a mass spectrometer. The amount of ^3He and its increase over time were used to determine when tritium had formed in the material. This date corresponded to when the sample was being electrolyzed and initially exposed to D_2.

Two studies stand out for having followed tritium production over time and explored variables that affect the process. In the first study, Bockris and students (Texas A&M)(*21, 36-42*) used an open electrolytic cell[9], the results from which were published in a series of papers that provided the data in Fig. 5. The cell produced no tritium for 40 hours of electrolysis, during which time the cell gained water from the atmosphere through a vent. Increased current started the process and made more tritium each

---

[9] An "open" electrolytic cell does not contain a recombiner catalyst so that the generated gases have to be vented to the atmosphere. This design contrasts to the "closed" cell containing a catalyst used to convert the gases back to D_2O, thereby allowing the cell to be sealed.

time more current was applied. Shaking the cell or adding D_2O caused production to stop and then return after a delay. Apparently, a critical condition present on the cathode surface was temporarily destroyed by violent motion of the electrolyte.

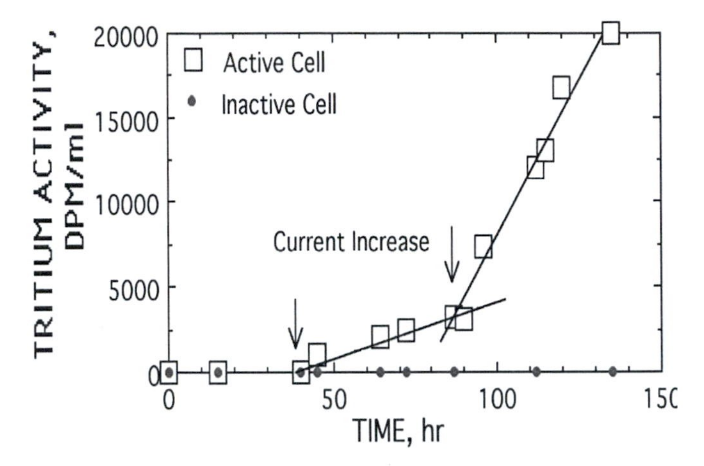

Figure 5. Tritium production in an open electrolytic cell containing D_2O+LiOD and a palladium cathode.(*42*)

After the study, the cathode was covered by fine dendrites of copper supplied by exposed wires. Another cell running at the same time and at the same location produced no tritium, showing absence of an environmental source and the usual typical lack of reproducibility.

In a similar study, Storms and Talcott(*43*) at the Los Alamos National Laboratory (LANL) observed similar behavior while using a cell in which tritium was measured in the electrolyte as well as in generated gas after converting the gas to water. Closed cells were used to prevent uptake of water from the atmosphere. In this case, only 3 days of electrolysis were required to start the process, as shown in Fig. 6. The tritium content increased over 15 days, followed by several bursts and then production stopped while the tritium content experienced a slow decline. Bockris et al. observed a similar slow decline once tritium production stopped. As can be seen in Fig. 6, another cell (#70) being electrolyzed at the same time at the same location using material from the same source showed no change in the tritium content, thus eliminating environmental contamination as an explanation.

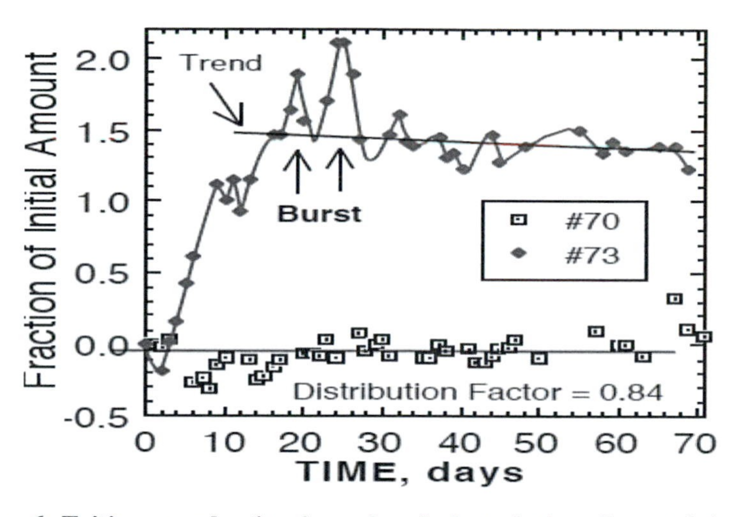

Figure 6. Tritium production in a closed electrolytic cell containing D_2O-LiOD and a palladium cathode. The apparent bursts may result from random scatter in the measurements. However, slow loss of tritium after production stopped is real.(*43*)

During the same study, contamination was explored by placing a cell in an environment known to contain tritium and watching the behavior shown in Fig. 7. In this case, the cell design is similar to one used in Fig. 6. Note that increase in the tritium content starts immediately after the cell is placed in the environment and continued until the concentration inside the cell equaled the concentration in the room. While the rate of uptake will depend on cell design and the amount of tritium in the environment, the general behavior differs significantly from that shown in Fig. 6.

Environmental tritium is easy to detect at levels that cause an obvious effect and is routinely monitored at LANL where such high levels might be possible, thereby providing further support for environmental tritium not being a source of contamination in this case.

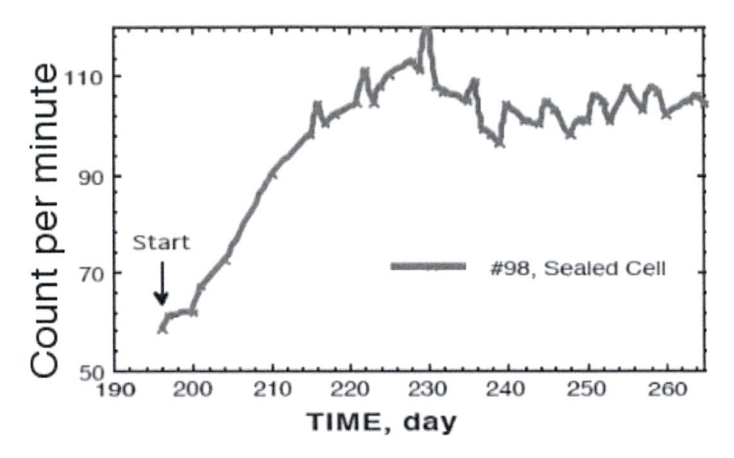

Figure 7. Cell similar to those used at LANL exposed to an environment containing tritium.(*43*) The apparent variations are caused by random scatter in the measurements.

What would happen if tritium had been added on purpose as claimed by Taubes?(*44*) To test the consequence of such action, tritium was added to cell #82, with the result shown in Fig. 8. Note the tritium concentration remained unchanged for 40 days before tritium is added. The addition caused an immediate increase, after which the tritium content again remained constant. Clearly, adding tritium to a cell cannot be mistaken for tritium production from an active cell because the behavior is different.

In addition, a new behavior is revealed. The increase in tritium content is gradual and slowly decreases once formation stops in an active cell. Apparently, a process is able to remove tritium from a cell that previously made tritium at active sites but not from an inactive cell to which tritium is added when active sites are not present.

Reifenschweiler(*45-49*) found tritium to disappear when it is reacted with fine particles of titanium (Ti). Perhaps tritium can be converted to a nonradioactive element by the LENR process by fusion with H or D when a NAE is present. This possibility is discussed in Chapter 5.

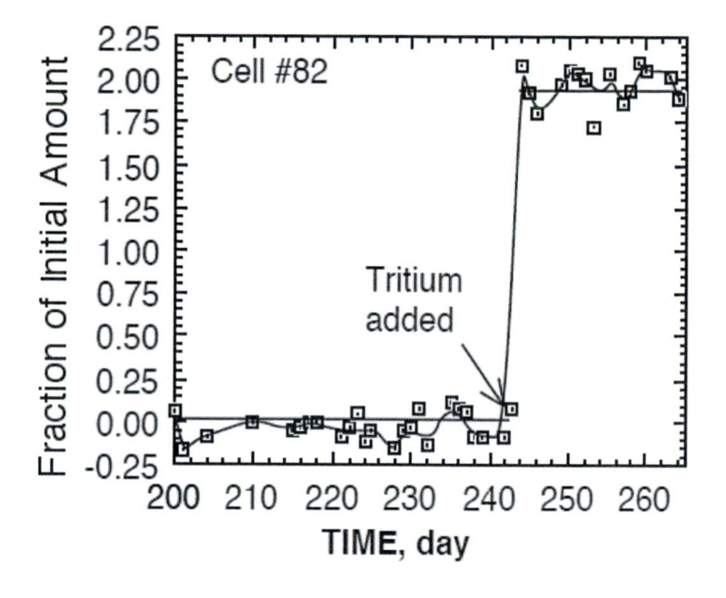

Figure 8. Effect of adding tritium to a closed electrolytic cell.(*50*)

Several other general behaviors of tritium are important. When tritium is made by LENR using electrolysis, it is found mostly in the electrolyte, not in the evolving gas. On the other hand, when a palladium cathode is used to which tritium has been added on purpose, tritium is found mostly in the evolving gas. This behavior is shown in Fig. 9, where the tritium content of the gas gradually increases until the gas contains all tritium previously in the Pd cathode. Very little of this tritium is retained in the electrolyte. This behavior shows that tritium made by LENR takes place only on or very near the surface from which the newly formed tritium ions can rapidly enter the electrolyte before they are able to react with surrounding deuterons and form TD gas. In contrast, tritium dissolved in palladium leaves only as TD gas, which is not able to dissolve in the electrolyte. This behavior also shows that tritium observed during LENR cannot result from tritium contamination that might have been present in the palladium, as some skeptics have suggested.(*51*)

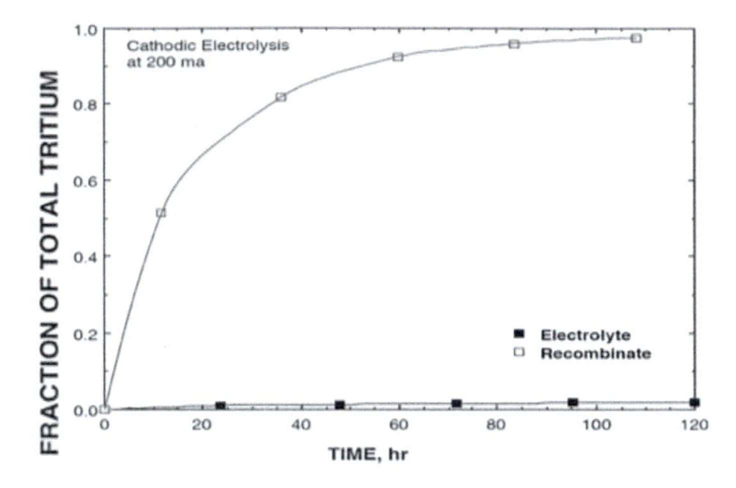

Figure 9. Effect of electrolyzing a cathode of palladium containing tritium. The recombinate results from collecting D_2 and DT gas that evolves during electrolysis and converting the gas to water, which is analyzed for tritium. The plot represents the total amount of tritium collected in the recombinate at each plotted time. The tritium content in the electrolyte was measured at the same time and represents the total tritium contained in the electrolyte up to that time.(*52*)

Conclusions about the effect of the D/H ratio must apply to the actual ratio in the NAE, which will be different from that in the electrolyte. As shown in Figure 10, the site where the nuclear reactions occur will contain a higher H/D ratio than in the electrolyte. This behavior can result in flawed interpretation when the measured H/D content in the electrolyte is used to explain behavior. This unexpected increase in H/D is expected to occur in the surface of any material exposed to a mixture of D and H in any chemical form.

2.1.4 How can tritium be produced by LENR?

It is safe to conclude hot fusion is not the source of tritium because a neutron does not accompany each tritium atom. In fact, although the typical n/T ratio is variable, the average value is near 10^{-6}. A possible source is discussed in Chapter 5. Conventional nuclear reactions known to produce tritium are listed in Table 1 to provide full understanding of how tritium might form.

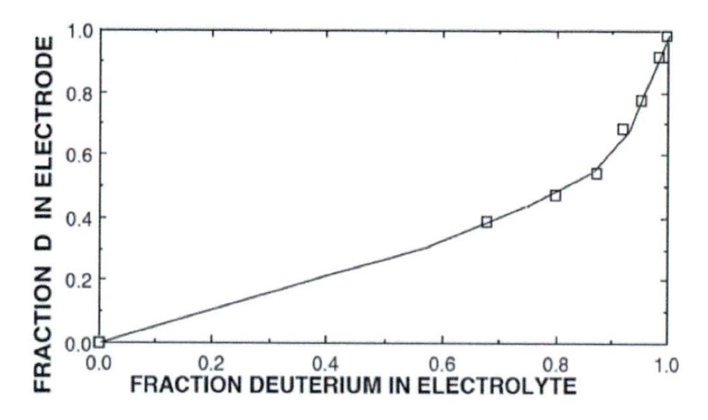

Figure 10. Comparison between the amounts of D in the electrolyte and in a palladium cathode. The hydrogen content of Pd is very sensitive to changes in the H content of the electrolyte when nearly pure D_2O is used.(*52*)

Table 1. Nuclear reactions that can make tritium (^{3}H) and the energy/event (Q).

	Reaction	Q
	$^6Li + n = {}^4He + {}^3H$	Q = 4.8 MeV
	$^{10}B + n = 2{}^4He + {}^3H$	Q = 0.3 MeV
	$d + n = {}^3H$ + photon	Q = 6.3 MeV
(hot fusion)	$d + d = {}^3H + p$	Q = 4.0 MeV
(cold fusion)	$d + p + e = {}^3H$ + photon	Q = 6.0 MeV

If neutrons were available to make tritium, the rate of such reactions would depend on the cross-section for the reaction, the concentration of neutrons, and the target isotope. The cross-section for reaction of ^{7}Li with neutrons as a function of energy is shown in Fig. 11. The cross-section for ^{6}Li is almost identical to ^{7}Li and the cross-section for reaction with ^{10}B is about 10 times greater over the energy range. Because all of these reactions show an increased cross-section as neutron energy is reduced, a corresponding increase in LENR is expected to result, as some theories have suggested. However, for the reaction to occur, the neutron must find a target as it moves through the material. The lower the energy, the slower it will move and the less often it would find a target even though the reaction will be more likely once the target is found. Consequently, neutron energy below a certain low energy would not cause increased reaction rate because the number of neutrons would be reduced by normal beta decay.

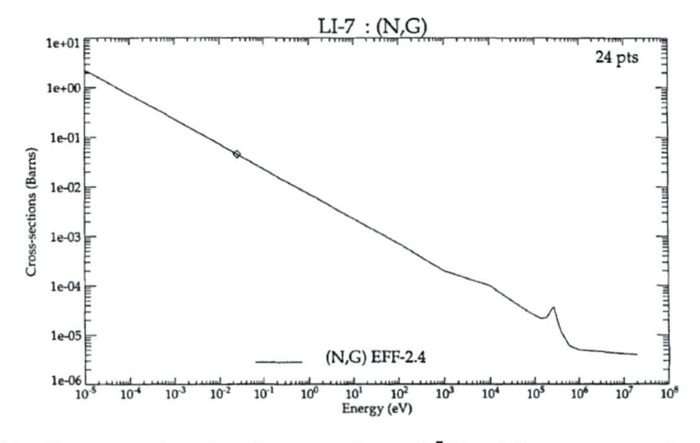

Figure 11. Cross-section for the reaction of ^{7}Li with a neutron having the energy shown.(*53*)

In addition, the thermal energy of each neutron will rapidly come to equilibrium with the thermal energy of atoms in the structure. As a result, no matter how low the initial energy might be, ambient temperature will determine the kinetic energy of the average neutron. These issues are examined in Section 3.2.4.

2.1.5 CONCLUSIONS ABOUT TRITIUM

After electrolysis begins, tritium will take time to appear, suggesting some special condition must form on the cathode surface. Tritium formation requires a critical current density, the production rate is erratic, and production is mainly on the cathode surface. Although neutrons and high-energy radiation are reported when tritium is produced, the rates are small compared to the amount of tritium. Consequently, most tritium does not result from the hot fusion process.

Tritium is sometimes produced when heat and helium are both detected, but not always. The rate of tritium production is always too small to produce detectable thermal power.

Studies using gas discharge indicate the rate of tritium production is sensitive to the D/H ratio in the surrounding gas. Tritium is produced in the absence of lithium or boron, but is produced on occasion when the electrolyte is normal H_2O containing the naturally occurring amount of D_2O. All of this suggests tritium results from fusion of D and H, which requires an electron be included in the nuclear product to avoid formation of ^3He.

2.2.0 HELIUM

Helium has two stable isotopes when the two protons combine with one neutron (^3He) or with two neutrons (^4He). When ^3He is found on rare occasions after LENR, its presence can be attributed to decay of tritium. Nevertheless, other sources of this isotope need to be considered. Several possible sources of ^4He also have to be considered.

Helium-4 is uniform in air at 0.00052% by volume[10] resulting from alpha decay of heavy elements during the life of the earth. Although this amount is small, it provides a potential source of helium when gas samples taken from operating cold fusion cells become contaminated with air. This potential error can be evaluated by measuring argon (Ar) in the generated gas as stand-in for He resulting from an air leak. Careful laboratory procedures generally eliminate this possible source of helium. After all, the investigators are aware of this potential error and take great pains to eliminate this error — just as skeptics are aware of the problem and take pains to point out the error.

Although helium is non-reactive under most conditions, to a small extent it will dissolve in materials and slowly diffuse under

---

[10] http://en.wikipedia.org/wiki/Helium

a concentration gradient like any other element.(*54, 55*) However, it can become trapped in a material when the diffusing atoms encounter a grain boundary where helium gas bubbles can form.(*54-58*) Once in the gas phase, the helium atoms cannot easily re-enter the structure. Removing this gas requires heating the material near its melting point. McKubre et al.(*59*) removed helium from a PdD-carbon catalyst by repeatedly reacting and removing D_2 in order to slowly sweep contained helium out of the material. This method of removal might result as changes in deuterium content cause expansion and contraction of cracks at the surface where helium is trapped, thus allowing its escape.

2.2.1 How is helium measured?

Although no recent attempts to measure helium have been published, the following summary repeats observations reported in the review published in 2010 by Storms.(*60*) Like tritium, helium in excess of a small background can only result from a nuclear reaction and the He/energy ratio can be used to identify the source.

Helium is measured using a mass spectrometer to separate $^4He^+$ from D_2^+. Because the atomic weights of $^4He^+$ (4.002602) and D_2^+ (4.028202) are very similar, the resolution required to separate these two ions has to be high, but easily accomplished using instruments available from several suppliers.[11] Additional sensitivity can be achieved by using chemicals to remove D_2 while leaving the inert helium unaffected. Although identification of helium at low concentration in a gas is relatively easy, determining its concentration in D_2 can be a challenge requiring careful calibration. This challenge is increased because helium can be lost from the gas by absorption on materials and then become a source of unexpected helium when it later desorbs.

2.2.2 Methods found to produce helium in LENR

Helium can result from several nuclear reactions as shown in Table 2. Indeed, some of these processes might take place at a

---

[11] For example, see http://www.mksinst.com/product/category.aspx?CategoryID=91

small rate along with the more active D+D fusion reaction. For example, Passell(*61*) suggested fusion of D with a small amount of ^{10}B might produce three ^{4}He rather than the expected ^{12}C (6.0MeV/He). He based this suggestion on the reduction in ^{10}B content in a Pd cathode after energy was detected during electrolysis. Such possible reactions are not likely to be a general source of helium because the required targets are not always present when heat is detected. Nevertheless, the measured amount of energy required to produce each helium atom listed in the table can be used to help identify the major source of heat and helium. The measurements are given as [He atoms]/[watt-sec] so that all values can be compared using the same units, remembering one Joule (J) equals one watt-sec.

Table 2. Possible nuclear reactions that produce helium.

Reaction	MeV/He	He atoms/J
$D+D={}^{4}He+energy$	23.8	2.6e11
$4D={}^{8}Be=2{}^{4}He$	23.8	2.6e11
$D+T=neutron+{}^{4}He$	17.5	3.6e11
$D+{}^{6}Li={}^{8}Be=2{}^{4}He$	11.2	5.6e11
$H+{}^{7}Li={}^{8}Be=2{}^{4}He$	8.4	7.4e11
$2H+{}^{7}Li={}^{9}B=2{}^{4}He+p$	8.4	7.4e11
$n+{}^{7}Li={}^{8}Li={}^{8}Be+beta$ (13 MeV)$=2{}^{4}He$	13.4	4.7e11
$n+{}^{6}Li=T+{}^{4}He$	4.3	14.5e11
Transmutation	3-6? from alpha	

From Storms(*1*)

Because the errors in measurement of heat and helium are independent of each other and sensitive to different variables, the probability that the He/energy ratio would fall near the expected value for D-D fusion by chance is very small, although impossible to calculate without more information. Nevertheless, the observations listed in this section clearly demonstrate a correlation between heat and helium when LENR occurs.

The basic question is whether the heat and helium result from a single nuclear reaction claimed to be D-D fusion or a chance combination of different reactions. This chance combination requires one reaction make energy without helium and another make helium, but produce less energy compared to D-D fusion. The two reactions would then have to combine with exactly the correct rates to produce a He/energy ratio equal to that produced by D-D fusion. This process has been suggested by Krivit(*62*) and challenged by Storms(*63*) in a recent exchange of short papers.

Krivit asks whether the observations can be interpreted to support a claim for reactions other than D-D fusion as the source of energy and helium, as proposed by Widom-Larsen.(*64*) An answer can be sought by asking two questions.

1. Are other nuclear reactions possible as a source of energy and/or helium besides D-D fusion?
2. Is the measured He/energy ratio sufficiently uncertain to allow an explanation other than D-D fusion?

An answer to question 1 starts by identifying the detected products. Besides helium, only tritium and transmutation products are normally detected. The energy obtained from tritium production can be calculated based on a generous evaluation of the few reported rates of production, which is estimated to be less than 10^{10} atoms/sec. If the maximum possible energy/reaction were assumed to be 6 MeV/tritium (Table 1), the maximum amount of power from this reaction would be 10^{-2} watts. Clearly, heating power generated by tritium production would be trivial.

The only other source of energy comes from transmutation. Calculation of power from transmutation is more complicated because many reactions are observed, the rates are seldom determined, and transmutation is not always observed when heat is measured. Transmutation of lithium is the most obvious possibility, which Widom and Larsen(*65*) suggest is the source of heat and helium as a result of reaction with neutrons. However, because this reaction produces too little energy to be consistent with measurement, it must be combined with an independent source of energy to produce the observed He/energy ratio. To accomplish this goal, neutrons are also proposed to react with other elements in the material to supply the missing energy. As discussed in Chapter 3, this proposed reaction would not explain observed heat and helium production in the frequent absence of lithium(*35, 66, 67*) and a role for neutrons in the process cannot be justified because free neutrons are not detected directly nor are their predicted reaction products detected. Consequently, this proposed explanation is not consistent with what is observed.

Krivit(*62*) goes on to cite the analytical studies of Passell(*61, 68*) and claims these studies show that transmutation could account for some observed power. Passell examined palladium-black power that had been exposed to D_2 gas by Arata(*69*), followed by production of significant energy and helium gas. The ^{110}Pd/^{108}Pd ratio was found to have increased slightly (1% to 8%), the implication being that either the ^{108}Pd experienced addition of two neutrons to produce ^{110}Pd or d was added, converting it to radioactive ^{110}Ag, which decayed to form stable ^{111}Cd. If dineutrons were added, each event would have to dissipate 14.97 MeV, presumably by photon emission. If the entire sample of about 3 gm were equally transmuted[12], about 42 MJ would be produced. Reaction with d would produce slightly more energy. Tens of MJ were claimed. Consequently, the energy matches the measurement within the uncertainty of the assumptions so that the conclusion must rest on explaining how the dineutrons were made in the first place, why they react only with ^{108}Pd, and how energy is dissipated without detection.

The concentrations of ^{109}Ag, ^{59}Co, and ^{64}Zn also were found significantly increased over the untreated palladium. This result is difficult to explain unless these elements were to result from fission after a deuteron was added to Pd. Bush and Lagowski(*70*)

---

[12] This assumption is very implausible but made to give an extreme value for the energy.

also found less ^{108}Pd as a result of LENR. These possible transmutation reactions are discussed in Section 2.3.2.

In conclusion, a plausible source of helium other than D-D fusion and a matching source of additional energy have not been identified. Nevertheless, changes in isotopic ratio and identification of transmutation products indicate that other nuclear reactions take place at low rates along with D-D fusion. Sorting this out will challenge graduate students for years.

Question 2 must be answered by judgment. If several independent sources of energy and helium were operating, would these reactions always combine to produce the observed values for the He/energy ratio even in absence of lithium? The reader has to decide how likely such a combination would match the measured values. In addition, many studies demonstrate correlation between no heat and no helium. If energy and helium were not produced by the same nuclear reaction, occasional production of energy without helium would be expected. Only one sample studied by Miles(*71*), consisting of Pd-Ce, is reported to produce heat without detected helium. Without measurement of helium captured in the alloy, this result is impossible to interpret correctly.

Rather than examining every potential error in individual measurements of helium, a more productive approach is to identify patterns shown by many measurements. If many studies show the same behavior, error can be ruled out. The various studies reporting correlation between heat and helium are listed below in approximate chronological order.

Wendt and Irion (1922)(*72*) claimed helium could be produced from an exploding tungsten wire in vacuum using electric discharge. Helium was detected using the spectroscopic method.

Paneth and Peters (1926)(*73*) placed hydrogen in contact with various metals, including palladium on asbestos, and exposed the material to electric discharge. Helium was produced when palladium was used and was detected using the spectroscopic method. The claim was retracted in 1927.(*74*)

Albagli et al. (1990)(*75*) (MIT) electrolyzed Pd in D_2O+Li and reported no heat and no helium.

Alessandrello et al. (1990)(*76*) (INFN) electrolyzed Pd in D_2O with external stress applied. No heat and no helium were detected,

Brudanin et al. (1990)(*77*) (USSR) electrolyzed Pd in D_2O+Na_2CO_3 and found no heat and no energetic alpha emission.

Matsuda et al. (1990)(*78*) electrolyzed Ti in D_2O+Li and found no heat and no helium.

Morrey et al. (1990)(*79*) analyzed five samples of Pd provided by Fleischmann and Pons (F-P) for helium using the double blind method involving six laboratories. Helium was found within 25 μm of the surface of small palladium rods. The active sample made 10.5-16.8 kJ and produced 0.52±0.33 ng He/cm^3 of Pd, which would be about $6x10^9$ atoms of He for each watt-sec of energy. Although this amount is far below the amount expected from D+D fusion probably because most helium was lost to the electrolyte, the presence of extra helium in the Pd cathode is clearly demonstrated. No ^{3}He was detected, which also rules out the presence of significant tritium in the sample supplied by F-P.

The effect of helium loss from a cathode surface into the gas was explored by implanting some samples with helium to a maximum depth of 1.2 μm. Because this helium was not removed by electrolysis, the authors assumed no helium was lost to the gas when heat and helium were made by LENR. Since it is now known that a large fraction of the helium does appear in the gas after LENR, this conclusion needs to be explained. Two explanations are plausible: either most helium made by LENR is produced closer to the surface than about 1 μm, or the cracks present when LENR is initiated provide a path for loss from a deeper location than is available when He is applied by bombardment.

Bush et al. (1991-1993)(*58, 80-84*) electrolyzed Pd in D_2O+Li using a glass system and captured the evolving gas. The D_2 was removed from the gas and the amount of He was determined using a mass spectrometer. The results are listed in Table 3 for samples electrolyzed in D_2O that gave energy and helium and samples electrolyzed in H_2O that produced neither. Radiation was detected from the cell using X-ray film while energy was made. Following this quantitative study, the method was improved and provided three more values listed in Table 4.

Table 3. Pd electrolyzed in D_2O and H_2O.

Sample	P_{ex}/W	$\Delta H_{out}/\Delta H_{in}$	Results [a]
(1) 12/14/90 A	0.52 [b]	1.20/1 [b]	^{4}He observed as large peak, long dwell; no ^{3}He [b]
(2) 05/05/75 B	0.46	1.27/1	^{4}He observed as large peak, long dwell [c]
(3) 11/25/90 B	0.36	1.15/1	^{4}He observed as large peak, long dwell; no ^{3}He
(4) 11/14/79 B	0.17	1.12/1	^{4}He observed at detection limit; no ^{3}He
(5) 04/29/65 A	0.24	1.10/1	^{4}He observed medium peak, some dwell; no ^{3}He
(6) 11/27/90 A	0.22	1.09/1	^{4}He observed as large peak, long dwell [c]
(7) 03/26/69 A	0.14	1.08/1	^{4}He observed at detection limit; no ^{3}He
(8) 01/18/37 A	0.07	1.03/1	No ^{4}He or ^{3}He observed
(9) 12/17/90 B	0.29 [d]	1.11/1 [d]	No ^{4}He or ^{3}He observed [d]

[a] Mass spectrometer, always at highest sensitivity.
[b] Current was 660 mA, all other experiments used 528 mA.
[c] No measurement of ^{3}He was made.
[d] The D_2O solution level of the cell was found to be excessively low resulting in an erroneou calorimetric result.

H_2O + LiOH electrolysis. Checking for ^{4}He in effluent gas

Sample	Results [a]
(1) 1/9/91 A-2	No ^{4}He or ^{3}He observed
(2) 1/16/91 A	No ^{4}He or ^{3}He observed
(3) 1/16/91 AA	No ^{4}He or ^{3}He observed
(4) 1/16/91 B	No ^{4}He or ^{3}He observed
(5) 1/17/91 A	No ^{4}He or ^{3}He observed
(6) 1/17/91 B	No ^{4}He or ^{3}He observed

[a] Mass spectrometer, always at highest sensitivity; any gas passing though the cryofilter was allowed time to accumulate and then surged into the mass spectrometer.

In response to critiques(*85-90*), Miles repeated the measurement using an apparatus made of stainless steel (see Table 6 from Miles et al. 1994) and even measured the diffusion rate of helium through glass(*82*) to further support the results when glass was used. Helium resulting from diffusion through glass was found to introduce only a negligible error.

Liaw et al. (1990-1993)(*91-96*) used the unique method of fused salt electrolysis. The electrolyte was molten KCl-LiCl-LiD heated to about 460°C containing a torch-melted Pd anode and an aluminum cathode. In this case, Li^+ goes to the cathode and D^- goes to the anode. Consequently, Li is not available to support a nuclear process with D, yet significant extra energy and helium were produced. The palladium was analyzed for contained ^4He and ^3He subsequent to heat production. Samples exposed to D showed excess helium and heat while those exposed to H showed neither. No sample contained excess ^3He. Less ^4He was found in the palladium than expected based on measured energy, presumably because most left the cathode with the D_2 gas, which was not measured.

Zywocinski et al. (1991)(*97*) electrolyzed a tube of Pd-Ag alloy in D_2O-Li. No ^4He was detected when the tube was subsequently heated to 870K. Gas was collected from the interior of the tube and no helium was detected. Unfortunately, this method would not be expected to release He even if some had been produced.

Bockris and students (1992)(*41, 42*) produced tritium and helium by electrolyzing Pd in D_2O+Li. Although a search for extra energy was not made, simultaneous production of tritium along with helium demonstrates that these two nuclear processes occasionally can occur at the same time in the same sample.

Table 4. Additional samples from Miles et al.(*71*)

Sample (Date)	P_{EX} (W)	P_{out}/P_{in} (X)	$^4He^a$ (Atoms/500 mL)	$^4He/sW^b$
12/30/91-B (Flask 5)	0.100[c]	1.08	1.34×10^{14}	2×10^{11}
12/30/91-A (Flask 3)	0.050[c]	1.02	1.05×10^{14}	2×10^{11}
01/03/92-B (Flask 4)	0.020[d]	1.01	0.97×10^{14}	5×10^{11}

a. Error range reported by Rockwell International was $\pm 0.01 \times 10^{14}$ atoms/500 mL($\pm 1\ \sigma$).
b. Corrected for a background level of 5.1×10^{13} atoms/500 mL.
c. I=525 mA. An anomalous GM count (27 σ) was measured during this period.
d. I=500 mA. The GM count rate was within the normal range.

Sakaguchi et al. (1992)(*98*) reacted $LiNi_5$ with either D_2 or H_2 and looked for ^3He and ^4He. Extra ^3He was produced only when D_2 was used along with perhaps some extra ^4He, but the study is incomplete. This study is worth repeating.

Yamaguchi and Nishioka (1992)(*99-101*) studied a sample of Pd with MnO_x deposited on one side and gold deposited on the other. The sample was reacted first with D_2 and then heated in vacuum while gas evolution was studied. They concluded that helium release was correlated with production of extra energy and "^4He production decreases with the amount of D_2 loaded in the Pd." Tritium production was also claimed. Neither ^4He nor tritium was found when H_2 was used even though excess energy was produced along with a gas having mass of 3. This ion peak might result from HD^+ rather than ^3He as suggested by the authors, with the D resulting from the proposed p-e-p = d nuclear reaction, as discussed in Chapter 5.

Zhang et al. (1992)(*102*) measured the negative ion spectrum resulting from applying SIMS to the surface of Ti after electrolysis in D_2O. Excess heat was observed along with ^4He. No ^4He was observed in the absence of excess heat.

Gozzi et al. (1993)(*103-107*) electrolyzed Pd rods in D_2O+Li and measured helium production in the evolving gas. Fig. 12 shows a delay in helium release after energy is produced in the case of one active sample. The amount of helium detected is somewhat larger than expected from the measured power, presumably because a slight air leak occurred before 650 hr, as the authors suggest. The sudden bursts of helium following energy production could not be explained this way. Tritium was detected but it did not occur in the same cells that produced helium. Unfortunately, the relative D and H concentrations in the cells were not reported.

Aoki et al. (1994)(*108*) Flow calorimetry was used to measure power produced by Pd electrolyzed in D_2O+Li and helium was measured in the gas using a gas chromatograph. Power of 4.8±0.5 watts produced $0.5\pm2.1 \times 10^{10}$ He atoms/sec, which gives 1.04×10^9 He/watt-sec. Tritium was also produced.

Miles et al. (1994)(*109*) electrolyzed Pd in D_2O+Li, first in 1993, and later collected the gas in stainless steel flasks to avoid helium from the air. This second study was made in 1994. Table 5(*110*) lists samples from both studies that did not make heat from which the helium background of 4.5±0.5 ppb can be determined. Table 6 lists the helium in excess of background when heat was made.

Botta et al. (1995)(*111, 112*) used loading to achieve an expected composition of D/Pd = 0.67. Current was caused to flow through the Pd to produce a local increase in concentration by electromigration.[13] Evidence for increased loading and heat

---

[13] Electromigration occurs when a current is passed through a material containing ions. The ions are caused to diffuse preferentially in response to the voltage applied, with positive ions moving to the negative electrode and

production near the cathode was reported. After running for 500 hr, the total amount of helium detected was 7.5×10^{16} atoms. The calorimeter was too insensitive to detect power resulting from so little helium.

Table 5. List of samples that produced neither energy nor helium.

The listed helium is background.

Electrode	Flask/cell, date	$^4He^a$, ppb	^4He, atoms/ 500 mL
Pd Rod[b] (4 mm x 1.6 cm)	1/C (2/24/93)	4.8 ±1.1	5.5×10^{13}
Pd-Ag Rod[b] (4 mm x 1.6 cm)	2/D (2/24/93	4.6 ±1.1	5.2×10^{13}
Pd Rod[b] (4 mm x 1.6 cm)	3/C (2/28/93)	4.9 ±1.1	5.6×10^{13}
Pd-Ag Rod[b] (4 mm x 1.6 cm)	4/D (2/28/93)	3.4 ±1.1	3.9×10^{13}
Pd Rod[c] (1 mm x 1.5 cm)	3/C (7/7/93)	4.5 ±1.5	5.1×10^{13}
Pd Rod[d] (4.1 mm x 1.9 cm)	3/D (3/30/94)	4.6 ±1.4	5.2×10^{13}
(Mean)		4.5 ±0.5	($5.1 \pm 0.6 \times 10^{13}$)

[a] Helium analysis by U.S. Bureau of Mines, Amarillo, Texas.
[b] D_2O + LiOD (I = 500 mA).
[c] H_2O + LiOH (I = 500 mA).
[d] D_2O + LiOD (I = 600 mA).

Figure 13 shows another later study where measured helium is compared to energy and plotted relative to the amount expected based on 23.8 MeV/He, shown as unity on the graph. The values before 220 hr are proposed to be excessively high because helium had not been fully flushed out of the system. Values after 220 hr show less helium than expected, presumably because the Pd cathode retained some helium. Radiation was measured using X-ray film, which showed an image of the cathode, and was assigned an average energy value of 89 keV.(*115*)

---

negative ions moving to the positive electrode. In the case of metal containing hydrogen, this atom moves toward the negative electrode where it accumulates. This process has been named the Coehn-Aharonov Effect(*113,114*) when hydrogen in palladium is subjected to electromigration.

Table 6. List of samples that produced both energy and helium.

Electrode	Flask/cell, date	$^4He^a$, ppb	Px, W	$^4He/s \cdot W^b$
Pd Sheet[c] (1.0 mm x 3.2 cm x 1.6 cm)	3/A (5/21/93)	9.0 ±1.1	0.055	1.6×10^{11}
Pd Rod[c] (1 mm x 2.0 cm)	4/B (5/21/93)	9.7 ±1.1	0.040	2.5×10^{11}
Pd Rod[c] (1 mm x 1.5 cm)	1/C (5/30/93)	7.4 ±1.1	0.040	1.4×10^{11}
Pd Rod[c] (2 mm x 1.2 cm)	2/D (5/30/93)	6.7 ±1.1	0.060	7.0×10^{10}
Pd Rod[d] (4 mm x 2.3 cm)	1/A (7/7/93)	5.4 ±1.5	0.030	7.5×10^{10}
Pd Rod[d] (6.35 mm x 2.1 cm)	2/A (9/13/94)	7.9 ±1.7	0.070	1.2×10^{11}
Pd-B Rod[d] (6 mm x 2.0 cm)	3/B (9/13/94)	9.4 ±1.8	0.120	1.0×10^{11}

[a] Helium analysis by U.S. Bureau of Mines, Amarillo, Texas.
[b] Corrected for background helium level of 5.1×10^{13} ^4He/500 mL.
[c] D_2O + LiOD (I = 400 mA).
[d] D_2O + LiOD (I = 500 mA).

Tables are from Miles et al. (1996) (*110*)

Arata and Zhang (1997)(*35, 66, 69, 116-123*) studied finely divided Pd-black contained in a Pd tube into which very pure D_2 gas was introduced by electrolyzing the tube as the cathode. They found the amount of helium in the contained gas increased as excess energy increased. A value for energy/He was not reported.

Bush and Lagowski (1998)(*124*) electrolyzed Pd rods in D_2O+Li+200 ppm Al, measured the energy using a Seebeck calorimeter, and collected the gas in stainless steel flasks. They detected helium when heat was produced in 8 samples and found no helium when no heat was detected six times. Three samples used to obtain values for He/watt-sec are listed in Table 7.

Qiao et al. (1998)(*125*) Palladium was reacted with D_2 gas in the presence of a hot tungsten filament.(*126, 127*) Helium was detected in the gas after excess energy was measured.

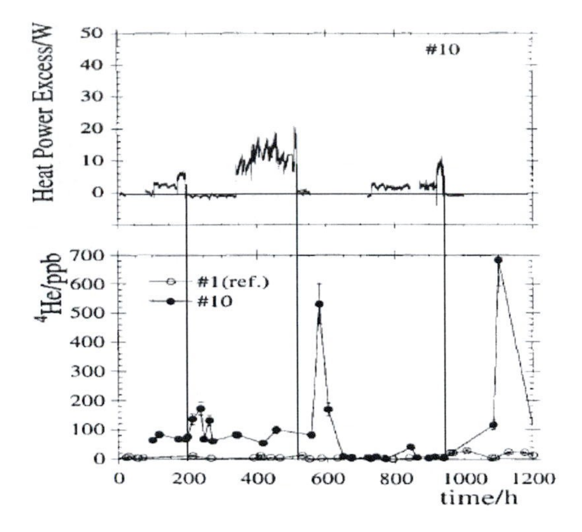

Figure 12. Comparison between excess power and helium concentration in the evolving gas. Note the delay in release of helium following production of excess energy. Helium concentration is compared to a dead cell running at the same time.(*103-107*)

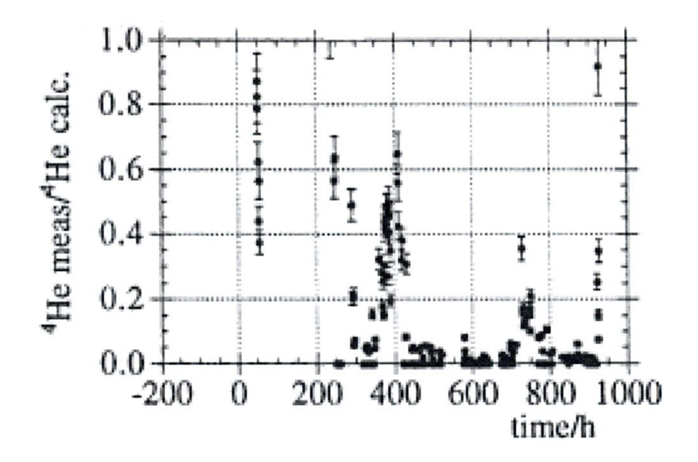

Figure 13. Measured helium compared to the amount of helium expected from the measured energy if the D+D reaction were the source. Helium produced at times between the periodic measurements was estimated and added. A value of 1.0 on the figure would result if all the helium produced by the reaction were measured.(*111, 112*)

Table 7. Excess heat and helium/energy for samples measured by Bush and Lagowski.

Excess (watt)	He atoms/watt-sec
0.047	$1.7x10^{11}$
0.035	$1.3x10^{11}$
0.055	$1.6x10^{11}$

Takahashi (1998)(*128*) found significant helium in four samples after heat was made using electrolysis of Pd in D_2O+Li. The amount of energy was not reported.

McKubre et al. (1999)(*34, 129, 130*) measured heat and helium produced by heating a catalyst supplied by L. Case in D_2 gas. The catalyst consisted of finely divided Pd deposited on coconut charcoal. The helium accumulated over time to give the behavior shown in Fig. 14. An average value of $2.0 \pm 0.8x10^{11}$ He/watt-sec is calculated from the slope of the lines through the measurements, which is consistent with values obtained using electrolysis. Repeated loading and deloading with D_2 was able to coax the remaining He from the sample to give a total amount of helium very near the amount expected from D+D fusion. As a result, this work shows independent confirmation for D-D fusion based on a comparison between the electrolytic and gas loading methods. Lithium cannot be a source of helium and heat because it is not expected to be present in the material.

Where in this complex structure the helium forms is not known. Although the Pd particles are assumed to be the NAE, as a result of their very small size, spill-over deuterium might also find active sites within the carbon structure.[14]

---

[14] Dried G75E catalyst provided by United Catalyst to Case consisted of highly fractured pieces of charcoal containing the following elements with the indicated atomic percent based on EDX analysis (Storms): C-95.1; O-5.2; Pd-0.5; K-0.4; Cl-0.4; Ni-0.12; Al-0.2; P-0.03; Si-0.02; S-0.02; with trace Fe and Na.

Isobe et al. (2000)(*131, 132*) electrolyzed Pd in D_2O+Li and measured helium in the gas and subsequently in the Pd cathode, as listed in Table 8. Although the study looks very well done, the relationship between excess energy and helium production is ambiguous, perhaps because the calorimeter was too insensitive to detect the small amount of energy produced.

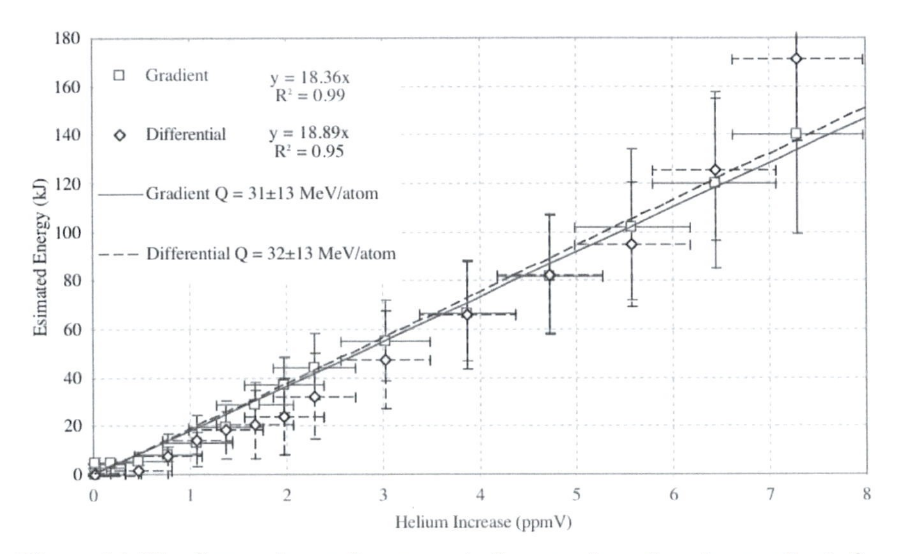

Figure 14. The figure shows the amount of energy based on two methods for its determination as a function of accumulated helium from Pd deposited on carbon.(*133*)

Miles et al. (2000)(*134*) produced excess energy in eight samples out of nine by electrolyzing a Pd-B alloy (0.5 wt. % B) in D_2O+Li.(*135*) One sample was found to produce 1.0×10^{11} He/watt-sec after generating an average excess power of 0.1 W.

Karabut and Karabut (2000) detected helium in samples of Pd subjected to glow discharge in D_2. The work is summarized recently(*136*) to show an increase in ^{4}He of up to 100 times and an increase in ^{3}He up to 10 times after excess energy was made. Tritium, a possible source of ^{3}He, is not reported.

DeNinno et al. (2000)(*137-140*) showed correlation between excess power and helium production, as seen in Fig. 15, but they could not obtain an accurate measure of excess energy. Consequently, a value of He/MeV cannot be calculated from the measurements. They report making about $7x10^{14}$ atoms of He during the study while about 20 mW of excess power was detected, which should have been about 10 times larger to be consistent with D+D fusion as the source.

Table 8. List of samples examined for heat and helium production by Isobe et al.

Exp. # (Duration)	D/Pd (maximum)	^{4}He detection: Inside the cell	^{4}He detection: Inside the cathode[a]	Neutron	Excess Heat
1 (163 h)	0.47	No	No	No	≦1.5 W
2 (201 h)	0.85	No	*Yes* (3.7×10^{14} atoms)	No	*2.6 W* (max.)
3 (264 h)	0.83	No	No	No	≦1.5 W
4 (167 h)	0.85	No	*Yes* (1.1×10^{15} atoms)	No	≦1.5 W
5 (243 h)	0.93	*Yes* (4.6×10^{16} atoms)	*Yes* (8.1×10^{16} atoms)	Not measured	≦1.5 W
6 (255 h)	0.96	*Yes* (3.3×10^{15} atoms)	*Yes* (8.8×10^{14} atoms)	Not measured	≦1.5 W
7 (740 h)	0.85	No	No	Not measured	≦ 1.5 W
8 (111 h)	0.87	*Yes* (2.2×10^{15} atoms)	No	No	≦1.5 W

a) Analysis of gases released from electrolyzed Pd cathodes under heating up conditions.
b) S.U.: Step-up mode L.H.: Low-high mode S.T.: Saw-tooth mode C.C.: Constant-current mode.

Stringham (2000)(*141-143*) used a novel sonic method to inject deuterium into palladium in contact with pure D_2O. The process is claimed to produce 60 watts of excess power for 19 hrs and to generate 10^{18} atoms of ^4He, which is equal to $2.4x10^{11}$ He/watt-sec, in good agreement with the other studies.

Apicella et al. (2005)(*144*) electrolyzed Pd in D_2O+Li while laser light is applied. Three experiments produced helium and heat to give the results shown in Fig. 16. The three results give $1.3x10^{11}$,

$4.4x10^{11}$, and $1.5x10^{11}$ He/watt-sec after the background of $0.6x10^{16}$ He atoms is subtracted.

Arata et al. (2009)(*145*) exposed an alloy of Zr-Ni-Pd to D_2 gas and measured excess energy and helium that had accumulated in the gas. Fig. 17 compares the amount of helium (based on measured current in the mass spectrometer) to the amount of energy. Absence of calibration allows only a qualitative correlation between energy and helium.

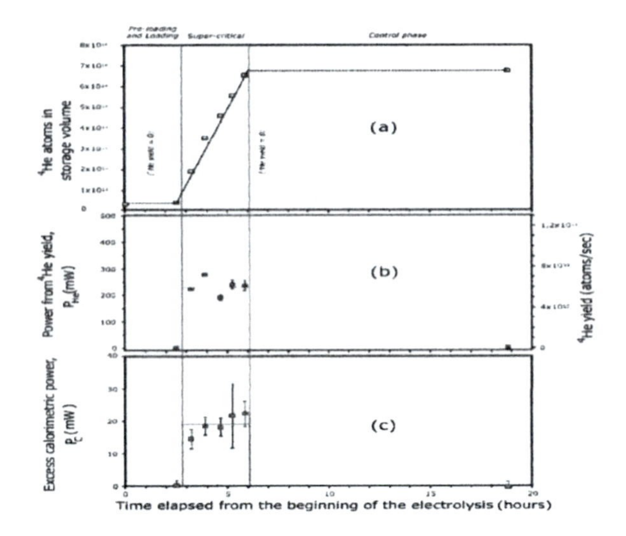

Figure 15. Relationship between measured excess power and helium production as a function of time as reported by DeNinno et al.(*139*) "Power from ^{4}He yield" on the figure is calculated based on expected power from the helium collected as if it had resulted from D-D fusion.(*137-140*)

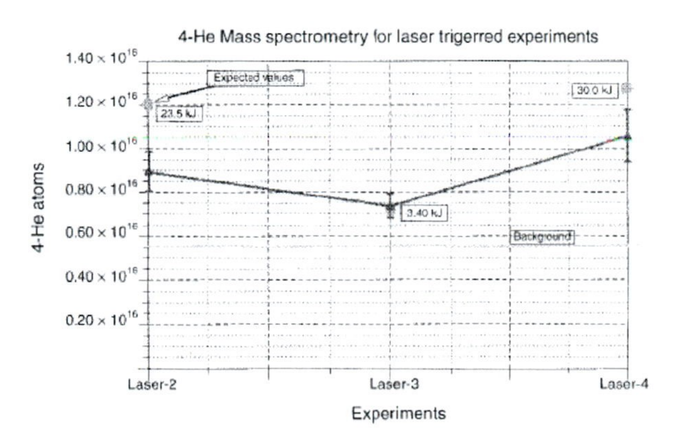

Figure 16. Three studies during which laser light was applied during electrolyses are compared to the helium and energy detected.(*144*)

2.2.4 CONCLUSIONS ABOUT HELIUM

The observations are all consistent: helium is detected when energy is produced and not found when extra energy is absent. Clearly, helium and heat are correlated; thereby demonstrating a nuclear process produces helium and energy. An average of 18 values based on six independent studies for He atoms/watt-sec result in $1.9 \pm 1.1 \times 10^{11}$ (Table 9). Note that 2.63×10^{11} He atoms/watt-sec would result if D-D fusion had occurred.

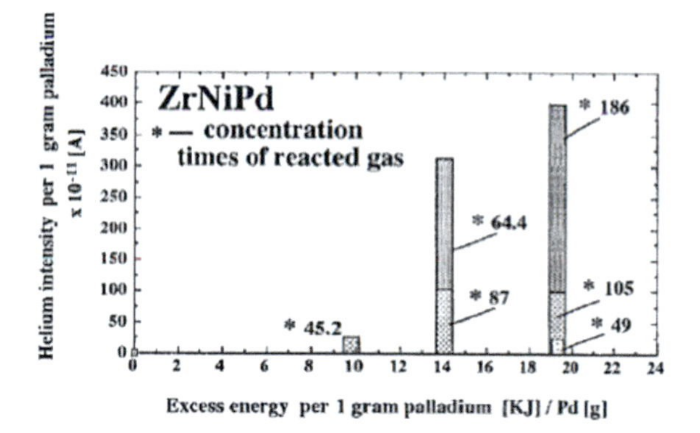

Figure 17. Comparison between current produced by the helium peak in a mass spectrometer and the energy produced. The interval (hrs) during which helium was collected is shown.(*145*)

Table 9. Summary of values for He/watt-sec.

Source	Power, W	He/watt-sec $x10^{-11}$	Method
Miles et al. (*71*)	0.1	2	electrolytic
	0.05	2	electrolytic
	0.02	5	electrolytic
(*110*)	0.055	1.6	electrolytic
	0.04	2.5	electrolytic
	0.04	1.4	electrolytic
	0.06	0.7	electrolytic
	0.03	0.75	electrolytic
	0.07	1.2	electrolytic
	0.12	1.0	electrolytic
(*124*)	0.047	1.7	electrolytic
	0.035	1.3	electrolytic
	0.055	1.6	electrolytic
(*133*)		2.0	gas loading
(*134*)	0.1	1.0	electrolytic
(*144*)		1.3	electrolytic
		4.4	Omitted as outlier
		1.5	electrolytic
(*141*)	60	2.4	sonic
AVERAGE		1.9	
STDEV		1.1	

Almost all reported values for helium measured in the gas are less than that expected based on 23.8 MeV/He because some helium is always retained by the palladium, with the amount being related to the nature of the material and its treatment. Perhaps helium is produced at different depths in different metals and alloys, resulting in the amount released to the gas being highly variable. Future studies need to measure helium in both the gas and metal. All of the expected helium could be removed from fine particles of palladium deposited on charcoal by repeated removal and addition of deuterium, which is a method worth exploring when bulk Pd is used. In the case of the sample using Pd on charcoal, unknown is whether helium is produced and retained in the Pd particles or in the carbon.

No correlation between heat production and presence of lithium has been found.

The helium must result from reactions within a few microns of the surface for any to be found in the gas. Therefore, the heat+helium producing reaction does not take place in bulk material. This means the LENR process must be sought in material existing in a thin and very impure surface layer, not in the majority of the chemical structure. As noted in Section 2.5.1, tritium is produced in this same region. This observation is essential to understanding and identifying the correct NAE.

2.3.0 TRANSMUTATION

Although hydrogen fusion and transmutation both produce a heavier nucleus by combining two smaller nuclei, the process is different. Fusion is normally used to describe the result of combining two nuclei of equal atomic number, usually hydrogen isotopes. Transmutation is used to describe the result of adding a lighter nucleus, usually hydrogen, to a heavier target. Both reactions share the problem of overcoming a Coulomb barrier and dissipating excess mass-energy from a single nuclear product.

Two different kinds of transmutation products are observed after one or more hydrogen nuclei are added to the target nuclei. One results in a single heavier nuclear product and another results in fragmentation of that product. A third kind is not discussed here because it apparently produces nuclei having much more mass than can be explained by the first two kinds. This issue is discussed in Sections 2.3.2, 3.2.6, and 5.4.0. Evidence for the various kinds of transmutation has been summarized(*146-155*) many times but with very little understanding of how the three kinds are related and can result from LENR.

Proof for transmutation is difficult to obtain because the amount of such material is comparable to normal impurities. Mechanisms that can concentrate impurities too rare to be otherwise detected amplify the problem. To give a few examples, diffusing hydrogen can carry with it impurity atoms and cause them to concentrate at chemically active sites where they can be detected and mistaken for a nuclear product. Hioki et al.(*156*) found that sulfur (S) in Pd is brought to its surface by annealing in vacuum. Electrolysis can dissolve impurities and deposit them in high concentration at chemically active sites. Gas discharge can sputter material and deposit it in high concentration at special locations on a surface. In addition, according to Afonichev(*157*), stress can cause rare impurities to move along the resulting dislocations to achieve detectable concentration at the surface.

To make the problem even more difficult, modern analytical methods are so sensitive, especially if the impurity is on a surface, that many elements not previously considered as possible contamination are frequently detected even without being further concentrated. Consequently, claims for transmutation need to be evaluated with caution.

2.3.1 How is transmutation measured?

EDX

"Electron dispersive X-rays" result when high-energy electrons bombard a material, generally using a scanning electron microscope (SEM). The X-rays are produced after electrons in the K and L levels of elements in the material return after being removed by the bombarding energetic electrons. Many elements can be identified because the resulting X-ray is unique to the element.

The method detects elements only within a few microns of the surface, with depth being determined by the energy of the X-ray and the energy of the bombarding electrons. Elements with atomic number below boron or those with overlapping X-ray energy are not easily detected by this method.

The method allows creation of a false-color image of the surface showing where each element is located and its apparent local concentration, making it a very useful and popular tool.

SIMS
"Secondary ion mass spectrometry" is undertaken by bombarding a surface with energetic ions in order to release elements as ions that are then detected using a mass spectrometer. The method is useful to determine isotopic composition as well as detecting elements within a few nm of the surface.

Composition of elements and isotopes can be determined as a function of depth when new material is exposed using ion milling.

Neutron Activation
Various elements and isotopes can be detected at very low concentrations by exposing the material to neutrons and detecting radiation emitted from those isotopes that become radioactive as a result.(*61, 158-162*) The method can determine the total number of atoms located anywhere in the sample. Sensitivity is limited by the cross-section for neutron activation, the type of radiation emitted by the product, its half-life, and the presence of interfering radiation. Although neutrons can react with most isotopes, an element cannot be detected if the resulting isotope is not radioactive. The method requires a high flux of neutrons, generally available from a nuclear reactor.

XRF
"X-ray fluorescence" results when an intense beam of energetic X-rays bombard a material causing emission of secondary X-rays that are characteristic of elements on the surface. Very small beams having large flux can be used to map very accurately the concentration and location of certain elements located only at the surface.

ICP-MS
"Inductively coupled plasma" mass spectrometry uses hot plasma to release materials from a solid and make them available for detection by a mass spectrometer. This allows the atomic

composition of the entire sample to be determined with high sensitivity by converting the entire sample to plasma.

AES

"Auger electron spectrometry" results when a surface is bombarded with electrons and the energy spectrum of emitted electrons is analyzed. Each element emits electrons of a characteristic energy depending on its chemical state. A false image of the surface can be made to show where the elements are located within a few nanometer of the surface.

XPS or ESCA

"X-ray photoelectron spectroscopy" applies X-rays to a surface and measures the energy of emitted electrons. This energy can be used to identify most elements and many compounds, but only on the surface.

2.3.2 What types of transmutation occur during LENR?

Transmutation of any kind is limited to a low rate because the Coulomb barrier is more than an order of magnitude greater than the barrier standing in the way of hydrogen fusion. Clearly, surmounting the barrier between H and Ni of 28 units or the barrier of 46 units between D and Pd would require a very unusual mechanism to provide sufficient energy or even greater shielding than required to make hydrogen fusion possible.

Once the barrier is overcome, the mass-energy must, on some occasions, be released by a single transmuted nucleus. How several MeV can be released without energetic radiation being emitted significantly challenges an explanation and makes a claim for significant energy production from transmutation very hard to justify.

At least two kinds of transmutation reactions can be identified; one that results in fission of the nucleus after addition of one or more hydrogen nuclei and another that does not result in fission. This description differs somewhat from that used by Kozima(*163*) based on fission and is significantly different from the process proposed by Miley(*164*) based on formation and fragmentation of a super-heavy nucleus, as described in Chapter 3. Regardless of which model is accepted, an effective explanation of LENR needs to account for two different kinds of transmutation products and how these different reactions relate to fusion of the hydrogen isotopes. These two transmutation reactions are discussed separately, starting with the fusion-fission type. A third type is not discussed here.

Transmutation produced by fusion-fission

Many examples of transmutation observed during LENR can be explained by one or more hydrogen nuclei adding to the nucleus of palladium or another element present on the surface of a nuclear-active material. Immediately following this fusion process, the resulting nucleus fragments (fissions) in various ways to produce a collection of lighter elements.(*165-175*) The fragment is seldom found to be radioactive, although such tests are not frequently made and elements having short half-lives might be missed. The suspected fragment is frequently reported to have an unconventional isotopic composition, but not always.

Most studies analyze part of a sample in which transmutation might have occurred and compare the analysis to what was found in the initial material. Unexpected elements are frequently detected in small, isolated regions on the surface. Even though this is where transmutation products are expected to form, attributing the detected elements to a transmutation reaction can lead to false conclusions unless many independent studies are examined and compared.

Figure 18 shows examples of craters where unanticipated elements are typically found. Many craters result from impurity deposits next to a crack. As seen in (B), impurities can accumulate and produce a volcano-like structure where convection caused by the bubble stream brings impurities to the site where they can deposit as a result of electrolytic action. This behavior shows clearly just how impure a surface can become and how easily these impurities can be concentrated at certain locations. Some craters appear to result from a local melting of the surface (C). Since the surface is very impure, the melting point can be much lower than pure palladium at certain sites. Other craters seem to form as a result of crystallographic reorganization of the surface as stress is relieved (A). These reorganization features are common in Pd and Ni after reacting with hydrogen but frequently show no evidence for LENR. Other examples are shown in Figs. 35 and 45 in a later section.

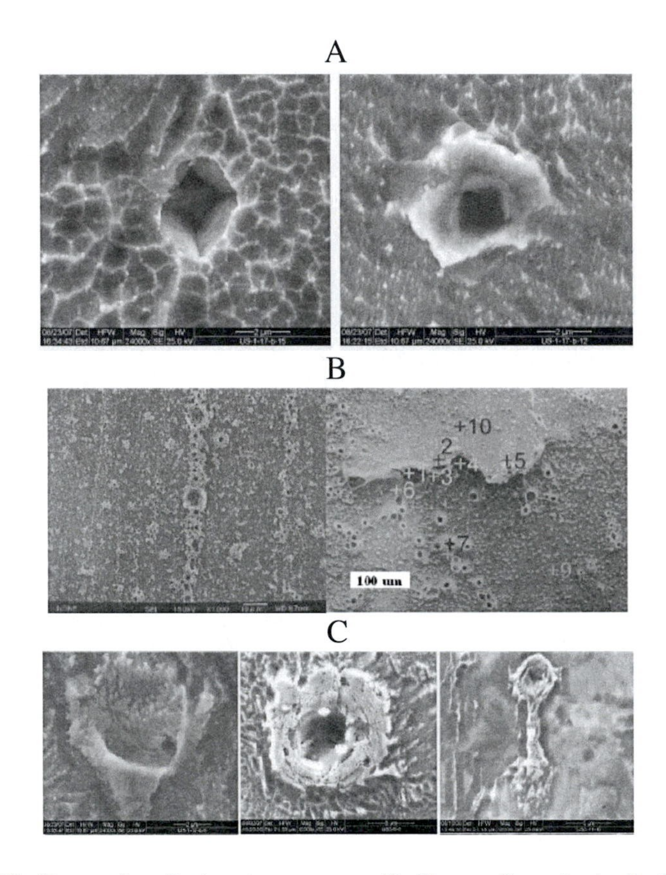

Figure 18. Example of structures on palladium after electrolysis in D_2O where localized concentration of assumed transmutation products are claimed to form.(*155, 176*)

Proper interpretation requires these different causes of structure formation be properly related to the potential source of the detected elements. In fact, many elements found on a surface can be attributed to local concentration of common impurities, not to LENR. Nevertheless, this fact must not be used to reject clear examples of LENR.

The frequency of detection for various elements, as assembled by different authors, is shown in Fig. 19. A few elements heavier than Pd are detected, as would result from addition of d to palladium (Pd) or to the platinum (Pt) impurity on the surface. However, many lighter elements are also found that can only result from the target nucleus splitting into two parts after d or p is added. This explanation is discussed by many researchers(*165-175*) because it would provide a mechanism for dissipating energy while conserving momentum. Fission is made even more plausible by the observations shown in Fig. 20, where two collections of elements having atomic numbers that approximately add to that of palladium are obvious. This bimodal effect is not as clear in Fig. 19. This bimodal distribution is consistent with conclusions reached by Miley and Shrestha(*177*) in their review. However, some elements cannot be justified as being the result of fusion-fission unless some of the fragments are radioactive and decay rapidly to produce the detected elements or unless some isotopes of palladium do not participate in the transmutation process. Passell(*161*) provided an indication of the latter possibility using neutron activation. This issue is discussed in Section 5.5.0.

A similar distribution (Fig. 21) was found when H_2O is used in the electrolyte. However, this figure is identified to result from Pd as the cathode as reported at ICCF-7(*178*) and from using Ni in the paper published in *Fusion Technology* a year later.(*179*) These results are not consistent with the expected behavior of Ni.

In contrast to the claim made by Swartz(*180*), the present evaluation could find no evidence for LENR occurring anywhere other than in the near surface region of a material. Of course, if the particle size were small, the near surface region would represent most of the material. Nevertheless, where a clear distinction can be made between surface and bulk, the surface is clearly the location of transmutation.

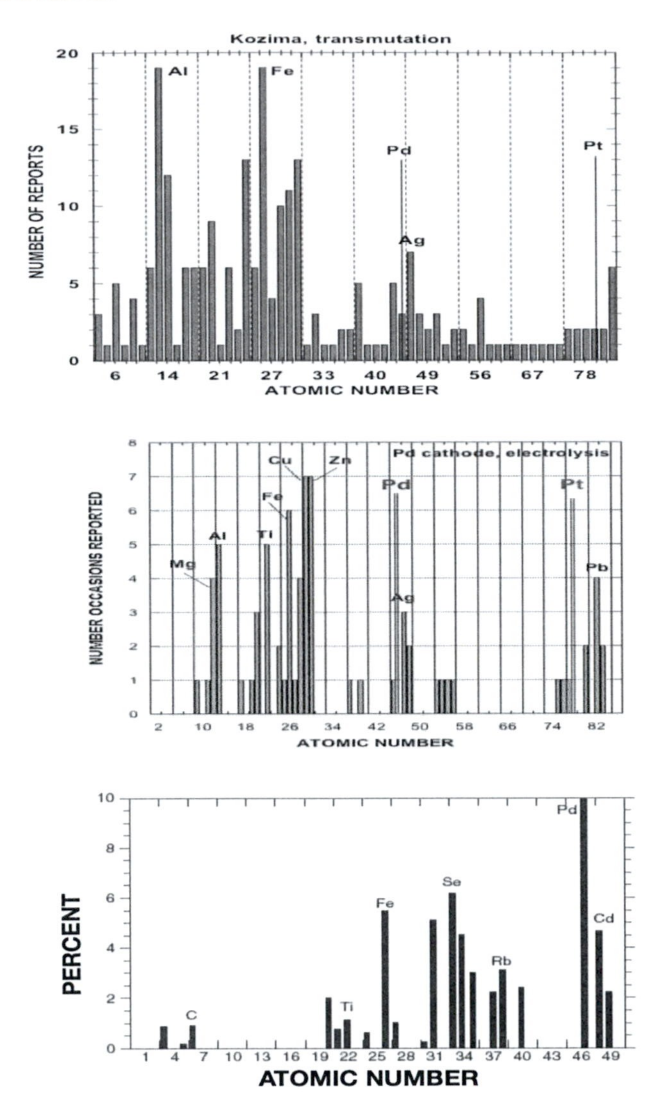

Figure 19. Number of reported detections of unexpected elements on the surface of a Pd cathode after electrolysis. The location of Pd and Pt are shown. Pt results from transfer from the anode to the cathode. #1 Kozima (*163*); #2 Storms(*1*); #3 Srinivasan et al.(*152*)

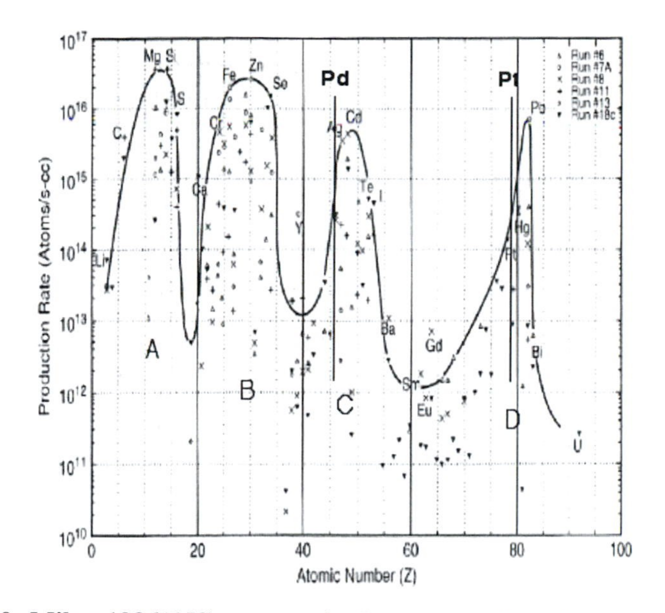

Figure 20. Miley 1996(*152*) summarized many studies of transmutation to show four regions of atomic number where large numbers of claimed transmuted elements are found. The region (C) forms next to palladium, from which the cathode is made, while the region (D) is near platinum (Pt), which is transferred to the cathode from the anode. Regions (A) and (B) might result from fragmentation of nuclei in regions (C) and (D).

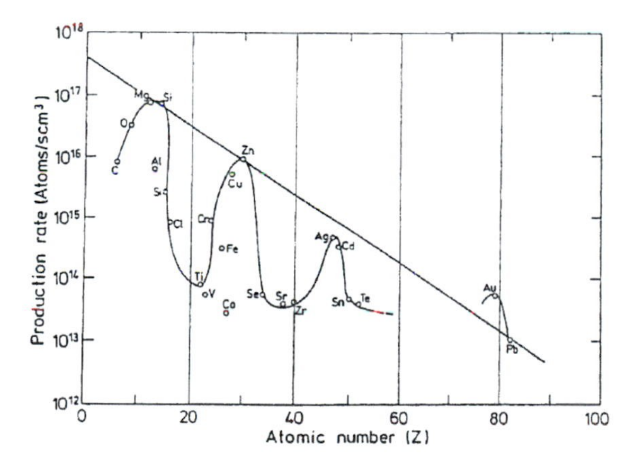

Figure 21. Element distribution obtained using H_2O in the electrolyte(*178, 179*) made between surface and bulk, the surface is clearly the location of transmutation.

Because the surface is very complex and non-uniform, exactly where the effect occurs within the surface region has not been identified. Clearly, these sites are many, randomly located, and variable in their reaction rate. In addition, this surface region can be the host to nuclear reactions involving many different elements and isotopes depending on their distribution and local conditions.

A few studies have detected some radioactive transmutation products.(*174, 175, 181-198*) If the half-life were short, radioactive isotopes produced early in the study might decay away before the material could be examined and remain undetected. For example, ^{107}Ag produced by transmutation might fragment into ^{27}Mg, which would decay by beta emission into stable ^{27}Al with a half-life of 9.5 m and ^{80}Br, which would decay by beta emission into stable ^{80}Kr with a half-life of 17.6 m. Krypton (Kr) being an inert gas would quickly leave the surface. Both reactions might be overlooked if the amount of material were small, if most of the radiation were stopped by the apparatus, and if too much time passed before the material were placed in a suitable detector. Thus, although a radioactive product is rarely reported, absence of radioactive isotopes resulting from transmutation has not been established.

The consequence of adding many protons, deuterons, or neutrons to a target can be easily visualized by studying Fig. 22. Here each stable isotope of elements above palladium in atomic number is shown by a dot. The consequence of adding the respective hydrogen nucleus can be found by following a line parallel to those shown on the figure that passes through the isotope of palladium of interest. Adding deuterons produces more elements without radioactive isotopes than does adding protons. Nevertheless, a very limited number of possibilities exist.

If neutrons were added, one isotope of palladium would be changed into another until a radioactive isotope is reached in the sequence, after which another element results from its slow decay. Consequently, production of new elements by neutron addition would be a slow process, generate radioactive isotopes, and not be consistent with observation.

Several researchers (*199, 200*), and a few people including Rossi(*201*) and Godes(*202*), have claimed transmutation is the source of energy produced by nickel saturated with H_2. To achieve this result, a significant fraction of the nickel lattice would have to allow a Coulomb barrier of 28 units be overcome at a significant rate, allow energy dissipation without producing significant radiation, and produce no detectable radioactive isotopes. Furthermore, the lifetime of the heat producing process would be limited by the amount of nickel available in each particle that is able to experience this process. These requirements seriously limit how much energy can come from this source.

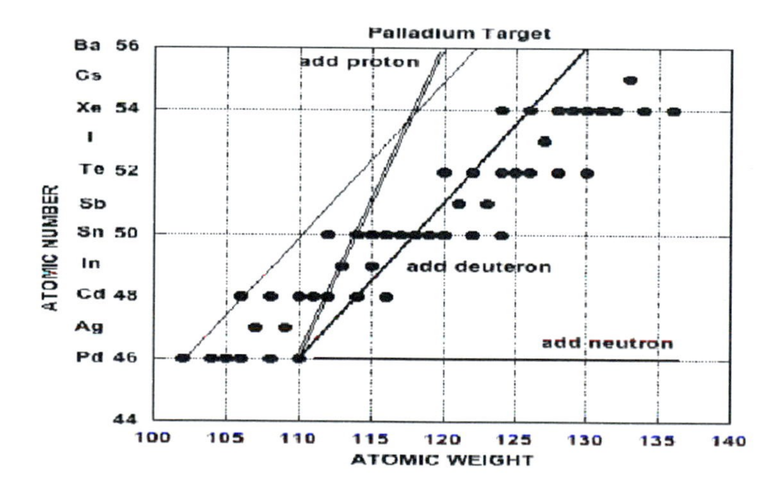

Figure 22. Effect of adding d, p, or n to isotopes of palladium. The dots represent stable isotopes. The lines show the consequence of adding p or d to ^{110}Pd. A stable isotope is formed where the line passes through a dot and a radioactive isotope results if a dot is not intersected at an atomic number. Addition of p or d to the other isotopes of Pd will have consequences shown by lines parallel to the ones shown.(*152*)

Given the fact that fusion takes place only in rare and isolated regions in a unique material and requires a very special process to overcome a barrier of only one unit, proposing that transmutation can take place at high rate throughout a material is hard to justify. Transmutation clearly occurs at a small rate, but no plausible evidence can be found to support a claim for a rate sufficient to produce detectable energy. More will be said about this issue in Chapters 3, 4, and 5.

Rossi(*203*) initially claimed copper resulted from addition of a proton to nickel. The observed normal isotopic ratio of copper challenges this conclusion, as does the absence of radioactive isotopes that would be expected to form. A mechanism explaining transmutation without these problems is proposed in Chapter 5.

Transmutation produced by fusion without fission

This method is discussed below in the section "Transmutation produced by gas loading."

2.3.3 What methods are found to produce transmutation?

Elements resulting from transmutation have been reported after electrolysis, gas discharge, gas loading, and plasma arcs formed in water. Both stable and radioactive transmuted products have been detected involving from 1 to 6 deuterons being added as well as addition of protons on occasion. The process is also found to occur in biological systems, but this method is not discussed here.(*153, 185, 204-226*)

Transmutation produced by gas loading

Iwamura et al.,(*227-229*) using the impressive tools available to Mitsubishi Heavy Industries Ltd., employed a very creative and well-studied method, published first in 1998. The process was found to occur on the surface of a sandwich composed of palladium (Pd) and calcium oxide (CaO) layers on Pd, with a final layer of Pd (400Å) on the top surface, as shown in Fig. 23. Use of yttrium oxide (Y_2O_3) instead of CaO was found to support the effect, but magnesium oxide (MgO) did not. The amount of target transmuted increased with increase in D_2 pressure and with increased rate at which D passed through the surface. In other words, the more D is available to the NAE, the greater the transmutation rate.

This composite was first used as a cathode in an electrolyte cell containing D_2O-Li with a platinum (Pt) anode.(*227-229*) For this use, the "Surface" was exposed to electrolysis as the cathode while the other side was exposed to vacuum, causing D to diffuse through the material from the top to the bottom of the figure. Even though this method does not allow a significant D/Pd ratio to build up, excess energy was measured along with X-ray emission, occasional neutrons, and unexpected elements on the "Surface." An unusual amount of titanium (Ti) was found on the "Surface" of one sample. Iron (Fe) was also detected on all surfaces, presumably resulting from ineffective protection of the steel cooling coils.

Nevertheless, an abnormal isotopic ratio resulted when D_2O-Li was used in the electrolyte, as shown in Fig. 24, with ^{57}Fe (2.1% natural abundance) and ^{58}Fe (0.3% natural abundance) present in abnormal amounts. A normal isotopic ratio was found when H_2O-Li is used. As can be seen in the figure, the ratio is highly variable with depth when D_2O is used. The two rare isotopes of Fe either are deposited in greater amount or are increased in amount by transmutation of palladium when D_2O is used. If these isotopes of Fe result from fusion-fission, the other element must be radioactive isotopes of scandium (Sc), depending on which isotope of Pd is transmuted. These Sc isotopes would rapidly decay by beta emission to give stable titanium, which is detected. The detected X-rays could be from Bremsstrahlung resulting from the beta-electron being stopped by the material. While not definitive, this observed behavior is consistent with fusion-fission of Pd. Excess energy would be expected from D-D fusion combined with energy from the fusion-fission reaction. The detected neutrons might result from occasional fractofusion.

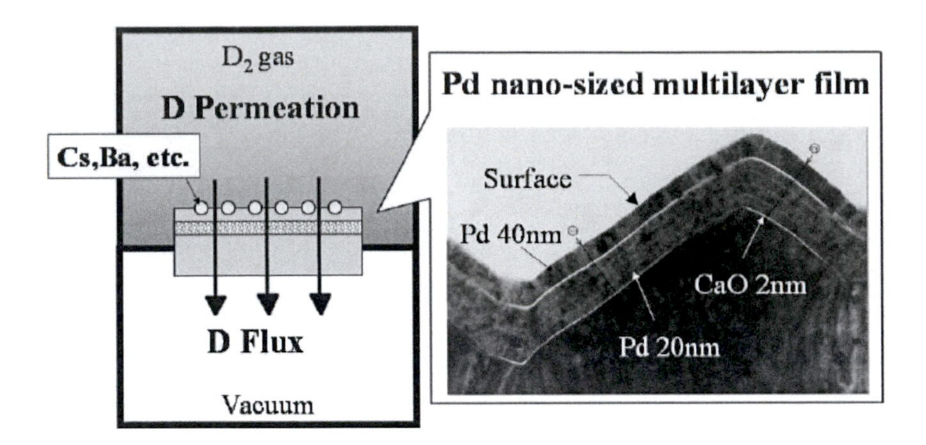

Figure 23. Cross-section of layers of CaO and Pd deposited on Pd. The lighter layers are CaO and the darker material is palladium.(*230*) The final deposit of Pd is 40 nm thick. Nuclear reactions were found to occur on the identified "Surface" where various elements were deposited.

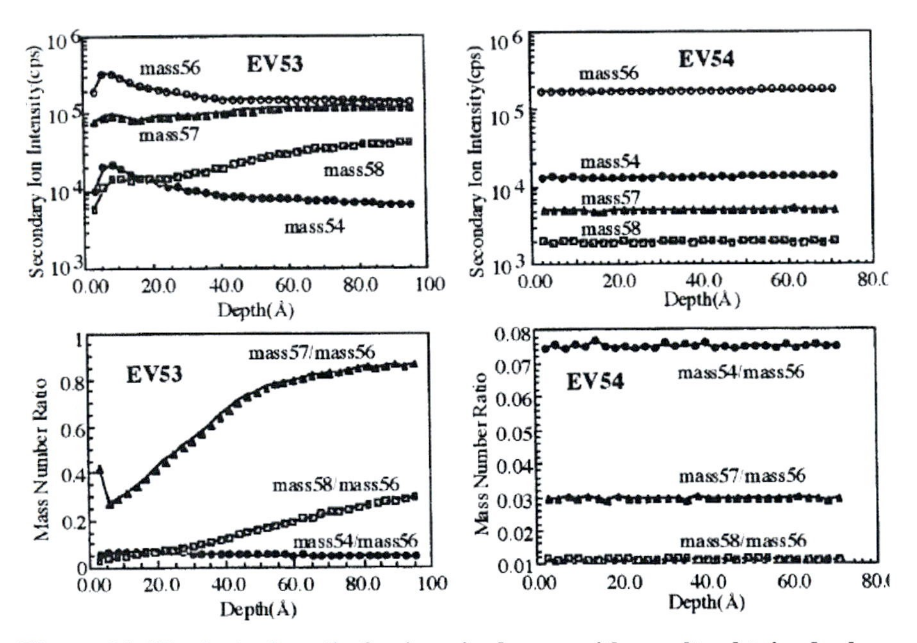

Figure 24. The isotopic ratio for iron is shown, with results obtained when D_2O is used as the electrolyte, shown on the left, and when H_2O is used shown on the right.(*227-229*)

Next, this universal Pd-CaO sandwich was placed in contact with D_2 gas on one side and vacuum on the other.(*231*) Three elements, Mg, Si, and S, were found to grow in abundance on the surface exposed to D_2. In addition, the sulfur (S) isotopic ratio, $^{33}S/^{32}S$, was found to be abnormally large. Because the $^{34}S/^{32}S$ ratio was within the error of the normal value, only the ^{33}S isotopic content had apparently increased. If ^{33}S resulted from adding 2D to Pd followed by fission, the other element could be both stable and radioactive isotopes of germanium (Ge). These would decay by rapid beta emission to arsenic (As) and selenium (Se), which were not reported. Formation of stable magnesium (Mg) from the same fusion-fission reaction could be paired with krypton (Kr), which would vaporize rapidly from the surface as non-reactive gas. The silicon (Si), which was found to have an abnormal isotopic ratio, would be paired with stable selenium (Se). Addition of Li to the surface resulted in appearance of fluorine (F) and aluminum (Al), which were not detected when Li was present during the previous electrolytic studies. This inconsistent behavior suggests another variable was operating, not just the presence of Li. The mechanism suggested by the authors, based on involvement of dineutrons produced by addition of an electron to the deuterium nucleus, is neither consistent with observation nor can it be justified, as explained in Section 3.2.4.

Iwamura et al.(*232-234*) repeated the above study in 2002, but this time they applied either cesium (Cs) or strontium (Sr) to the surface, after which deuterium was caused to diffuse through the structure. The surface composition was analyzed using *in-situ* XPS without the sample being removed from the apparatus. This analysis was done over time, shown in Fig. 25, while the concentration of cesium (Cs) on the surface decreased and the amount of praseodymium (Pr) increased. This change in concentration did not occur when an ordinary piece of Pd was used instead of the Pd-CaO complex or when the gas was H_2 rather than D_2.

Formation of Pr from Cs requires simultaneous addition of four deuterons to the nucleus without subsequent fission. This process has to take place as one event to avoid formation of intermediate nuclei (Ba, La, Ce), which were not detected and would be radioactive. The conversion rate is roughly a linear function of the rate at which D passes through the palladium complex and the resulting Pr is located in many very small sites having random distribution within a region about 10 nm below the surface(*235*), as shown in Fig. 26.

How four deuterons could simultaneously penetrate a Coulomb barrier of 55 units and dissipate the resulting energy of 50.5 MeV from a single nucleus without producing detected radiation remains a serious challenge to theory. This challenge is answered in Chapter 5. The observed reaction rate was too small to produce detectable energy.

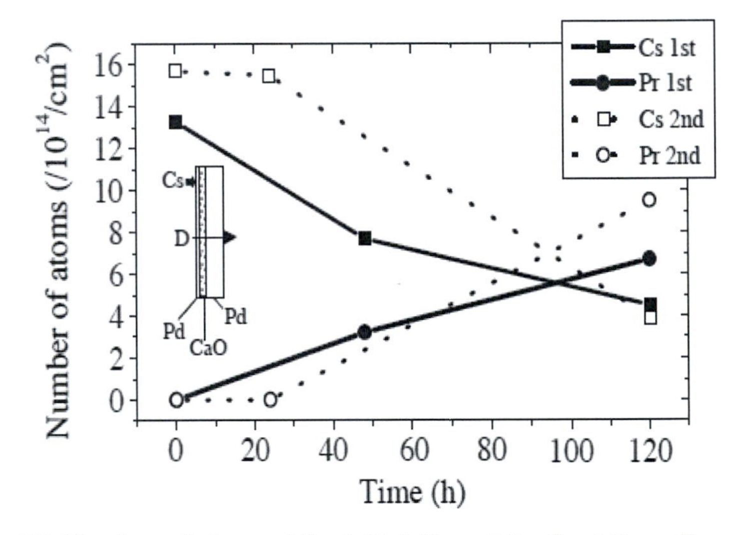

Figure 25. Number of atoms of the initial Cs and the final Pr on the surface of the Pd complex as a function of time while exposed to D_2.(*235*)

A similar study was made by applying strontium (Sr) to the surface. In this case, the amount of molybdenum (Mo) on the surface increased as the Sr content decreased, as seen in Fig. 27. In addition, ^{96}Mo increased, which would result by addition of four deuterons to the major isotope of Sr (82.6% NA), which is ^{88}Sr.

Again this addition had to occur as a single event and the energy had to be dissipated without detected radiation. No evidence for fusion-fission was found.

In 2005, Iwamura et al.(*236*) extended their study by applying barium (Ba) to the Pd-CaO complex and watched samarium (Sm) form. In this case, they used two samples, one containing natural Ba, which is mostly ^{138}Ba, and another sample enriched in ^{137}Ba. Addition of six d to the isotope of Ba produced the corresponding isotope of Sm, as shown in Fig. 28.

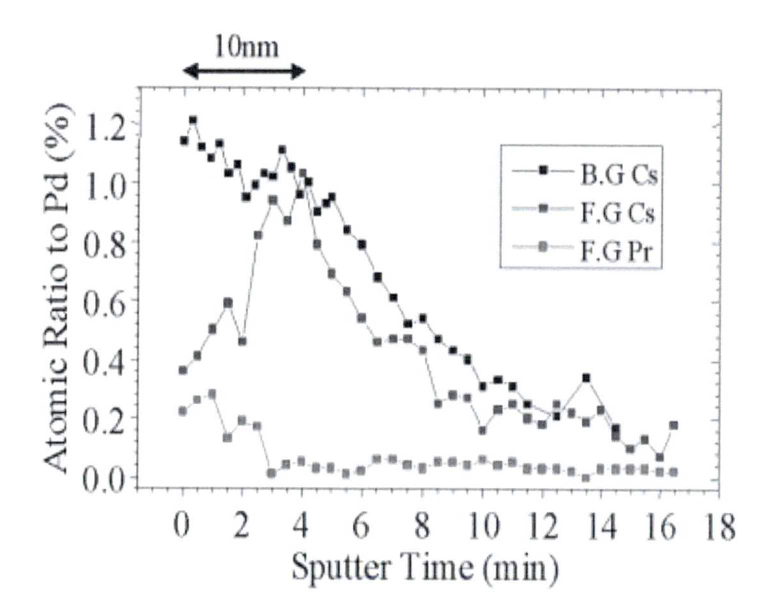

Figure 26. Depth profile of Cs and Pr in the Pd-CaO complex surface. B.G. is the initial Cs profile and F.G. shows the profile after exposure to D_2.(*235*)

Even though addition of two d to ^{138}Ba would produce stable ^{142}Ce and addition of four d would produce stable ^{146}Nd, these elements were not detected. Also, addition of two d to ^{137}Ba would produce radioactive isotopes of Ce, which also were not reported. In fact, all isotopes of Ba were found to produce stable isotopes of Sm except ^{146}Sm, which has a very long half-life of 10^8 years. The process would appear to add deuterons until stable isotopes are formed.

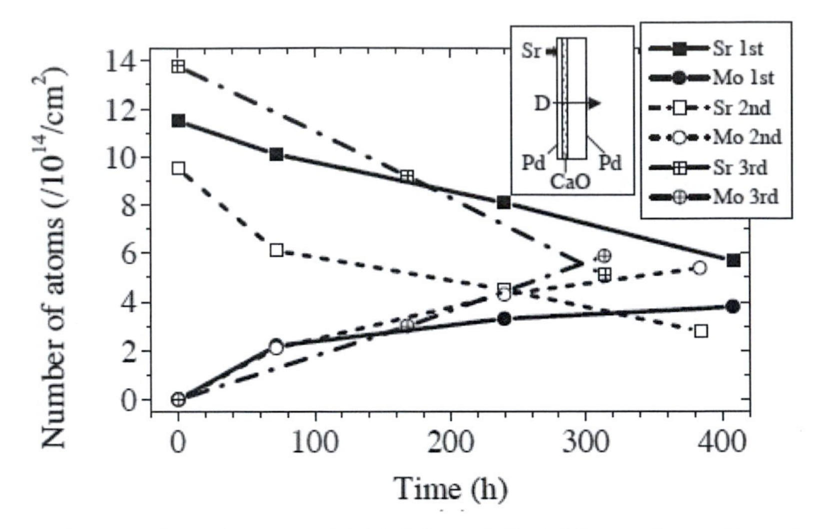

Figure 27. Number of atoms of initial Sr and final Mo on the surface of the Pd-CaO complex as a function of time while being exposed to D_2.(*237*)

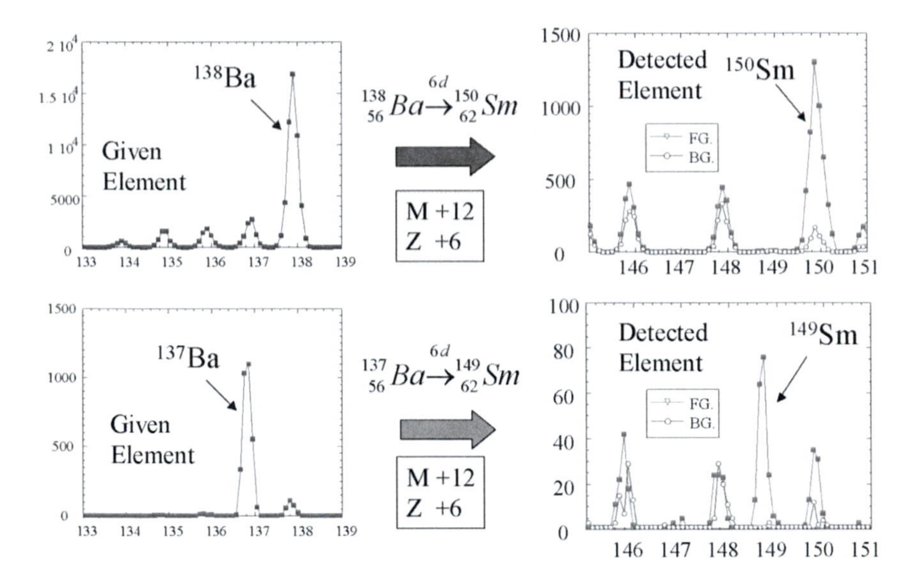

Figure 28. Isotopic distribution of Ba in samples applied to the Pd complex compared to the resulting isotopic distribution of Sm after exposure to D_2. (*237*)

Why six deuterons are added in this case and a fewer number when other elements are present on the surface; how the deuterons can get over a barrier having a magnitude of 56; and why fusion-fission does not occur are questions requiring answers provided in Chapter 5. This work has continued and resulted in a summary published in 2013(*230*), shown in Fig. 29. In each case, even numbers of deuterons are always added to produce stable isotopes.

$$^{133}_{55}Cs \xrightarrow{4d(2\alpha)} {}^{141}_{59}Pr$$

$$^{88}_{38}Sr \xrightarrow{4d(2\alpha)} {}^{96}_{42}Mo$$

$$^{138}_{56}Ba \xrightarrow{6d(3\alpha)} {}^{150}_{62}Sm$$

$$^{137}_{56}Ba \xrightarrow{6d(3\alpha)} {}^{149}_{62}Sm$$

$$^{44}_{20}Ca \xrightarrow{2d(\alpha)} {}^{48}_{22}Ti$$

$$^{184}_{74}W \xrightarrow{2d(\alpha)} {}^{188}_{76}Os$$

$$^{182}_{74}W \xrightarrow{4d(2\alpha)} {}^{190}_{78}Pt$$

Figure 29. Summary of nuclear reactions observed by Iwamura et al. to occur on the surface of the Pd-CaO complex.(*230*)

In the case of calcium (Ca) and tungsten (W), not all isotopes of the target can produce stable isotopes no matter how many deuterons are added. Although no radioactivity was reported, the reported behavior is not consistent with the other studies. Clearly the inconsistent data requires additional measurements be made before rules governing behavior can be proposed. Nevertheless, basic replication of the effect has been achieved.(*238*)

In all of this work, the β-PdD phase cannot form under the conditions used. Heating the complex to 70°C, exposing it to only one atmosphere of D_2 gas on one side, and to vacuum on the other side limits the deuterium concentration to a very small value. Consequently, a flux must be used to make deuterons available to the active sites instead of a high concentration, in contrast to the situation when electrolysis is used. More will be said about the relationship between flux and concentration in Chapter 5.

Transmutation produced by the glow discharge method

This method subjects a material to electric discharge as a cathode in low-pressure D_2. Ions of D^+ are injected into the surface with significant energy (100-1000 V), but not enough to initiate hot fusion or conventional nuclear processes at a detectable rate. The process significantly changes the surface morphology, as shown in Fig. 30, where material is transferred to the cathode from any nearby insulator. Many transmuted products, both stable and radioactive, as well as extra heat and radiation are reported.

This process can bring unexpected impurities to the surface where they can experience transmutation. Pulsed current is frequently used to provide a brief high current without melting the material. Nevertheless, local temperatures can be significant.

Researchers at the Scientific Industrial Association "Luch" in the Russian Federation have been exploring transmutation using glow discharge for years using voltages too small to produce conventional nuclear interaction.(*136, 165, 171, 174, 175, 182, 183, 188, 195, 239-257*) This work has revealed a wide variety of transmutation reactions, many of which produce radioactive isotopes that are consistent with fusion-fission of palladium. Many of the elements are found in the near-surface region, within about 1 μm of the Pd surface and have abnormal isotopic ratios.

Some ^{4}He is also detected in samples after excess energy is produced. Low intensity charged particle emission is detected after the discharge is turned off, with most energy in the 2-4 MeV region with a maximum near 18 MeV.(*240*) This delayed radiation is emitted only from a few isolated regions, shown in Fig. 31. In addition, the distribution of elemental and isotopic composition is not uniform, which complicates conclusions based on analysis made at a few locations. This behavior challenges the assumption that the entire lattice is involved in the nuclear process. Indeed, D-D fusion and Pd-D fusion-fission appear to be taking place only at a few special locations with some of the fission products being radioactive.

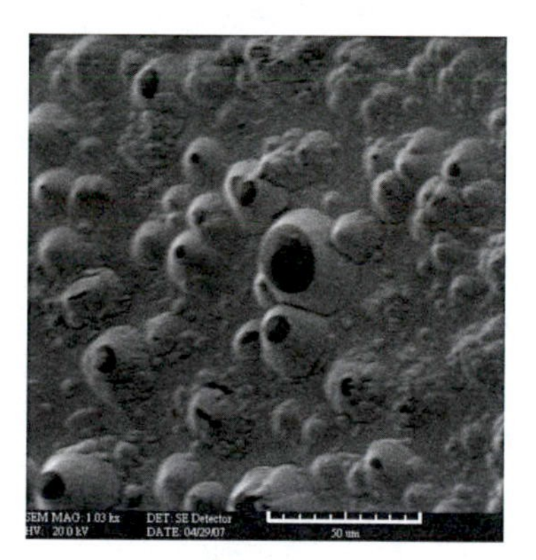

Figure 30. A common appearance of a Pd surface after glow discharge.(*258*) The dark regions are deposited material from a nearby insulator. The shape of the surface reveals complex interaction occurs between the discharge and the surface, with surrounding materials, and within the cathode surface.

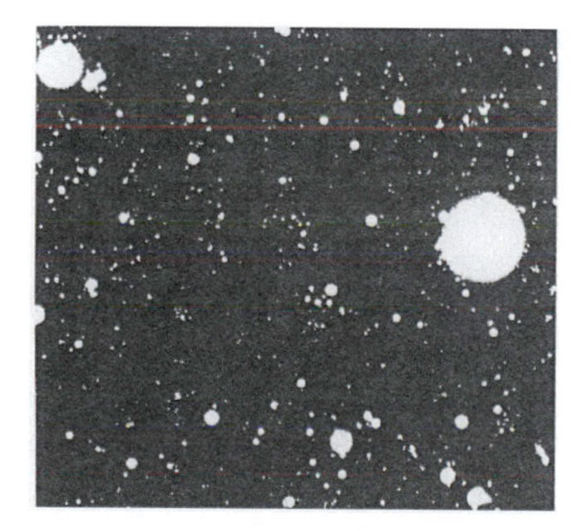

Figure 31. Autoradiograph of a Pd cathode after gas discharge using X-ray film with a 2 mm lead (Pb) absorber. Intense local photon decay occurs at the white regions.(*240*)

Transmutation produced by electrolytic method

A large number of studies, too many to list here, have reported finding elements on the cathode surface after electrolysis. These elements appear to result from fusion-fission of palladium, as well as elements resulting from addition of a deuteron or proton to another element without fission occurring. For example, Bush(*259*) found excess strontium (Sr) with an abnormal isotopic ratio in the electrolyte after electrolyzing H_2O+Rb_2CO_3. Bush and then Notoya and Enyo(*260*) found excess calcium (Ca) in a cell after electrolyzing H_2O+K_2CO_3. In both cases, a proton is proposed to be added without producing fission and the resulting element is dissolved in the electrolyte rather than remaining on the cathode surface. A different explanation is offered in Chapter 5.

Transmutation produced by plasma electrolysis and arc-plasma in water

If applied voltage is increased to over about 100 V during electrolysis, plasma forms around the cathode. This extra energy increases the rate and energy of ions bombarding the cathode surface and the cathode temperature, resulting in increased transmutation and excess energy production. In a similar fashion, formation of a DC arc between two electrodes immersed in water also produces transmutation. While these two conditions have much in common, they are best discussed separately, starting with arc formation.

Oshawa first used the arc method to make iron by transmutation in 1964, followed by Oguta et al. in 1990 and 1992, but the papers are not readily available. In 1994, Sundaresan et al.(*261*) repeated the study using DC arc in H_2O between ultra-pure carbon electrodes and observed formation of a magnetic powder that contained iron (Fe). Independently and at the same time, Singh et al.(*262*) produced iron using this method, but with a normal isotopic ratio. Grotz(*263*) emphasized the need to account for the frequently present iron impurity, which Sundaresan et al. were careful to do. Production of a powder containing iron (Fe) using AC instead of DC between carbon electrodes was replicated by

Jiang et al.(*264*) In this study, the ^{58}Fe isotope was enhanced and other elements including Cr, Co, and Zn were detected by neutron activation and spectroscopic analysis. Ransford(*265*) also replicated production of iron. Hanawa(*266*) produced Fe using this method along with a variety of other elements consisting of Mn, Co, Ni, Cu, and Zn in lesser amounts. These elements all lie between an atomic number of 25 and 30, with ^{26}Fe located in the middle. In this case, the site of nuclear activity was found to be located on the anode. This result is important because people do not normally examine the anode for transmutation products, but instead collect material present in the liquid or located on the cathode.

Matsumoto(*267, 268*) extended the method by using electrodes of various metals and found many of the expected transmutation products at isolated locations, again on the anode when Pd was used. When he(*269*) used lead (Pb) and copper (Cu) electrodes with AC, many unusual particles were observed when the lead melted, which contained some apparent transmutation products. Arc formation between Ni (-) and graphite (+) in a solution of $(NH_4)_2MoO_4$ in H_2O by Nakamura et al.(*191*) generated radioactive products that decayed slowly by photon emission (~0.13 MeV).

In this case, the nuclear products appear to result from several heavy nuclei, not hydrogen, combining to produce a still heavier nucleus that does not fragment, which is much different from what happens when transmutation occurs any other way. Apparently, this is an example of a third mechanism causing transmutation — one that is very difficult to explain because it is not consistent with fusion-fission.

Rather than creating an arc, Mizuno et al. increased the voltage (~110 V) applied to a normal electrolytic cell until plasma formed around the cathode. Using Pt as both cathode and anode with sodium or potassium carbonate in the electrolyte, he detected neutron emission that correlated with applied current and was 10 times greater when using D_2O compared to H_2O.(*270*) Elements near iron (Mn, Fe, Ni, and Cu) were detected, but this time on the cathode surface, to which they might have transferred from the anode. In this case, these elements might also result from fusion-fission of platinum. Lifetime of the cathode was improved by changing the cathode material to tungsten. This study produced transmutation products listed in Fig. 32 when excess energy was detected.(*271, 272*) Although the concentrations of transmutation products were nonuniform within the cell, production of several elements having a lower atomic number than tungsten, such as germanium (Ge) and indium (In), are unexpected and could be fusion-fission products of tungsten. The complexity of sorting out such studies requires much more information.

Emission of neutrons from a Mizuno-type cell containing a W cathode and K_2CO_3 in H_2O was replicated by Cirillo et al.(*273*) The neutrons were detected using CR-39 to detect alpha particles resulting from H_3BO_3 placed on the CR-39 surface in which the $^{10}B(n, a)^7Li$ reaction took place. The flux was reported as 720 n/sec-mm^2.

Element	Before (mg)	After electrolysis Large excess heat	After electrolysis Small excess heat
Al	0 .006	1.2	0.5
Si	1.4	2.0	1.5
P	0.5	0.5	0.5
S	0.4	0. 6	0.6
Cl	0.8	1.0	0.85
Ca	0.06	0.45	0.12
Ti	0.001	0.8	0.15
Cr	0.001	1.4	0.005
Fe	0.024	2.5	0.055
Ni	0.001	0.02	0.004
Cu	0.14	0.03	0.16
Zn	0.001	0.75	0.001
Ge	0.01	1.2	0.01
Pd	0.01	0. 5	0.01
Ag	0.01	0. 12	0.01
In	0.01	1.1	0.01

Figure 32. Transmutation products produced during plasma electrolysis using a tungsten cathode, Pt anode, and K_2CO_3 in H_2O.(*271*)

2.3.4 Transmutation using applied laser radiation

Energy applied as laser radiation can cause transmutation and increase excess energy production. This section focuses only on claims for producing transmutation products. Claims for stimulating energy production are examined in Section 2.5.2. These two different effects are related because even though too little transmutation is produced to make detectable energy, the location of the transmutation products indicate where on the surface nuclear reactions are expected, the assumption being that fusion of hydrogen isotopes and transmutation occur in the same NAE.

The group at the University of Lecce, in Italy, studied the effect of irradiating PdD with UV laser starting in 1997. Nassisi(*274, 275*) used an XeCl laser (308 nm, 4.02 eV, 0.5 J/cm^2, 2 Hz rate) to radiate $PdD_{0.77}$ formed by exposing the treated Pd metal to D_2 gas. On several occasions, the surface of the PdD became sufficiently chemically active to react with air afterward, as is frequently observed when PdD is made other ways and exposed to air. Pits were produced in the surface typical of Pd after being repeatedly heated in D_2 gas, as seen in Fig. 18A. This behavior asks the question, “Is the laser important because this radiation excites a fusion process already underway or does the heat pulse produced by the pulsed laser trigger formation of NAE?” An answer is suggested in Section 2.5.2 and Chapter 5.

Significant quantities of Al, C, Ca, Fe, Mg, Na, O, and Si were detected by Nassisi in certain locations using EDX. The elements are expected to result from fusion-fission of Pd with aluminum being the major product. The largest rate of transmutation resulted by applying 308 nm radiation to $PdD_{0.7}$ in D_2 gas for 20-30 days, although some products were also found when H_2 was used. A small delayed neutron flux was detected afterwards. The transmutation products were only found in a few local regions within a few microns of the surface and were generally associated with small pits. Similar pits are frequently reported when other methods are used and are frequently found to contain transmutation products. Radiation was apparently produced that caused the quartz window to crack, after which vanadium (V) and zinc (Zn) were detected on its surface.

This study was continued by Castellano et al.(*276, 277*) using PdD films (16-141 nm) created by depositing Pd on silicon (Si) and exposing this to D_2 gas. Many cracks and bubbles formed as a result of hydriding, an example of which is shown in Fig. 33.

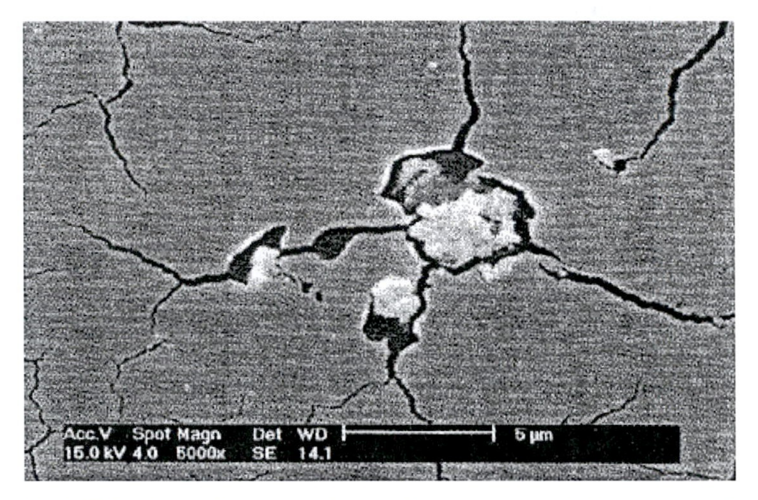

Figure 33. SEM view of the surface where transmutation products are detected in Pd thin films.(*276*)

Some samples were exposed to 1 sec pulses of 308 nm laser radiation for five weeks after which the surface was examined using EDX. Extra elements are found in the walls of cracks both with and without using the laser. An especially large number of elements are found in the thicker layers even though the PdD was not exposed to the laser. This observation suggests reaction with D_2 alone produced cracks in which transmutation occurred without additional laser energy being required. A typical bimodal collection of elements below the atomic number of Pd and a few elements above Pd were found as shown in Fig. 34.

Di Giulio et al.(*277*) also used PdD applied to Si. They found that a rough surface did not cause cracks to form in the PdD and did not produce transmutation products. In contrast, smooth Si produced the crack structure shown in Fig. 35 with Ca and Fe found as transmutation products in the cracks whether the laser was used or not. The same treatment of PdH resulted in Ca, Fe, S, and Ti when the laser was not used, with addition of Zn, Cu, and Cr to the products when the laser was applied.

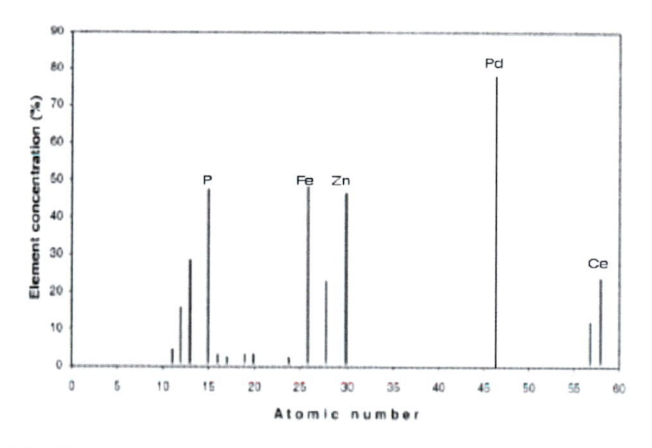

Figure 34. Transmutation produced without laser using 125 nm layer of PdD on Si. Atomic number of Pd is noted for reference.(*276*)

This work was continued by Nassisi et al.(*278*) by implanting boron (B) in the PdD film on Si. As before, transmutation products were found in what appeared to be irregular holes rather than in a cracked surface. These products did not seem to involve boron and many were present regardless of whether the laser was used or not.

Table 10 lists the detected elements in the Si film. The concentration and distribution of elements were variable and related to film thickness. In each case, the result is consistent with the fusion-fission transmutation reaction. Further study(*279*) by these authors found the laser pulse they used caused wire samples and presumably all other samples to heat and then cool before the next pulse was applied. Presumably this heating resulted from energy applied by the laser and not because LENR energy was produced. This rapid heating would be expected to produce stress, reorganization of the material, and cracks. Typically, the D/Pd ratio reached the expected value near 0.7, which means the samples they studied converted completely to the beta phase, but with a low composition. The conditions used this time produced zinc (Zn) as the main transmutation product, while the expected companion element is not reported.

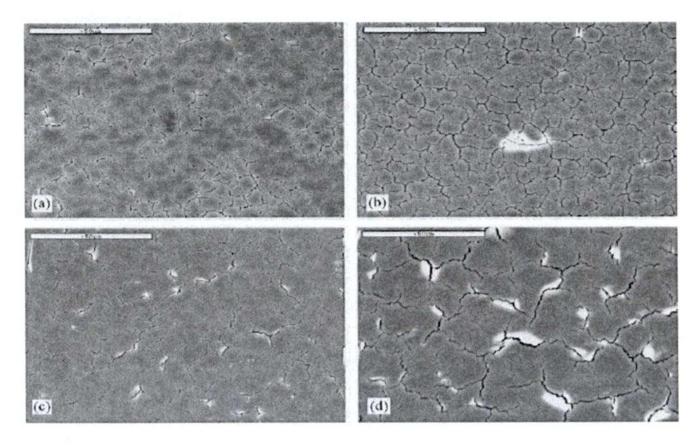

Figure 35. Typical aspects of the surfaces of Pd film deposited on polished silicon wafers, after the completion of the gas-loading phase. (a) D_2 without laser treatment; (b) D_2 and laser treatment; (c) H_2 without laser treatment; and (d) H_2 and laser treatment.(*277*)

Nassisi et al.(*278*) exposed Pd implanted with boron (B) to laser radiation with 648 nm wave-length and 2 mW power while held in either D_2 or H_2 gas. Pits were formed on the surface and many new elements were found on the walls of the pits after lengthy exposure. Use of deuterium produced extra elements without requiring the laser, while H_2 produced transmutation products shown by EDX in Fig. 36 only when the laser was used. The products are those expected from fusion-fission of Pd.

Violante et al.(*280*) provide a tutorial about surface plasmons and how a laser might interact with them. This mathematical description is applied to examining thin films of NiH, both smooth and "black," using a He-Ne laser. Amazingly, the amount of ^{65}Cu relative to ^{63}Cu present as contamination in the "black" sample was increased by laser treatment.

Table 10. List of elements detected using EDX in holes in a thin layer of PdD or PdH containing boron on a smooth Si substrate.(*278*)

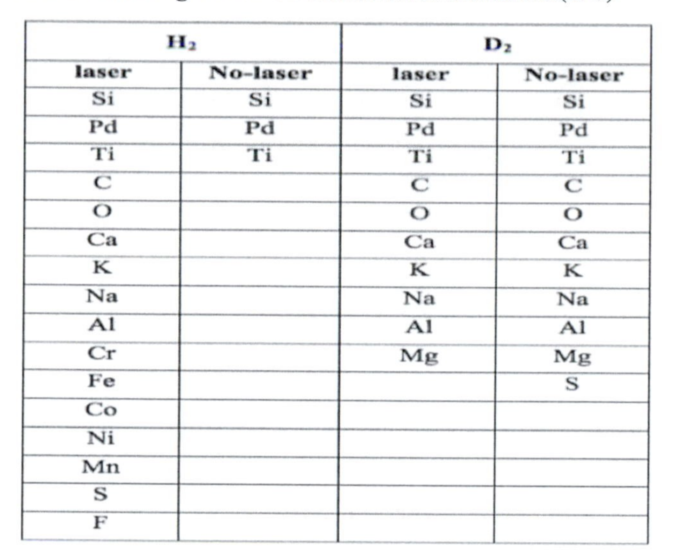

H_2		D_2	
laser	No-laser	laser	No-laser
Si	Si	Si	Si
Pd	Pd	Pd	Pd
Ti	Ti	Ti	Ti
C		C	C
O		O	O
Ca		Ca	Ca
K		K	K
Na		Na	Na
Al		Al	Al
Cr		Mg	Mg
Fe			S
Co			
Ni			
Mn			
S			
F			

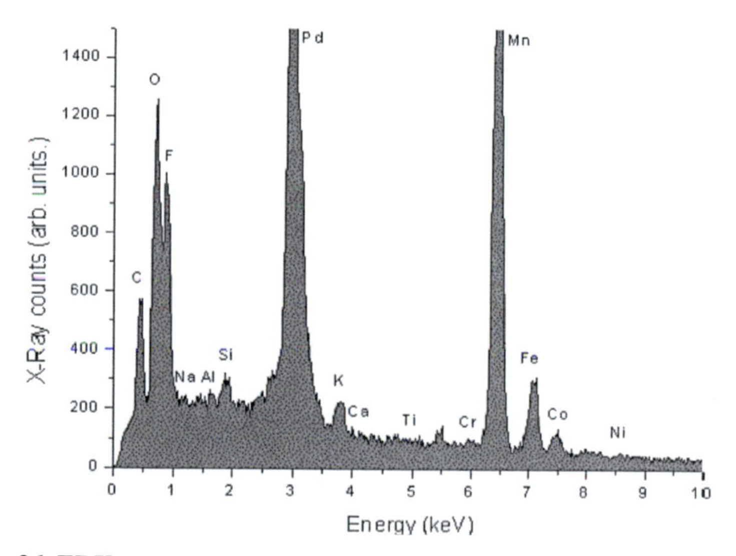

Figure 36. EDX spectrum after exposing Pd in H_2 to laser radiation.(*278*)

Tian et al.(*281*) used a YAG laser (532 nm) in a failed attempt to stimulate heat production from gas loaded PdH with a

H/Pd ratio from 0 to 0.8. Instead, LENR was revealed by the surface showing many cracks and fractures containing silver (Ag) and cadmium (Cd) as transmutation products.

Iwamura et al.(*282*) radiated the surface of their typical composite (Section 2.3.3) with laser light (355 nm). Enhanced conversion of cesium (Cs) to praseodymium (Pr) was not detected. However, uptake of deuterium by the composite from the surrounding D_2 gas was reduced by laser radiation.

2.3.5 CONCLUSIONS ABOUT TRANSMUTATION

Two kinds of transmutation are possible under LENR conditions. Elements applied to the surface of a material containing NAE can apparently fuse with an even number of deuterons to produce new elements on the surface without significant detected radiation being emitted or the final nucleus being broken into smaller parts. On the other hand, when transmutation involves elements in the normal chemical structure or perhaps when protons are involved, transmutation is followed by fragmentation (fission) into two smaller nuclei. These smaller nuclei can be stable or radioactive with a variety of atomic weight combinations. Like the fusion reaction between hydrogen isotopes, transmutation apparently takes place only near the surface and in isolated locations associated with cracks.

A third and more complex transmutation process occurs when high energy is applied with high density, such as in an electric arc. Here, elements are created with more neutrons and protons in their nucleus than present initially. An electric arc in water produced elements surrounding iron on the anode (+) where hydrogen would not be available to cause the reactions typical of LENR. On the other hand, as discussed in Section 2.5.1, bombarding a material by electrons can apparently initiate similar nuclear reactions.

How the large Coulomb barrier can be overcome during transmutation and how the resulting energy is dissipated are serious challenges faced by an explanation. While a few transmuted products might result from a few random occasions when the barrier is overcome, a claim for production of significant power from this process is very hard to justify. Regardless of the method, the demonstrated rate of transmutation has so far been much too small to make detectable energy.

Lithium deposited on the surface during electrolysis is expected to experience transmutation. As the radiation measurements discussed below reveal, the expected energetic alpha emission (8.7 MeV/He and 11.4 MeV/He) resulting from adding an isotope of hydrogen to an isotope of lithium, shown below in bold, may have been detected using the CR-39 method described in Section 2.4.3.

$^{6}Li + d = {}^{8}Be = 2\,{}^{4}He$ Q = 22.4 MeV = 11.4 MeV/He
$^{7}Li + d = {}^{9}Be$ (stable)
$^{6}Li + p = {}^{7}Be$ (electron capture) = ^{7}Li + gamma
$^{7}Li + p = {}^{8}Be = 2\,{}^{4}He$ Q = 17.3 MeV = 8.7 MeV/He

This reaction also is consistent with the observed change in isotopic ratio of lithium observed in the surface region of nuclear-active material.(*283*) In this case, transmutation results in a fusion-fission type-reaction, with the energy released into the two fragments. The resulting alpha particles are expected to interact with the accumulating ^{9}Be in the NAE as a result of the initial nuclear reaction, thereby generating neutrons by forming ^{12}C. This source might account for the neutrons frequently detected along with alphas. Clearly, many questions remain unanswered. A path to these answers is suggested in Chapter 5.

In many studies, transmutation products appear in both PdD and PdH films whether a laser is used or not. This experience demonstrates the importance of the commonly observed cracks rather than the hydrogen isotope present. Of course, visible cracks are too big to be the proposed NAE. Instead, the NAE is expected to be nanogaps present in the walls of visible cracks.

2.4.0 RADIATION

Radiation is not observed when LENR occurs unless it is sought using sensitive detectors. In no case is the amount of detected radiation of any kind consistent with the amount of power produced based on conventional theory. That the required dissipation of nuclear energy does not produce significant detectable energetic radiation is one of the deep mysteries of the LENR process.

The detected radiation consists mostly of photons—consisting of X-ray, gamma ray, IR, RF, and annihilation radiation, along with an occasional alpha (^{4}He) and neutron. In addition, a new and unusual kind of radiation is reported that can cause nuclear reactions in material through which it passes, but not like a neutron. This "strange" radiation is discussed in Section 2.4.6.

2.4.1 How is radiation detected?

Radiation is detected several different ways, with each method having different limitations. Radiation detectors are well understood and very reliable in showing the presence of energetic emissions. However, each kind of detector has a different sensitivity to the different kinds of radiation, requiring care in their choice and use. The following detectors have been used to study LENR.

CR-39

The plastic called CR-39 is sensitive to being bombarded by energetic particles such as alpha and neutrons while being insensitive to photons, electrons, and electronic "noise." The appearance and size of pits produced in the plastic after "development" can be used to estimate the type and energy of the particle. However, the pits assigned to alpha particles could also result from energetic ^{3}He as well as from the larger and very energetic products emitted by fusion-fission transmutation. Because this is an accumulating detector, it can detect a very small amount of radiation, but it cannot determine at what time during the study it was emitted or for how long.

Geiger-Müller Detector
The Geiger-Müller detector (GM) is sensitive to most types of radiation with a wide range of energy. The type of radiation and its energy are not measured directly but must be inferred by measuring the effect of absorbers on the intensity of radiation reaching the GM.

Proportional Detector
Radiation passing through a gas called P-10 (90% Ar + 10% CH_4) generates a current that is related to the particle energy deposited in the gas volume. The total energy of a particle can only be measured if it is stopped completely within the detector. The flux can be determined by counting the rate at which voltage pulses are produced.

Silicon Barrier Detector (SBD) and Germanium Detector (Ge)
These are called semiconductor detectors in which radiation changes the effective resistance of the material by releasing charge carriers. Most energy is dissipated in the material because the material has high density, which allows a wide range of energy to be accurately measured with high resolution. Photons and charged particles can be detected.

Scintillation Detectors
A material, either solid or liquid, is used that emits light when radiation passes through it. The light is detected and measured by a photomultiplier to give values for flux and energy. Widely used solid photo-emitting materials are sodium iodide (NaI), cesium iodide (CsI), or zinc sulphide (ZnS).

Neutron Detection
Neutrons are detected only when they cause a nuclear reaction. These reactions can be detected by measuring the ionization current that results from reaction with ^3He gas or ^{11}B in BF_3. A photo-emitting material, such as $LiCaAlF_6$ or NE-213, can also be

used to detect a resulting nuclear reaction and determine the neutron energy.

2.4.2 Source of radiation

As Hagelstein(*284-286*) has evaluated, radiation can be emitted from several sources. Unfortunately, the source is very difficult to identify and can be incorrectly assigned. For example, hot fusion can be initiated when cracks form (fractofusion), radiation can result from the cold fusion reaction itself, and transmutation can produce both prompt radiation from the event and delayed radiation from radioactive products. Many secondary sources are also possible as the primary radiation interacts with material through which it passes. Of significance is the failure of radiation intensity resulting from LENR to correlate with energy production or the amount of other nuclear products. This lack of apparent correlation might result because the energy of generated radiation is too low to leave the apparatus.

Although the presence of detected radiation demonstrates a nuclear reaction is underway, it gives very little understanding about the source. The many different nuclear events occurring at the same time make interpretation very complex. In addition, H and D appear to produce radiation having different energy, resulting in different rates of escape and detection, further complicating interpretation. In other words, many important clues are being missed. Nevertheless, high-energy radiation can demonstrate without a doubt the presence of a nuclear reaction.

2.4.3 Detected radiation using CR-39

Workers at the SPAWAR Space and Naval Weapons Research Laboratory in San Diego used this method for many years before the work was terminated.(*287-290*) Much of the work is described by Mosier-Boss et al.(*291-305*) in many papers.

Evidence shows several kinds of energetic particles are produced at low rate when palladium is electroplated on a cathode using an electrolyte containing D_2O+LiCl+$PdCl_2$. Presence of a magnetic field is found to affect the morphology of the deposit and the ability to produce radiation. These emissions consist of neutrons having energy typical of D-D hot fusion (2.54 MeV) and 14 MeV neutrons possibly from D-T hot fusion; 3-10 MeV protons; and 2-7 MeV and 7-15 MeV alpha. Although these nuclear products can be assigned to various hot fusion reactions, they might also be produced by transmutation reactions of palladium, which the authors suggest as a possibility. Unexpected presence of Fe, Al, Cr, and Ni were also found on the surface, perhaps as result of a proposed fusion-fission reaction between Pd and D.

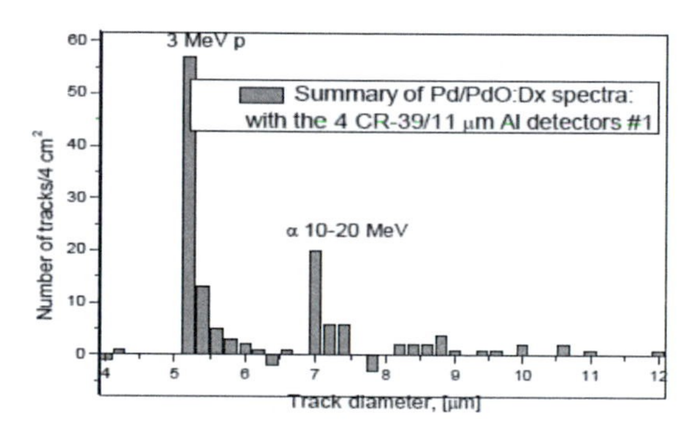

Figure 37. Typical result from a CR-39 sample showing distribution of track (pit) diameters and the calculated energy based on calibration using particles of known energy.(*306*)

Lipson et al.(*306-308*) applied this method using activated Pd made by flame-heating the metal in air, which cleans the surface and produces a blue-colored oxide identified as Pd/PdO. This oxide is easily reduced by hydrogen to produce a very reactive surface containing very little oxygen. Titanium (Ti) hydride was also studied. The CR-39 detector was used either during electrolysis or when the materials were treated in different ways afterwards. Steady improvements in the ability to accurately determine energy has reduced the uncertainty for the two commonly observed energetic particle emissions, one near 3 MeV for the protons and the other in the range 10-20 MeV for the alphas, as shown in Fig. 37. Their studies are summarized in chronological order below, from which steady progress characteristic of this field can be seen.

2000 (*309*) A sample was made by applying gold to Pd after it had been oxidized in air, identified as Au/Pd/PdO. This was electrolyzed in D_2O-Na, after which D_2 was rapidly removed under vacuum while heat and strain were applied. A NE-213 detector detected weak neutron radiation with energy of 2.45 MeV and 3 MeV protons were detected using a SBD. This result was confirmed using CR-39, which apparently also detected energetic tritons. This is radiation characteristic of hot fusion.

The work(*310*) also showed evidence for weak 8-14 MeV alpha emission. This study was motivated by the belief that multi-phonon excitations combined with some screening of the Coulomb barrier could initiate hot fusion. The behavior is consistent with fractofusion being the source of the radiation, not LENR.

2002 (*246*) Pulsed glow-discharge using D_2 and voltage of 0.8-2.45 kV were applied to a Ti cathode, which produced tracks identified as 3 MeV protons, presumably from the hot fusion reaction. Other tracks were identified as 13.5±2.0 MeV alpha particles, and a few tracks suggested either 1.7 MeV photons or 2.3 MeV deuterons. Based on the yield of 3 MeV protons as a function of applied voltage, a screening factor of 620±140 eV is calculated for initiating the hot fusion reaction in Ti.

2002 (*311, 312*) Pd plated on glass or Al_2O_3 was electrolyzed in Li_2SO_4-H_2O and charged particle emission was detected using CR-39. Alpha particles having 11-16 MeV and protons near 1.7 MeV were detected.

2003 (*313*) A study using gas discharge with H_2 revealed X-ray emission (photon) with an energy of 1.2-1.5 keV based on using plastic scintillation detectors. This behavior is consistent with the observations reported by Storms and Scanlan(*314*) for gas discharge using D_2.

2003 (*315*) Alpha emission is explored in more detail by attaching CR-39 chips to both sides of Pd plated on glass after which the assembly is electrolyzed in D_2O-Li. In this case, radiation originates in material that is shielded from electrolytic action, hence has low deuterium content. Nevertheless, 1.5-1.7 MeV protons and 11-16 MeV alphas are found.

A sample of Ti was exposed to gas discharge in D_2, which also produced ~3 MeV protons with the expected energetic alphas. Another particle identified either as 1.4 MeV proton or 2.8 MeV deuteron, called 'd?p', was detected. The flux of 3 MeV protons increased with applied voltage while the alpha and 'd?p' fluxes remained relatively unchanged, but did appear to increase slightly when applied power is increased.

Laser radiation (1054 nm) was applied apparently in vacuum as a very small spot (<15 μm) with high intensity (1-2 10^{18} W/cm^2) to 30 μm thick foils of TiD and TiH previously prepared using electrolysis.(*316*) Alpha radiation (13±2 MeV) similar to that found during the electrolytic study was detected when TiD was used. Alphas and other energetic particles were also emitted from TiH, but with a different energy profile.

2004 (*317*) Foils of Pd/PdO or Ti implanted by helium on one surface increased the rate of energetic alpha emission from the implanted surface when electrolyzed as the cathode in H_2O-Li or D_2O-Li. Such implantation would produce local stress in a region 20-30 nm thick, which would perhaps have effects not anticipated by the authors.

2005 (*307*) The Pd/PdO was loaded to $PdD_{0.7}$ by electrolysis in D_2O-Li. The sample was placed under stress and allowed to deload in air. The observed temperature increase is typical of reaction with air, which is not the explanation provided by the authors. Loss of D_2 from PdD is an endothermic reaction, not exothermic as assumed. Consequently, this sample cannot be said to have made anomalous energy. The usual 3 MeV protons and 11-16 MeV alphas were detected using CR-39. Soft X-ray emission near 1.30±0.15 keV was also detected.

2005 (*318*) The usual Pd/PdO structure was reacted with light hydrogen during electrolysis in H_2O-Li to produce PdH, after which the sample was allowed to react with air. Once again, CR-39 was used to reveal emission of alphas with 10-13 MeV and 15-17.5 MeV. Protons with 1.7-1.9 MeV were also detected. This result is similar to previous studies of PdH but once again with a different energy compared to the same type of radiation from PdD. Apparently, both isotopes of hydrogen can produce the same nuclear products, but with different energy.

2007 (*319*) The authors examined the relationship between their measurements and the concept of screening potential, which is sometimes called the enhancement factor (Section 2.8.0). While the occasional proton, triton and neutron emissions clearly result from hot fusion, explaining the observed rare alpha emission is a challenge.

Soft X-rays have been observed during several studies, which the authors summarize. They conclude both glow discharge and other treatments of PdD produce low-energy X-rays below 1.5 keV. Because photons of such a low energy would suffer significant reduction in intensity while passing through the walls of a typical apparatus, the flux at the source would have been much greater than detected radiation.

2009 (*306, 320, 321*) Electron bombardment (0.6 mA, 30 keV) of the Pd/PdO+D and TiD produced the frequently detected 3 MeV protons and 11-20 MeV alpha emissions. Bombardment of deuterium containing material by electrons has a long history discussed in Section 2.5.1.

A project to replicate particle emission while Pd is electroplated by co-deposition, called the Galileo Project, was initiated by Krivit[15] after which many resulting CR-39 samples were analyzed by Lipson or Roussetski.(*322-324*) Mixed success was obtained. Although the behavior has been independently replicated, success apparently requires skill and some luck, as is characteristic of LENR in general.

Oriani and Fisher (*325-327*) suspended CR-39 chips in the electrolyte and in the vapor above the surface of either H_2O-Li or D_2O-Li electrolyte. Energetic particles were detected during electrolysis that were claimed to support the polyneutron theory of Fisher.(*328*) The polyneutron theory is discussed in Sections 3.2.4 and 4.3.0.

2.4.4 SUMMARY OF CR-39 STUDIES

Studies using CR-39 clearly revealed evidence for hot fusion taking place when samples are subjected to conditions expected to produce LENR. Not revealed is when during the study the low-rate nuclear reaction took place and the instantaneous rate at that time. Nevertheless, the measured flux is always very small compared to the flux required to account for detected energy production from a source considered LENR.

Determination of the alpha and proton energy is difficult using CR-39 because emission from any location other than the true surface of the source would cause reduction in energy of the particle. This behavior suggests the values in Fig. 37 are lower limits to the actual initial energy.

A similar range of energy produced when PdH is examined eliminates deuterium as the unique nuclear reactant. Similar behavior observed when Ti is used eliminates Pd as the unique source. What nuclear reaction can produce alpha emission with energy near perhaps 20 MeV that can be initiated when either isotope of hydrogen is present in different chemical structures? An answer is proposed in Chapter 5.

---

[15] See Nov. 10, 2006, issue of *New Energy Times* (http://www.newenergytimes.com).

2.4.5 Real-time detection of radiation

Storms and Scanlan(*314*) studied energetic particle emission from a cathode made from various metals using steady glow-discharge in D_2 with applied DC voltage of 500-900 V. What appeared to be deuteron emission was detected using a SBD and showed energy between 0.5 MeV and 3 MeV, with a regular series of peak energy about 0.5 MeV apart. Intensity of the peaks decreased as the energy of each peak increased, as can be seen in Fig. 38. Intensity of emission and energy of individual peaks are sensitive to gas pressure, type of gas, and applied voltage, but the basic shape of the spectrum remained unchanged. This shape suggests the generation process is influenced by a quantized feature in the source — perhaps by the number and kind of atoms in the structure where a nuclear reaction takes place. In addition, such a regular series of energy values with an exponential increase in intensity at lower energy suggests a resonance process as the source. Radiation of such high energy can only result from a process involving mass-energy conversion. No indication of alpha emission was seen even though alpha radiation could be detected. X-rays were detected having energy below or above applied voltage, depending on the voltage used, with a greater applied voltage producing greater X-ray energy. None of this radiation could be detected outside of the apparatus.

Iwamura et al.(*329*) observed X-radiation near 75 keV during electrolysis in D_2O-Li, after which lead (Pb) was found on the surface of palladium. The authors attributed this radiation to the K-a X-ray of Pb, which would only result if an equal source of energy were available to eject an electron from the Pb atom. The source of lead contamination and the required energy to eject the electron is unknown. A few neutron bursts were also detected without correlation to X-ray production.

Piantelli(*199, 200, 330-341*), with many co-authors, has published a series of papers over the years showing radiation emitted from nickel when exposed to hydrogen gas (H_2), which is summarized in Fig. 39. A large photon emission was found with energy near 0.66 MeV, smaller emissions were detected near 1.5

MeV and 2.6 MeV, and a few delayed energetic particles were observed using a cloud chamber. The cloud chamber result is ambiguous because radon and other natural radioactive elements will be present to provide occasional particle emission. Evidence for excess energy and transmutation products is also reported.

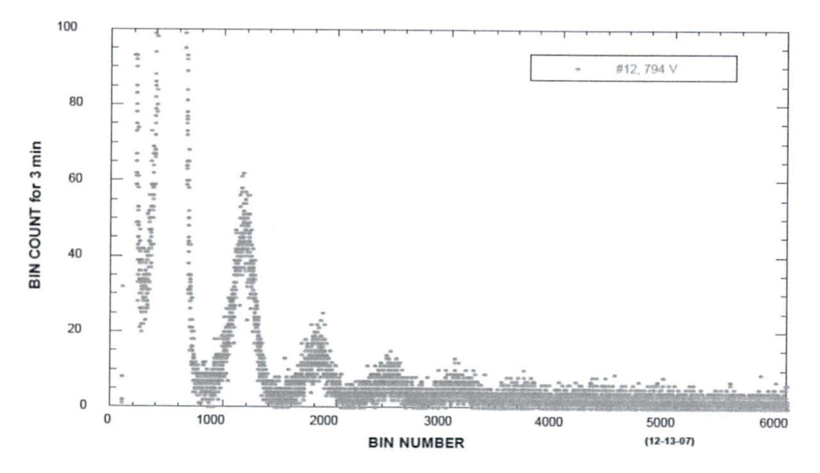

Figure 38. Spectrum of deuterons emitted from a metal target during DC glow discharge in D_2. The BIN numbers correspond to energy values using a silicon barrier detector with BIN 1600 equal to approximately 1 MeV.(*314*)

Violante et al.(*342*) used a Ge detector to measure radiation from thin films of various metals consisting of Cu(45 nm), Ni(45 nm), Cu(25 nm)/Ni(45 nm), and Cu(25 nm)/Pd(45 nm). These films were deposited on polyethylene and subjected to electrolysis in H_2O-Li while radiation was measured. In addition to radiation, the total inventory of potential transmutation products was determined. The double layer of Cu/Pd (Fig. 40) and Ni (Fig. 41) showed evidence for radiation. Note the tendency to produce radiation with regularly spaced energy, similar to the behavior reported by Storms and Scanlan (Fig. 38), but at a much lower energy and with smaller intervals. Tritium was found in one cell.

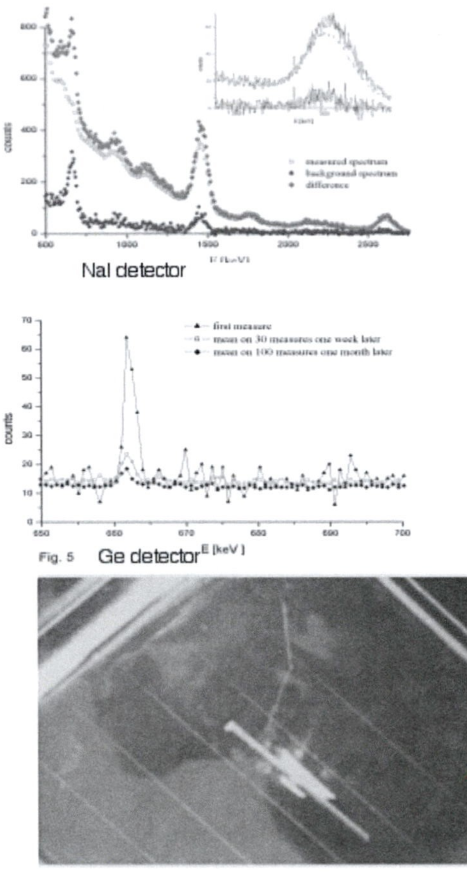

Figure 39. Examples of radiation reported by Piantelli et al.(*339*) using specially treated Ni exposed to H_2. The cloud chamber picture shows a single energetic particle being emitted and suffering deflection when it encountered a nucleus in the gas.

Bush and Eagleton(*343*) used a NaI detector to study a Pd+Ag alloy cathode in D_2O-Li and a Ni cathode in H_2O-Li using electrolysis. Unfortunately, the energy of detected radiation is too poorly resolved to allow unambiguous interpretation. The Pd+Ag cathode apparently showed photon radiation near 76 keV that produced what may be a series of peaks between 65 keV and 67 keV, and between 21 keV and 24 keV. When Ni was used as the cathode in H_2O-Li, the total radiation above background was found to roughly correlate with excess power production, as shown in Fig. 42.

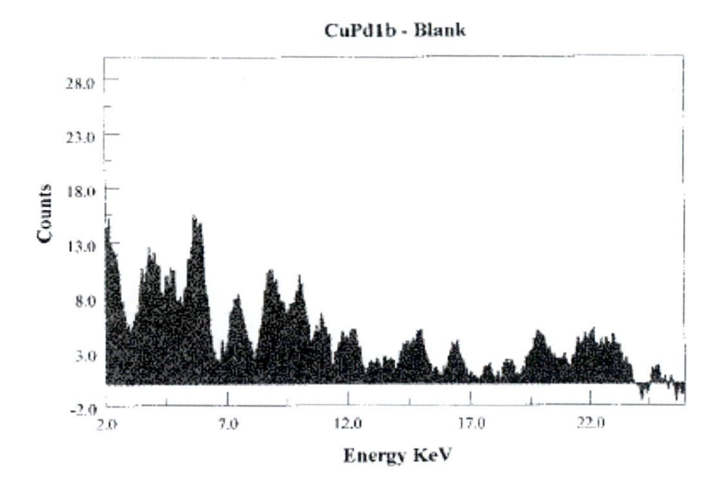

Figure 40. Count rate in excess of background for active Cu/Pd cathode electrolyzed in H_2O-Li.(*342*)

Storms and Scanlan(*344*) exposed coatings of various metals on Pd to H_2 gas and detected radiation using GM detectors located outside the apparatus. The flux was observed to increase slowly over several hours. Insertion of a lead (Pb) absorber between the source and the GM detector caused an immediate reduction in flux followed by slow decay, shown in Fig. 43.

Apparently, some part of the material in the detector[16] was activated by radiation being emitted by the sample. This activation appeared to produce two radiation sources with slightly different decay rates, shown in Fig. 44. Rapid decay was followed by a longer decay with a half-life of about 76 min. Once the lead (Pb) absorber was removed, the primary flux immediately increased, thereby causing a slow increase in radiation as the activation

---

[16] The mica window was analyzed and found to contain potassium and carbon as the only elements not common in the shell of the detector or in the apparatus itself.

process resumed. The sample showed cracking and pits typical of all samples found to emit similar radiation, as can be seen in Fig. 45. Eventually, radiation from all samples slowly dropped to zero for no apparent reason.

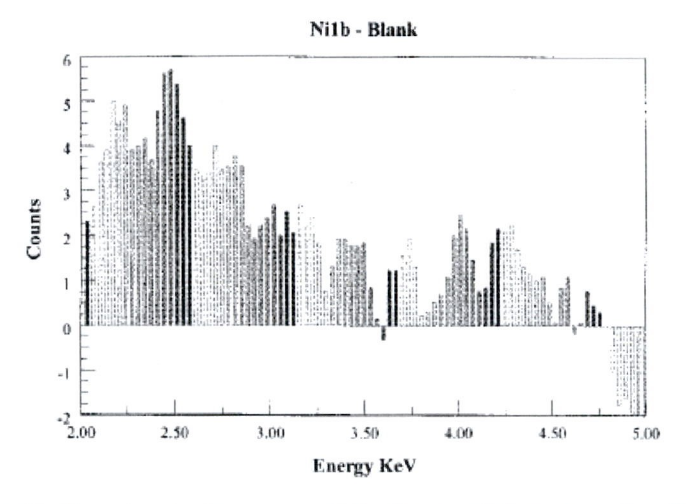

Figure 41. Count rate in excess of background for an active Ni cathode electrolyzed in H_2O-Li.(*342*)

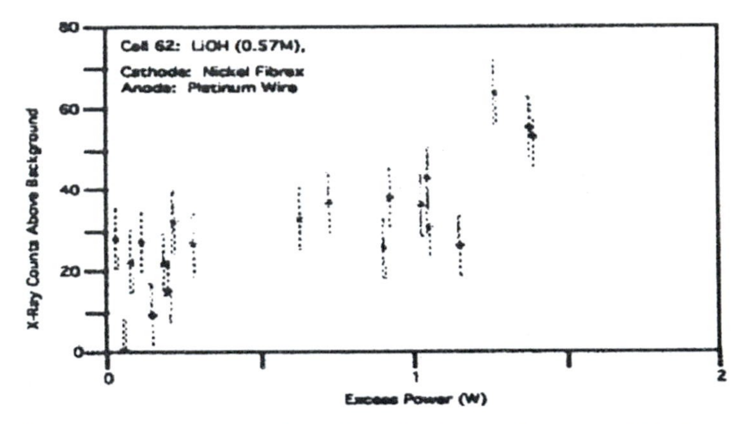

Figure 42. Photon radiation detected by Bush and Eagleton(*343*) from a Ni cathode in H_2O-Li electrolyte while making excess power.

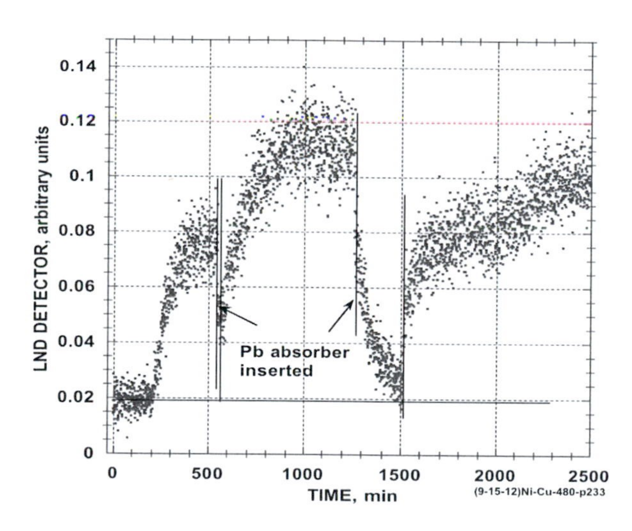

Figure 43. Ni coated by Cu and exposed to H_2 gas. Reaction caused cracks and radiation.(*344*)

A nickel cathode studied by Bush(*345*) showed radioactive decay with a half-life of about 3.8 days (Fig. 46) after being electrolyzed as the cathode in H_2O-Rb_2CO_3. To further complicate interpretation, the half-life appeared to be different for the different energies of radiation, suggesting the activity resulted from a mixture of radioactive isotopes. Unfortunately, the energy was not reported and the claimed radioactive isotopes cannot be identified because the flux is too small.

Notoya and Ohnishi (1996)(*192, 196*) electrolyzed H_2O-Na_2CO_3 with a nickel cathode and claimed to detect gamma emission from the electrolyte with energy near 1368 keV while excess heat was made. The conclusion about the source of radiation provided in the paper is best ignored until more measurements are available.

Bernardini et al. (2000)(*190*) electrolyzed titanium (Ti) in D_2O-K_2CO_3 and reported heat production as well as weak gamma radiation from the cathode, with values for two energy ranges of 890-982 keV and 1036-1123 keV. The source cannot be identified based on the information provided.

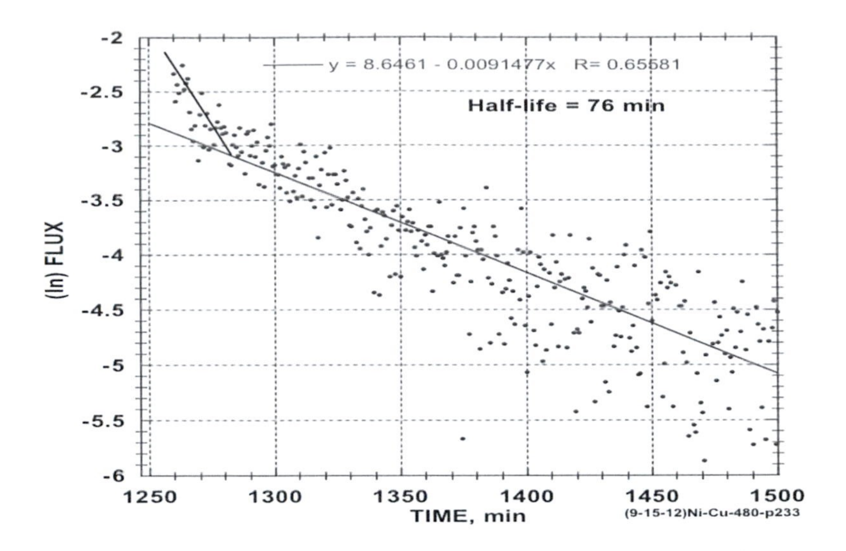

Figure 44. ln flux vs time for the decay in Fig. 43 between 1250 and 1500 minutes. Two different decay rates are revealed, with the slower having a half-life of about 76 min.(*344*)

Iyengar et al.(1990) reported occasional gamma ray bursts of 14-20 min duration at 1186 keV to result from Pd while being electrolyzed in D_2O-Li. The peaks plotted in Fig. 47 showed slow decay consistent with a short-lived radioactive isotope after sudden production. These bursts were not correlated with occasional low intensity neutron bursts.

Afonichev(*347*) mechanically deformed a piece of titanium alloy while in deuterium gas at 710°C. RF radiation in the radio frequency range was emitted as bursts during the process without neutrons being detected. Tritium was detected in the surface region after the process. Titanium is noted for producing hot fusion (fractofusion) as it cracks, but in this case, the expected neutrons were not observed. Detection of RF radiation is unique and apparently associated with production of tritium initiated by stress cracks.

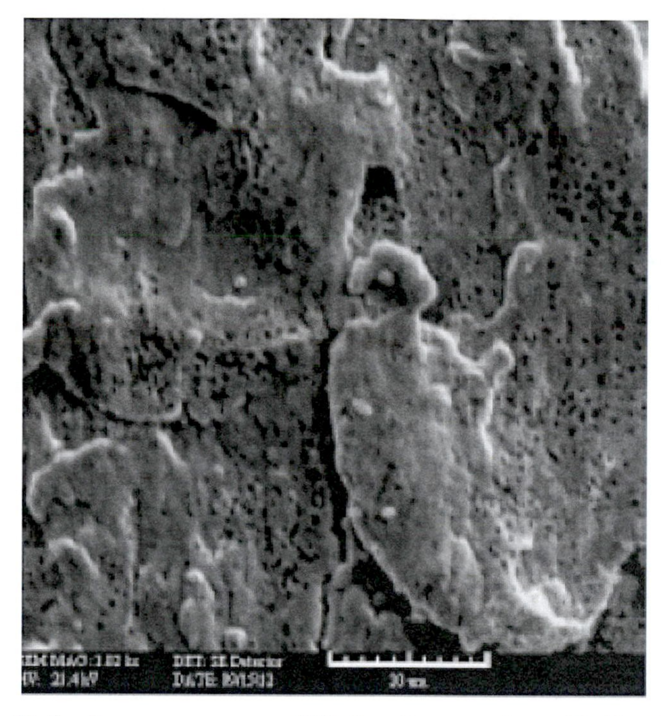

Figure 45. Surface appearance of Ni-Cu after reacting with H_2.(*344*)

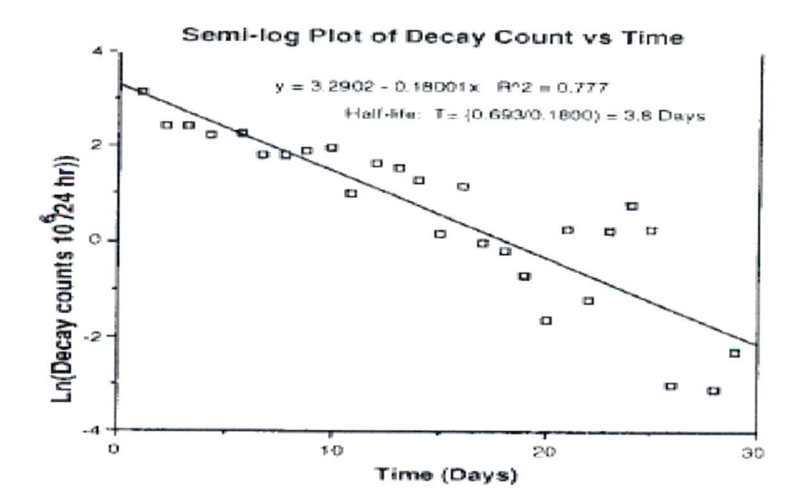

Figure 46. Decay plot of a nickel cathode after electrolyzing using H_2O-Rb_2CO_3.(*345*)

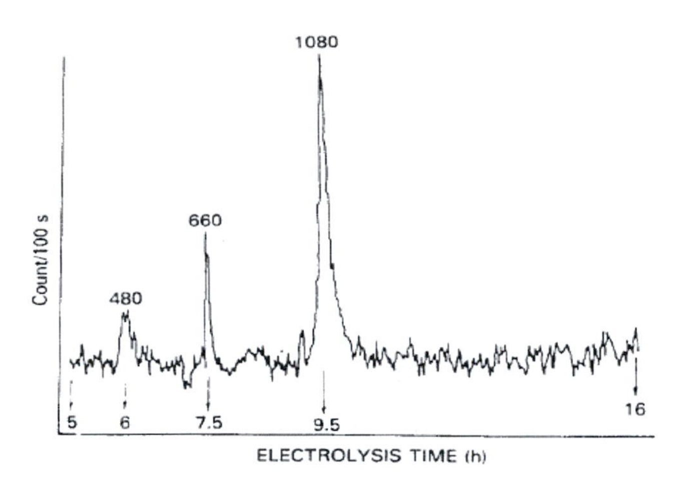

Figure 47. Gamma events as a function of time while Pd is electrolyzed in D_2O-Li.(*26, 346*)

Karabut(*348*) using gas discharge, observed a cathode to produce photons in the range 1.5-2.0 keV (0.6-0.8 nm wavelength) with coherent behavior similar to a laser and emitted perpendicular to the cathode surface. The intensity was sufficient to cause damage in nearby organic material. Normal incoherent radiation was also emitted in all directions when energy applied to the discharge was reduced.

In a summary of their work, Karabut et al.(*349*) measured the energy of X-radiation and found a few bursts up to 10 keV when the discharge voltage was 1-2 keV. Some radiation continued for many hours after the discharge was turned off. Some photon radiation was collimated as observed previously, which is produced using all gases and studied cathode materials. Radiation flux is increased by increased current, longer duration of discharge, and increased voltage. Radiation occasionally continued after discharge was terminated.

Keeney et al. (2003)(*350*) studied the fractofusion reaction known to occur in TiD. Charged particles and neutrons typical of hot fusion were detected at low level at rates consistent with previous studies of this material.(*351-359*)

Various forms of secondary radiation can be generated when energetic particles, such as beta radiation, are stopped in a material, which results in X-rays called Bremsstrahlung as noted by Swartz and Verner.(*360*) Because this kind of secondary radiation always has a much lower intensity than the primary flux, detectable flux would not be expected unless the primary beta radiation from LENR had a large flux with a large energy. Absence of significant Bremsstrahlung indicates absence of significant beta flux from a source within the apparatus.

2.4.6 "Strange" radiation

Existence of "strange" radiation has been proposed independently of and before LENR was discovered. Nevertheless, LENR has encouraged a search for its presence. While some kind of unusual radiation might be involved either as a catalyst for or emitted as consequence of the LENR reaction, its role in the LENR process is unknown. Consequently, the reported observations described here are provided only to give information, not to suggest a relationship to LENR. Nevertheless, radiation can be emitted under certain conditions that can initiate nuclear reactions in nearby material, but not in the way neutrons produce nuclear reactions. Clearly, more study is required, but with greater willingness to accept evidence for such novel radiation.

Even before LENR was discovered, McKibben(*361*) speculated privately about a primordial particle having a fractional charge and the ability to cause nuclear reactions. Bazhutov et al.(*362-365*) propose existence of a heavy negative particle arriving from outer space called an Erzion. Rafelski et al.(*366*) suggest a similar particle they call the X-particle. In both cases, the particle is proposed to catalyze a nuclear process.

Matsumoto(*267, 268, 367-372*) describes what he calls the Iton, which can carry energy from the LENR reaction and is occasionally detected as unexplained tracks in X-ray film. Savvatimova(*373, 374*) also reported seeing unusual tracks in X-ray film when transmutation occurred while using deuterium during glow discharge.

Lochak and Urutskoev(*375*) detected novel tracks in film after a thin wire of Ti was exploded in water by passing high current. Changes in the isotopic ratio of the Ti and transmutation products were also produced. Tanzella et al.(*376-378*) found this method produced excess energy.

Biological effects of radiation produced by a similar process was studied by Pryakhin et al.(*379*) The process used by the author would be expected to produce the large clusters of electrons called EVO by Shoulders(*380-382*), which can pass through materials and cause unusual nuclear reactions. The so-called "ball lightning" identified by Lewis(*383*) might be related to the EVO.

Oriani and Fisher(*384*) detected radiation well away from an operating electrolytic cell using CR-39. The results can be explained just as well as "strange" radiation being emitted instead of polyneutron emission, as favored by Fisher.

Vysotskii and Adamenko(*385*) (Section 2.5.1) found that intense bombardment of materials by electrons produced unusual radiation that caused nuclear transmutation in nearby material. Nuclear activation of material contained in a GM detector reported by Storms and Scanlan (Fig. 44) might also result from such radiation.

2.4.7 CONCLUSIONS ABOUT EMITTED RADIATION

Because required detectors are seldom present during most LENR studies, many occasions of radiation production might be missed. Nevertheless, low intensity radiation is often detected when a search is made during heat production. The possibility of many kinds of nuclear reactions occurring makes the source of this radiation difficult to determine. For example, this radiation might be produced by transmutation reactions as well as by D-D fusion, both the hot and cold types. Emission of coherent radiation having a particular direction relative to the cathode is especially important but easy to miss unless the detector is properly located. Emission of novel radiation, with its unexpected interaction with materials and detectors, adds further complexity. Some detected photon radiation is found to have coherent properties.

Creation of radioactive products would add decay radiation as they rapidly form stable isotopes later observed on the surface. This conclusion will be discussed in more detail in Chapter 5. Many studies of TiD show radiation from fractofusion, which is not related to LENR because hot fusion is the source.

Absence of significant Bremsstrahlung radiation helps limit the kind, intensity, and energy of radiation being generated at the site of LENR.

While CR-39 is useful in showing the presence of a small radiation flux, the calculated energy is much less certain, as Fig. 48 makes clear. Careful calibration is required, absorbers must be used to identify the kind of particle being detected, and the very low flux adds statistical uncertainty.

Apparently, no matter which method for initiating LENR is used, PdD and TiD emit a very small amount of energetic radiation during the process with proton energy near 3 MeV and alpha energy near 10-20 MeV, along with particle radiation with greater energy. PdH produces the same emissions but with slightly different energy.

Gas discharge produces a significant amount of photon radiation with energy equal to or greater than applied voltage, some of which leaves the cathode in a particular direction relative to the surface. This preferred direction would cause some radiation to be missed by a detector located in the wrong position. On the other hand, useful knowledge about the process would result if the location from which the different radiations were emitted could be determined.

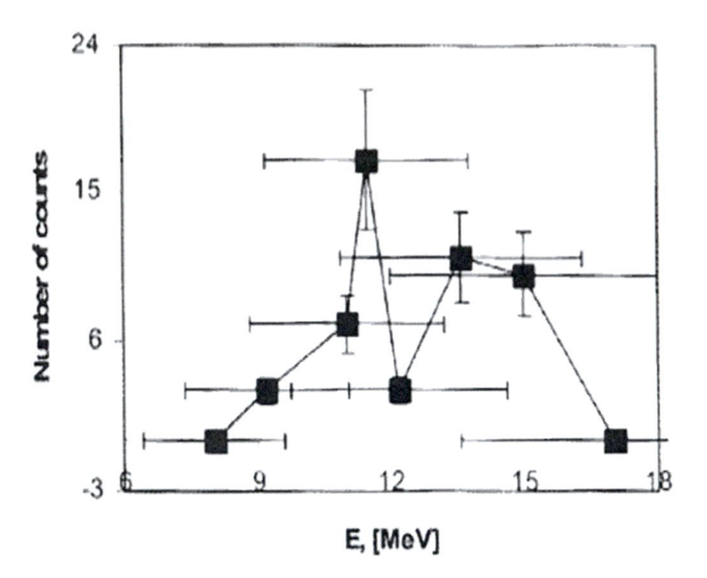

Figure 48. Alpha spectrum resulting from PdD using CR-39 to which error limits are applied.(*319*)

The neutron and proton energies indicate occurrence of hot fusion at low rates in PdD or TiD when subjected to a variety of treatments.

Evidence for fusion-fission of Pd is reported by many studies. This process would not be expected to produce alpha emission unless helium was one of the fragments. This possibility is discussed in Chapter 5. Some observed soft X-rays might result as Bremsstrahlung from these fusion-fission reactions.

Although radiation clearly demonstrates nuclear activity, the measurement and interpretation are too complex to be useful in identifying their source and explaining LENR at present.

2.5.0 APPLIED ENERGY

Applied energy can either initiate a nuclear process or increase its rate once it starts. Which of these effects is operating is frequently not clear, although in many cases the radiation energy obviously increased the rate of a reaction already underway.

2.5.1 Applied electron flux

Applied electron flux has been observed to produce nuclear reactions in some materials. This treatment has two plausible effects; it can partially reduce a positive charge in local regions or it can produce large local magnetic fields. However, large currents are routinely caused to flow in materials without any evidence for nuclear reactions being initiated. In addition, electron bombardment of materials is also well known to be ineffective in causing nuclear reactions. Perhaps, a novel condition must be present for electrons to cause nuclear behavior similar to the special NAE required to initiate LENR by other methods.

Kamada (1992)(*386, 387*) bombarded aluminum (Al) containing D or H with 200-400 keV electrons in a SEM and detected energetic particle emission identified as alpha emission using CR-39. The hydrogen in the aluminum was found contained in bubbles or tunnel-like structure that forms in the region 50-100 nm from the surface as a result of hydrogen ion implantation. During later studies(*388, 389*), 175 keV electrons were applied to similar material. Local melting was observed only in regions containing deuterons. Success in producing melting depended on the amount of D loaded into the Al and energy of bombarding electrons. The authors describe how such melting can only result from a nuclear reaction.

Matsunaka et al.(2002)(*131, 132*) used 3 keV electrons to bombard PdD coated with copper and uncoated TiD. Slight photon radiation in the 10-20 keV range was detected.

Chernov et al.(2007)(*390, 391*) applied 20 keV electrons to PdD with a current between 50 μA and 150μA. Organic dye was used to detect emission of D atoms from the surface. The electron flux was found to cause loss of D_2 from the material and to cause emission of atomic D from all surfaces of the sample, even from where electrons did not impact.

Adamenko and Vysotskii(*385, 392-399*) applied intense and short duration bursts of electrons to materials used as the anode while looking for radiation and transmutation products. Photon radiation between 2 keV and 10 MeV was detected with a maximum energy near 30 keV. Some of this radiation would be normal X-rays and some might result from gamma emission as new elements are created. This high concentration of electrons was found to create many stable elements and to even convert radioactive ^{60}Co to a stable isotope. In addition, "strange" radiation was emitted from the anode that caused transmutation when it interacted with nearby materials.

Although all of the conditions described above are far removed from those present during LENR, the observations reveal the unexpected ability of energetic electrons to cause nuclear reactions in some materials. Once again, a unique NAE is expected to be required before a nuclear reaction can occur.

2.5.2 Applied laser radiation to produce fusion

Laser radiation, when applied to a nuclear active material, is found to affect the rate at which energy is produced as well as cause emission of radiation and produce transmutation products. As discussed in Section 2.3.4, location of transmutation products suggests a nuclear reaction takes place in the walls of cracks located within the surface region of active material. In contrast, fusion is not so easy to locate. This section focuses mainly on how laser radiation affects energy production and explores the question, "Are heat and tritium production affected by laser radiation in the same way as transmutation and are each of these nuclear products formed at the same location?"

Laser radiation was suggested as a stimulant for LENR as early as 1990.(*400*) In 1991, Beltyukov et al.(*401*) report being able to initiate the commonly observed hot fusion reaction produced in Ti-D by exposing the material to laser radiation. Simply exposing liquid D_2O to laser had no effect.(*402*)

Violante et al.(*403*) and Apicella et al.(*144, 404*) applied 632 nm (33 mW) laser radiation to a cathode loaded to $PdD_{0.95}$ while it produced excess energy during electrolysis. Presumably, the entire surface was the source of fusion energy while only a small spot on the surface of one side was stimulated by the laser where additional extra energy might have been produced. Use of p-polarization produced excess energy while s-polarization did not, suggesting a structure having special orientation to the surface is important. Such a structure could be provided by cracks that generally form perpendicular to the surface.

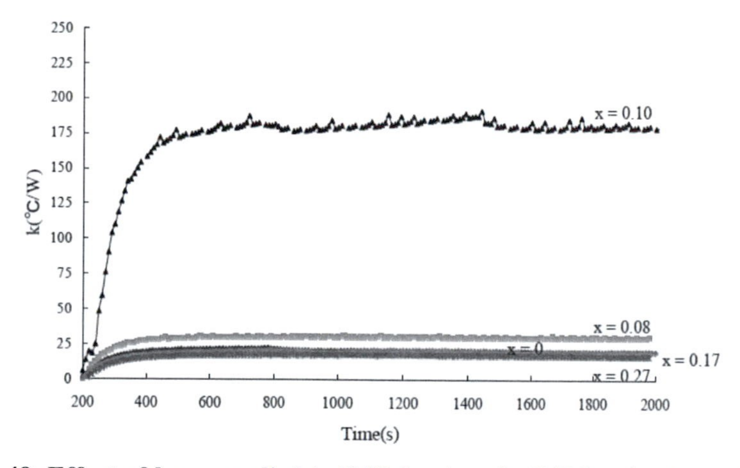

Figure 49. Effect of laser applied to PdD having the D/Pd ratio shown. The laser had a wave length of 532 nm (green) with a pulse width of 50 μsec.(*405*)

Tian et al.(*405*) applied laser light of 532 nm and 2 watt to Pd wire containing different amounts of deuterium. Excess energy was related to the loading ratio and how energy was applied, *i.e.* whether it was continuous or pulsed at different durations and rates. The most impressive power was produced by applying a pulse width of 50 μsec to $PdD_{0.10}$, shown as the upper curve in Fig. 49. The sample was mostly α-PdD with only a little β-PdD present, based on the composition. This treatment is expected to produce stress-cracking, which was not considered by the authors. Obviously, a large composition is not required and the effect is not directly related to composition when the laser is used. Apparently, other variables were influencing energy production during this study.

Barmina et al.(*406*) note that nuclear reactions have been initiated under various conditions when laser intensity is near 10^{10}-

10^{13} W/cm^2. With this experience in mind, they attempted to form tritium by exposing Ti or Au targets immersed in D_2O (98.8% D) to laser radiation. Power density of 5x10^{11} W/cm^2 was applied at 532 nm in one case and at 10^{13} W/cm^2 using 1064 nm radiation in another case, with the former wave-length having the greater effect even at lower intensity. Although tritium could be made simply by exposing the target while immersed in pure D_2O, applying electrolysis in D_2O-Na greatly increased tritium production. The process abraded the surface to create a complex morphology.

Badiei et al.(*407, 408*) irradiated vapor from the hot surface of Fe_2O_3+K in D_2 gas using a pulsed laser (2.2 eV, 564 nm) and detected production of energetic D^+ ions (630 eV) using the time-of-flight method for energy measurement. While vaporization of D+ is expected, the reported high energy of the ion is a surprise. The authors propose the laser causes bonding electrons in the gas molecule to be ejected, as shown in Fig. 50, after which Coulomb repulsion pushes the two nuclei apart with great energy. They propose the amount of energy reveals that the two deuterons were initially much closer than in a normal D2 molecule and perhaps close enough to fuse after application of only a little extra energy. The authors assumed this material, which they call Rydberg matter, consists of metallic hydrogen. The possibility that this large energy might result from breaking a bond between D and O or between D and Fe is not considered. This possibility could account for the large ejection energy without having to assume a small bond distance between deuterons. Whether this energy state has any relationship to LENR, Rydberg molecules, or metallic hydrogen is not clear. Nevertheless, the method provides a useful tool to identify the energy state of hydrogen present in the surface region. Raman spectroscopy(409), laser induced fluorescence(410), and X-ray imaging induced by laser radiation also have been used(411) to explore the energy state of hydrogen on a surface.

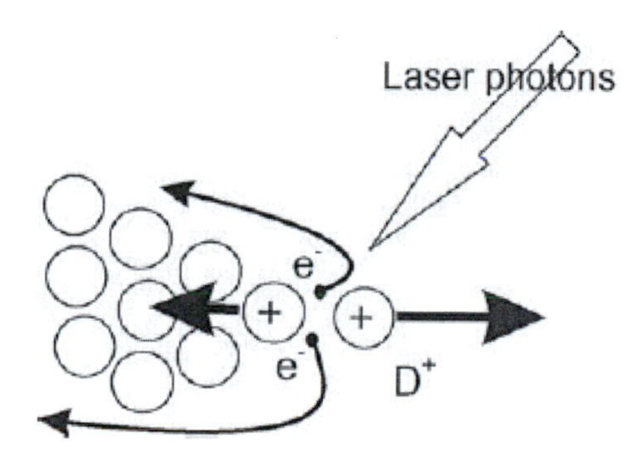

Figure 50. Proposed effect of laser photons on electrons holding d nuclei together in a metallic-like bond near the surface of a material.(*407*)

Arata et al.(*412*) and independently by Letts and Cravens(*413, 414*) stimulated production of energy by applying laser radiation during electrolysis of D_2O-Li. Letts found a coating of gold (Au) is required on the Pd cathode before the laser has any effect and that a weak magnetic field apparently affects behavior. Figure 51 shows the typical complex appearance of such a surface along with EDX analysis of a sample studied by Letts. The surface is covered unevenly by a rich assortment of elements, all of which apparently came from cell components. Where exactly the extra energy is created is not known, but it is assumed by the author to be produced in the PdD rather than in the gold. The laser effect was then replicated by Storms(*415*) and McKubre et al.(*416*), in collaboration with Letts, and by Swartz(*417*), who used very pure D_2O without Li. As described shortly, Letts et al.(*418*) used a dual laser during electrolysis with interesting results.

Storms(*415*) used a platinum (Pt) cathode that eventually became active as Pd transferred from the anode and deposited on the cathode. Excess power produced without the laser is compared in Fig. 52 to excess power generated by the activated Pt cathode when a single laser is used. The laser only had an effect once excess power was produced without the laser. The greater amount of excess produced without the laser, the greater the amount of excess power added by the laser.

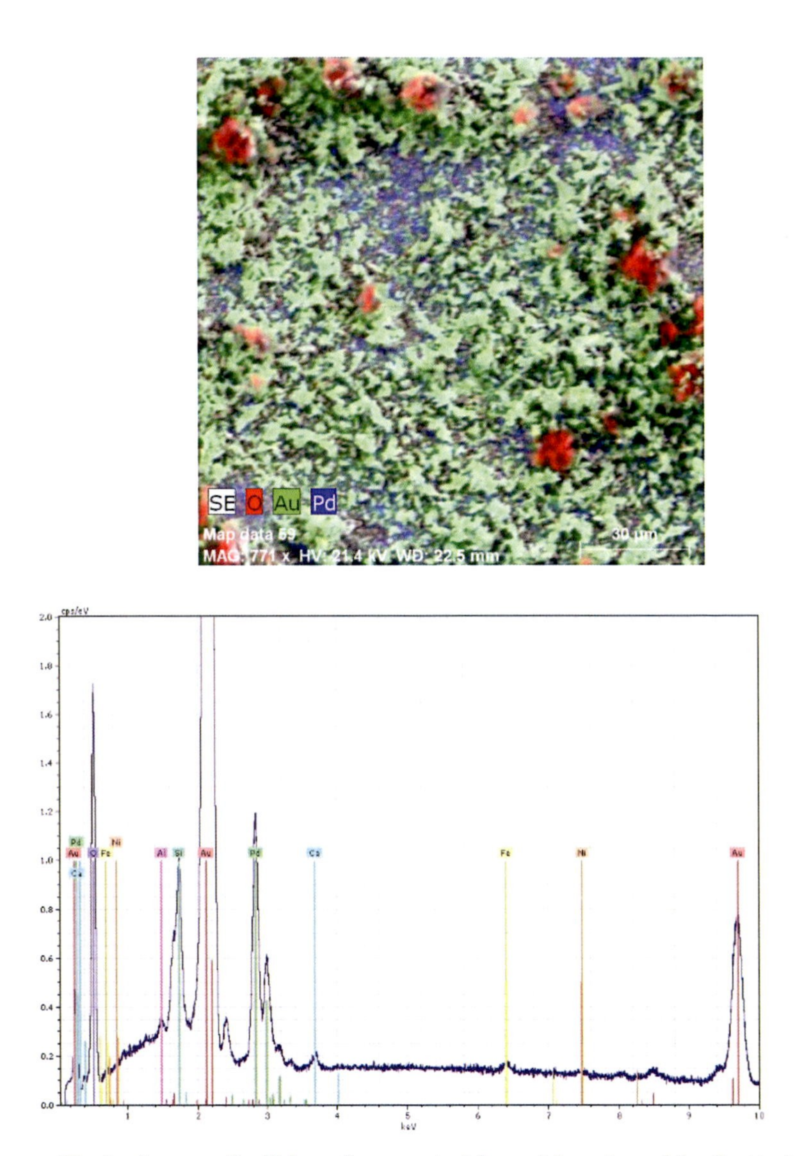

Figure 51. An image of a Pd surface coated by gold and used by Letts.(*413*) An EDX analysis of the surface is shown.

On the other hand, the conditions on the cathode surface used by Letts produced no detectable power until the laser light was applied, which caused detectable power until the laser was turned off, as shown in Fig. 53. Polarization of the laser radiation with respect to the magnetic field was found important. How this polarization is related to surface morphology or to the nuclear mechanism is unclear.

Letts and Cravens found application of certain elements to the surface (Fig. 54) to be beneficial. These elements are expected to deposit in a thin layer having very uneven distribution and concentrations. Where on the surface and how the elements can affect the nuclear reaction is not known. Nevertheless, once again the LENR effect is shown to be sensitive to conditions in the near surface region where pure PdD is not present. In addition, the laser light is expected to penetrate only a few nm into the surface and to have an effect only in this very limited region.

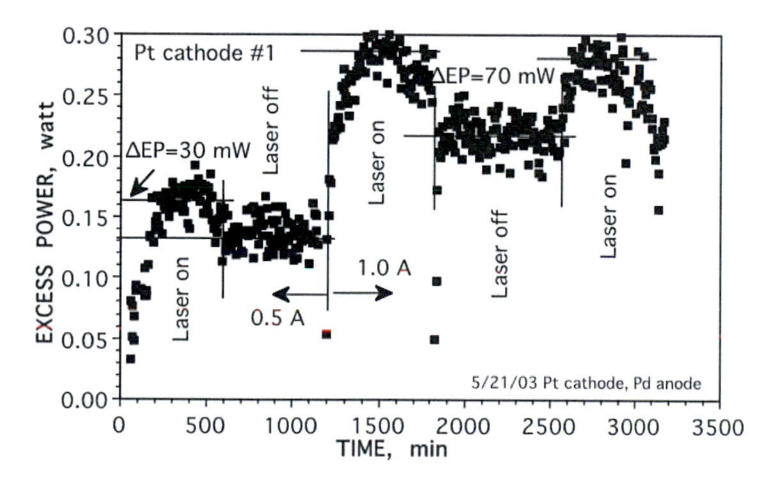

Figure 52. Effect of applying 680-686 nm laser light (35 mW) to an activated Pt cathode during electrolysis in D_2O+Li. Excess power is corrected for the 35 mW added by the laser itself while measured by a Seebeck calorimeter.(*415*)

The dual laser method was used by Letts to create a beat frequency as a method to explore a theory proposed by Hagelstein.(*419*) Excess power was produced at certain frequencies, as can be seen in Fig. 55, where the difference between the two laser frequencies is plotted on the X axis. The effect of relative polarization between the two beams shows that a beat frequency is important. The only unknown is whether the required beat frequency is the sum or difference between the two lasers.

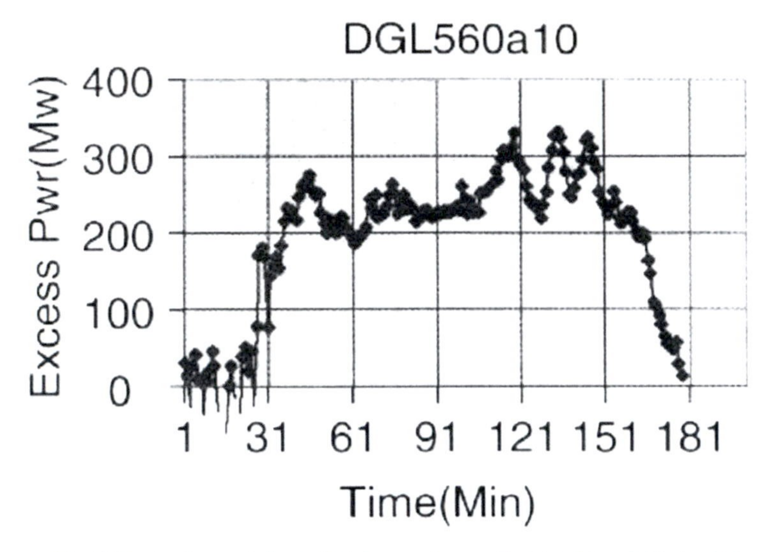

Figure 53. Effect of applying 661.5 nm laser light (30 mW) to Pd coated with gold while electrolyzing in D_2O+Li containing a rare earth additive.(*413, 414*)

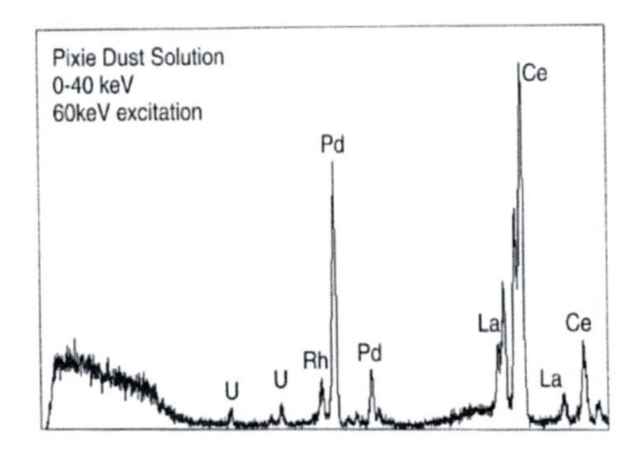

Figure 54. Surface analysis of an active cathode to which certain elements were added to the electrolyte.(*413, 414*)

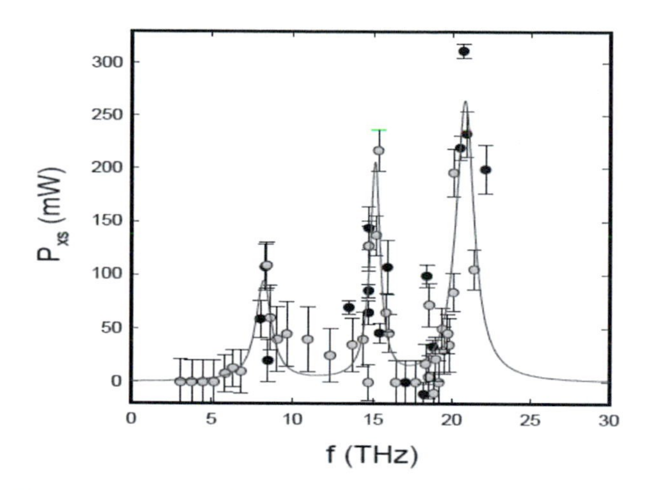

Figure 55. Excess power produced by two lasers superimposed on the same spot on a Pd-Au cathode and with the same polarization. The frequency difference between the two lasers is plotted.(*418, 419*) The study was done by holding one laser frequency constant while the other frequency is changed.

The beat frequency difference is chosen because Hagelstein believes this unique frequency stimulates an energy state in the density-of-states of the hydride, thereby causing the fusion process to start.(*418-420*) This explanation is hard to relate to the known effect of a single laser in stimulating heat production because a beat frequency difference cannot occur. On the other hand, suppose the active frequency actually resulted from a beat frequency created by the sum of the two lasers. This frequency could be produced by both a single and dual laser with similar values with the only difference being that a dual laser allows the resulting beat frequencies to be slightly changed.

A comparison between the primary wavelength, the sum, and the difference are compared to a region of the electromagnetic spectrum in Fig. 56. The primary wavelength used by Letts falls in

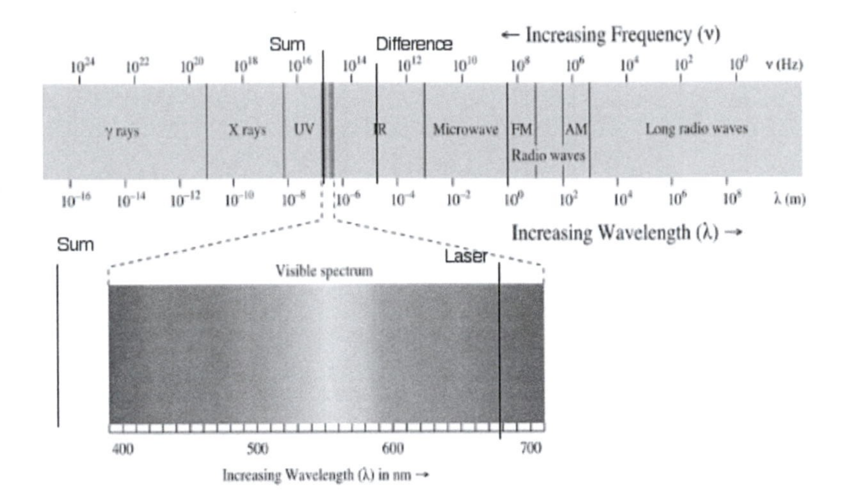

Figure 56. Relationship between frequency and wavelength in the electromagnetic spectrum (from Wikipedia). Approximate positions of the primary laser, the difference between the wavelengths, and the sum of wavelengths used by Letts are shown.

the red region of the spectrum, the difference is in the far infrared, while the sum is in the ultraviolet (UV) region, with a wave-length near 343 nm and a frequency near 8.7×10^{14} cycle/sec. The resulting radiation would be complex because the fixed laser would produce a fixed beat frequency while the variable laser would combine its frequency with itself and with the other laser to produce two variable frequencies. If the sum were the important variable, the plot shown in Fig. 55 would have the same shape except the X-axis would go from about 8.7×10^{14} to 11.7×10^{14} cycles/sec. This analysis suggests use of lasers in the UV end of the spectrum, as found effective in other studies, would be more convenient than attempting to generate high frequency by creating a weak beat frequency using lasers in the red region of the spectrum. Bass(*421*) predicts the most effective wave-length is 40 nm, which is in the deep UV.

2.5.3 CONCLUSIONS ABOUT APPLIED RADIATION

Use of a laser is important because it allows the near-surface region to be explored and location of the NAE to be found. Future studies might explore the entire surface, rather than a few square mm, to identify active areas so that they can be analyzed using other methods to determine the unique conditions causing the nuclear reactions.

Production of excess energy by many studies without use of laser light shows a laser is not required to trigger the process. Application of this energy apparently only increases the rate of a nuclear reaction already underway. The sensitivity of the calorimeter would determine how much, if any, of this initial power could be detected and how much enhancement would be required to produce detectable power.

Application of laser light, especially in the UV end of the spectrum, increases the rate of the nuclear reactions in the region being radiated over that produced by the normal F-P effect. Both excess energy and transmutation products have been produced. Why this is true has not been satisfactorily explained. The effect seems to be related to the polarization angle of the laser radiation relative to the surface. Presence of a magnetic field also seems to affect the rate of power production. Not clear is whether this field affects formation of the NAE or instead accelerates the process taking place in the NAE once it forms.

The wave-length of applied radiation is important, with shorter wave-lengths having a larger effect. The process is not influenced by the deuterium content in the target in a clear way. Various surfaces, even ones covered by gold and containing either D or H, have their nuclear behavior changed by laser radiation.

Measurement of induced energetic radiation while laser radiation is applied can give insight into the nature of the LENR reaction.

2.6.0 HEAT ENERGY

The discussion now focuses on conditions required to produce significant energy using LENR. Heat energy gives LENR commercial value, but it is not useful to reveal the mechanism or even to prove the energy has a nuclear source. Nevertheless, energy production in excess of a prosaic source combined with correlation to nuclear products provides overwhelming support to claims of energy produced by LENR.

Seven different methods have been used to produce energy from LENR, with the electrolytic method being the most studied and the most successful. The ability to initiate the effect several different ways adds credibility to the claims. Surprisingly, simply exposing any hydrogen isotope to nuclear active material is sufficient to initiate a fusion reaction.

For one watt to be produced, a reaction rate near 10^{11} events/sec is required when D-D fusion is the source and about 10^{9} events/sec are needed for power to be detected with confidence. Although rare and difficult to initiate, once started the nuclear reactions can be made to take place at a significant rate.

Several kinds of responses are observed with one being universal. Typically, the power is unstable and occurs in bursts from local regions on the surface. In addition, the active materials seem to have a limitation to how long this energy can be produced. This behavior suggests a chaotic process operating from independent and unstable locations. Measured power would result from the sum of these individual sources. Consequently, understanding what happens at each independent reaction site is critical to explaining the observed average behavior. Unfortunately, study of heat production from individual reaction sites is not presently possible.

The discussion here is not focused on proving the claimed heat energy is real. That goal was accomplished by Storms(*1*) in a previous book. The focus here is to identify features of energy production an explanation needs to address.

2.6.1 How is energy measured?

Heat energy is measured most often by using a calorimeter in which temperature change is used to determine the rate of heat flow from the source to a heat-sink. Energy is calculated by multiplying power and time to give watt-sec or joules. Most experiments require some power be applied to the experiment to initiate the reaction. This power needs to be identified and subtracted from the measured value to obtain power generated by the nuclear process.

The following general calorimeter types have been used with many complex variations in design and accuracy. Accuracy has two components: absolute accuracy based on calibration and relative accuracy based on random variations in the measured value. In general, absolute accuracy of ±50 mW can be obtained and is common. A measurement of change in generated power is limited mainly by random variations and can usually be determined with sensitivity greater than absolute accuracy. As a result, patterns produced by variation in power as a result of changes in applied conditions can be determined with greater certainty than the absolute amount of power being measured.

Isoperibolic calorimeter

This calorimeter type relies on a temperature gradient being created across a thermal barrier located between the source of energy and a heat-sink having constant temperature. This barrier can be the wall of the glass container or a second wall located outside of the container. Correct power is only obtained when the temperature is constant and the temperature gradient is the same as that present during calibration. These requirements are difficult to achieve and can cause unacknowledged error.

Single wall- Early measurements of power were based on a temperature measured at one location in the electrolyte and compared to a uniform and constant temperature outside the glass cell, using the glass wall of the cell as the thermal barrier. A serious error can result from temperature gradients within the fluid causing the measured temperature not to correctly represent the average fluid temperature. Calibration is best done by using a dead cathode rather than an electric heater because bubbles generated during electrolysis will reduce this temperature gradient in the fluid and produce a gradient similar to that created when an active cathode is studied that also produces bubbles. A cell having a tall-narrow configuration was used by Fleischmann and Pons to reduce the effect of potential temperature gradients in the electrolyte.

Double wall- A second thermal barrier is located outside of the cell. This method can be very accurate and convenient because it eliminates the uncertain gradient in the cell.

Adiabatic calorimeter

The amount of power is determined by how rapidly temperature changes in a material having a known heat capacity. Generally, the electrolytic cell itself is used as the heat-absorbing material, which permits power to be estimated while change takes place if the total thermal capacity of the cell is taken into account.

Seebeck calorimeter

The cell is placed in a container having walls in which many thermal-electric converters are located, with the outside of the wall held at constant temperature. The average temperature gradient across the wall generates a voltage that is related to the average energy flux regardless of where heat passes through the wall. The calorimeter can also be designed with most of the surface being a good thermal insulator so that all heat is lost through only a few thermoelectric converters. Because this is a version of the isoperibolic calorimeter, accurate values are only obtained when generated power is constant.

Flow calorimeter

A flowing fluid, generally water, is used to carry energy away from the source. The source must be well insulated so that energy can only leave with the fluid. The amount of power is calculated from the flow rate, the heat capacity of the fluid, and the temperature difference created between the fluid going into and coming out of the calorimeter. Accurate values require power production to be constant.

Phase change calorimeter

This type of calorimeter measures energy rather than power. The energy is determined by measuring the amount of a material that has changed phase. This phase change can be from solid to liquid, such as melting ice or mercury, or from liquid to gas, such as when liquid air or liquid Freon is used. Constant temperature at the energy source is not required to obtain accurate values.

IR radiation calorimeter

The rate at which energy is produced can be determined by measuring the amount of power lost based on emitted radiation, generally in the infra-red (IR) region of the spectrum. This method is best used when the source has a high temperature.

2.6.2 What methods are found to produce heat?

Heat production, as well as the normally detected nuclear products, has been produced several different ways. In each of these methods, a NAE is created by applying different conditions before or during the study. The challenge is to identify the common and universal nature of the NAE.

Electrolysis

Current is passed between two electrodes in an electrolyte containing a soluble ion, generally lithium. The anode (+) is usually platinum and the cathode (-) can be any metal, with palladium being most frequently used. The electrolyte is usually D_2O although H_2O and other liquids have been explored. The process generates gases from decomposition of the electrolyte that have to be either vented or combined to reform water. When vented into the air, the cell is called "open." Often a catalyst is placed in the cell to recombine the gases back to water, which is called "closed." Corrections have to be made for energy lost by the exiting gases when an open system is used.[17] A closed system can

---

[17] This correction is applied using the "neutral potential." The method of calculation and values for D_2O and H_2O are provided by Storms.(1)

be sealed so that nothing enters or leaves the cell and no corrections are required. Most modern studies use a closed system.

Heat, radiation, transmutation, tritium, and helium have all been produced using this method.

Plasma electrolysis

When the voltage applied to an electrolytic cell is increased generally to over 100 V, energetic plasma forms near the cathode. As a result, hydrogen ions are injected into the cathode more rapidly and with greater energy than during normal electrolysis. This method increases the rate of LENR. Corrections for energy loss are complex because the generated component gases are mixed with vapor.

Electroplating

When salts of palladium are added to the electrolyte, palladium can be plated on the cathode by a process called co-deposition. Under normal conditions, this process can take place slowly as metal is transferred from the anode to the cathode by normal electrolysis. The morphology of the deposit is influenced by a magnetic field.(*292*) Electrodeposition of metals besides palladium has been explored.

Gas (glow) discharge

Application of a voltage to low-pressure D_2 or H_2 gas will cause plasma to form that can be used to bombard the cathode with ions. Pulsed DC is frequently used to apply high current without melting the cathode material. Heat, radiation, transmutation, and tritium are produced depending on the cathode material and the hydrogen isotope.

Gas loading

Materials made nuclear active can produce heat, tritium, helium, and transmutation simply by being exposed to H_2 or D_2. Because this method is so simple, it is being explored as a source of useful energy using both D_2 and H_2.

Sonic

Acoustic waves generated by a transducer are caused to pass through D_2O to a target. The distance between the target and transducer is adjusted so that bubbles produced by the acoustic wave collapse precisely on the target surface. As a result, D^+ is introduced into the target by bubble collapse. The injected D^+ finds a NAE and initiates LENR. As a result, a steady supply of hydrogen fuel is available while the surface is subjected to stress by the acoustic wave. This method has produced all of the nuclear products expected from LENR.

Electromigration and diffusion

Forced motion of deuterium through a material either by imposing a voltage or a concentration gradient can increase excess energy production. Both methods make the fuel more available to the NAE. In addition, applied voltage causes hydrogen to concentrate near the negative end of the wire where the higher concentration can increase the fusion rate in this region.

2.6.3 Production of energy

Too many reports of energy production are available to discuss each in detail. Many studies can now be ignored because they were designed to prove the claim to skeptics or to simply replicate previous success, which provides very little understanding about the mechanism. Because the electrolytic method using D_2O has provided most information about energy production, this method is examined first.

Often overlooked is the endothermic nature of the overall reaction when water is used as a source of hydrogen to react with Pd. As a result, a known amount of energy is absorbed as deuterium reacts with Pd and an active composition is achieved. This behavior allows calibration using this initial apparent loss of energy, as described by Storms.(*1*) Otherwise, calibration can be done using a non-reactive cathode, usually platinum, or an internal resistor. Correct application of the neutral potential to correct for loss of gas from an open cell is essential.(*1*)

2.6.4 Energy produced by electrolysis using D_2O

Location of some nuclear products has been identified in previous sections. Obviously, all of these reactions make energy and some might make enough power to be detected. The first question is, "Are all nuclear reactions and the resulting energy produced in the same location?" A visual location of energy production can be provided by IR images taken of a Pd surface while Pd is electrodeposited from D_2O. The process, called co-deposition in 1991 by Szpak et al.(*292, 293, 303, 305, 422-431*), was found to produce flashes of light at random locations, shown at one instant of time in Fig. 57. This observation reveals the local nature of the heat-producing process.

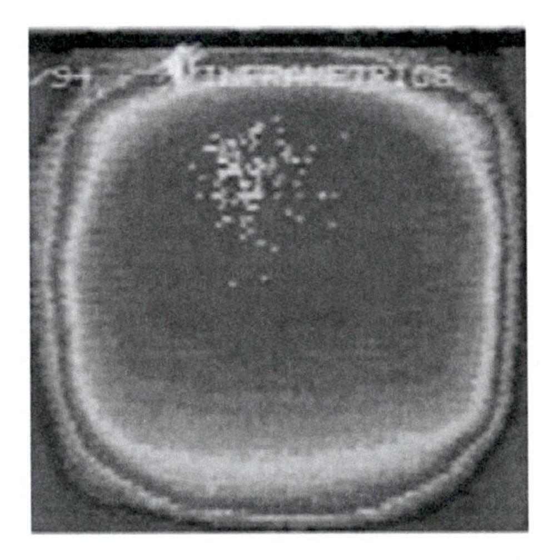

Figure 57. IR picture of the Pd surface during co-deposition of Pd from an electrolyte containing D_2O + $PdCl_2$ + LiCl.(*427*) The bright spots are regions where temperature has briefly increased.

Swartz et al.(*432*) report the same effect. These flashes originate from small regions on the surface where temperature has briefly increased. The total measured power results from the sum of these random events.

Co-deposition was studied independently by Shirai et al. in 1991(*433*) and used by numerous people since then,(*21, 434-436*) with excess energy and nuclear radiation being frequently reported.

As an aside, in view of chlorine being a poison for reaction of Pd with hydrogen, the method might be more successful if this element were eliminated from the electrolyte. Tanzella et al.(*376*) report successful co-deposition using D_2O-$PdSO_4$ containing various additives.

The rate of energy production is sensitive to the average D/Pd ratio, as can be seen in Fig. 58 where the classic relationship reported by McKubre et al. (*59, 437*) is plotted. This relationship between D/Pd and excess power is universal but the details depend on the shape of the cathode, with wires showing a different behavior compared to plates of Pd. Tanzella et al.(*438*) found the surface to have a higher composition than the bulk during loading and when bulk composition was large. Indications of nuclear activity are detected at lower compositions depending on the nature of the cathode and how deuterium is applied. Sometimes LENR might occur with a rate too small to make detectable power.

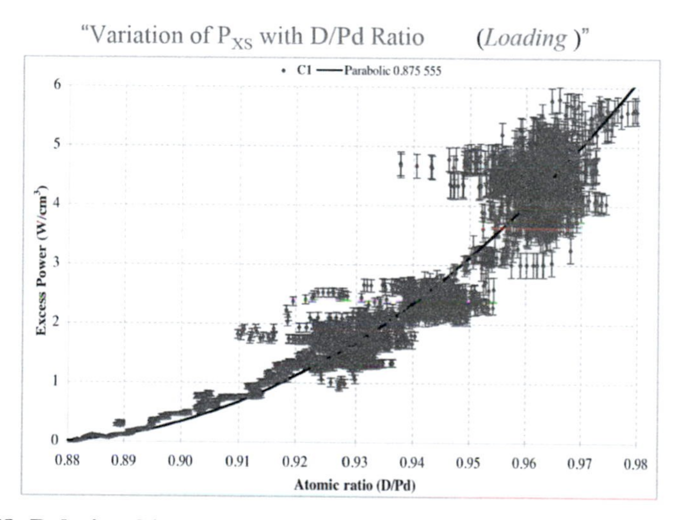

Figure 58. Relationship during electrolysis of Pd between excess power and deuterium composition of the average bulk cathode wire based on resistance change.(*59, 437*)

Storms(*439*) showed that the rate of energy production is related to the surface composition, with a critical value being required to produce detectable power. Values as high as D/Pd = 1.5

have been reported.(*440-442*) Because this ratio exceeds full saturation of the face-centered-cubic (fcc) beta-phase, a change in structure apparently takes place at the upper limit of the beta phase. This issue is discussed in Section 2.7.1.

The ability to achieve a nuclear-active surface composition is related to the amount of current (A/cm^2) applied to the cathode, with a critical onset value for power production being required. This critical current depends on the kind of cathode used, with wire cathodes(*443*) having a typical behavior shown in Fig. 59 while thin coatings of active material plated on an inert substrate show less critical onset current, as shown in Fig. 60. This difference can be explained by a bulk sample needing extra current to compensate for the loss from many large cracks — a loss that is much reduced when thin films are used.

Thin layers of Pd have been applied to the cathode several different ways, resulting in increased success in generating excess energy. First Oriani et al.(*444*) and then Storms used palladium instead of platinum as the anode, which caused palladium to transfer to the cathode where it deposited as a thin active layer. It is perhaps ironic that simply reversing the anode and cathode materials would improve reproducibility. Success of such thin deposits adds emphasis to the conclusion that the effect does not occur in the bulk, but instead in a region perhaps within a few microns of the surface.

Production of energy can be affected by many variables in addition to composition. McKubre et al.(*445*) propose flux of D through the material is another factor. Although flux is expected to have an effect, an exact correlation between flux and power has not been established. Such a correlation is clouded by the chaotic processes operating on a cathode in local regions, such as bubble formation, and changes in local surface structure as a result of reaction with materials in the electrolyte. As a result, the behavior of NAE and the conditions it experiences in the complex surface cannot be clearly related to average behavior.

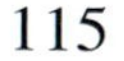

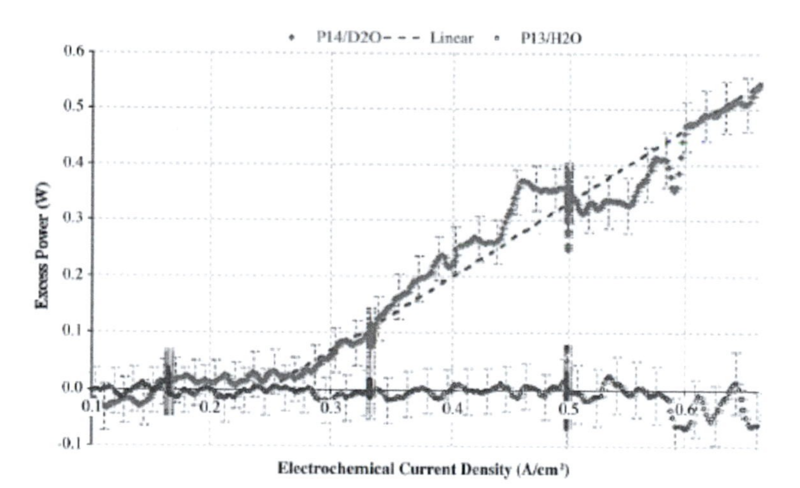

Figure 59. One of many examples of how applied current changes excess heat production.(*443*)

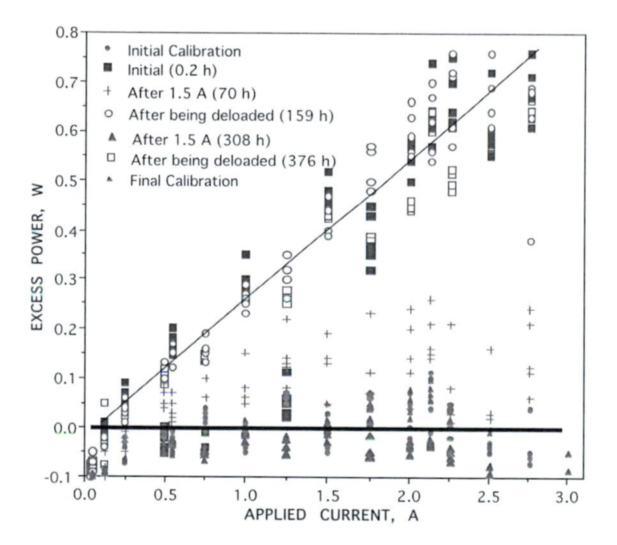

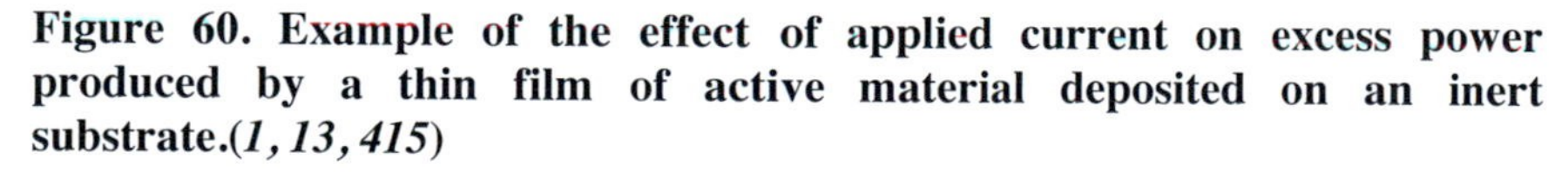
Figure 60. Example of the effect of applied current on excess power produced by a thin film of active material deposited on an inert substrate.(*1, 13, 415*)

As Fig. 61 shows, excess power and deuterium content are highly variable during electrolysis. In fact, this variation is probably greater than data in the figure indicate because thermal inertia of the calorimeter would damp-out the more rapid variations. The random and local nuclear events shown in Fig. 57 would be expected to cause such behavior as the active sites flashed off and on. Steady power would result only when a great many of these events occurred in a truly random sequence.

Even local temperature is not constant thanks to local nuclear and chemical reactions. However, some cathodes are more stable than others, with part of the instability being hidden by the thermal inertia of the calorimeter itself. In addition, no calorimeter can give accurate values for power as long as power is not constant, which adds some uncertainty to all measurements as power is being absorbed or released by the heat capacity of materials in the calorimeter. Although corrections can be made, they are not frequently applied.

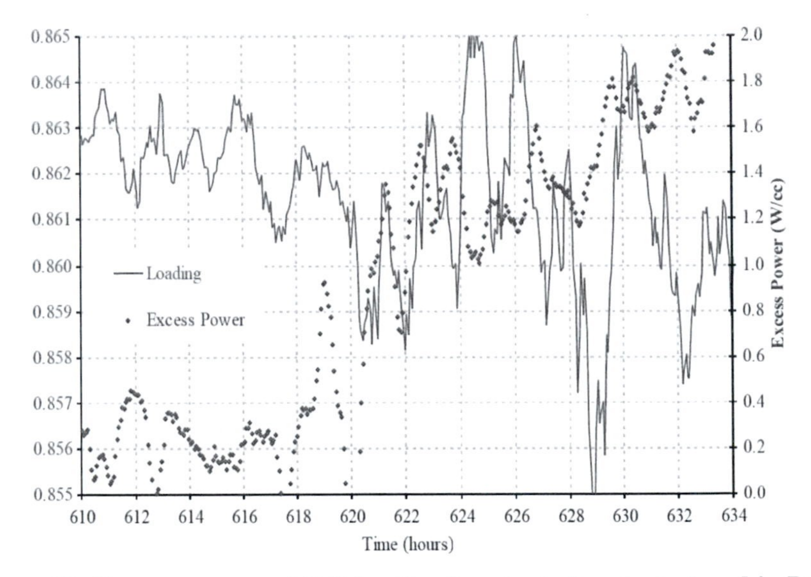

Figure 61. Excess power and D/Pd ratio of a wire being electrolyzed in D_2O compared over time.(*443*)

All active cathodes eventually stop producing excess power as the surface becomes modified by further deposition of material, causing the cathode to lose deuterium or destroying the NAE. In fact, after excess energy has been produced, the surface is frequently found highly contaminated by Li, O, B, Pt, and Si.(*446-448*) Occasional Cu is deposited when wires are exposed to the electrolyte. Which part of this complex surface produces anomalous energy is unknown and is generally ignored in proposed theories.

Swartz and Verner(*449*) avoid deterioration produced by deposited lithium and various other impurities by using very pure D_2O. High voltage must be applied between the anode and cathode to obtain useful current. This method produces excess energy, which demonstrates lithium and other impurity elements are not required to cause LENR.

Other variations of the electrolytic method have been successful. Dardik et al.(*450-456*) add current having a complex form called a super-wave to the normal DC current. As a result, the Pd acquires a greater concentration and the resulting excess has been increased to significant levels. Presumably, certain frequencies and waveforms within the super-wave are more effective than others, but the most effective combination has not been revealed. Whether the super-wave increases power production simply by increasing the concentration of D in the surface, increasing flux, or increasing the amount of NAE is not yet known.

If applied current is increased until applied voltage approaches 100 V, plasma forms around the cathode. This plasma allows ions to be applied to the cathode surface with greater rate and energy than available from normal electrolysis. As a result, more excess power results. Mizuno et al.(*270*) used this process initially with platinum electrodes but they found a tungsten cathode(*271, 272, 457-459*) to last longer. Mizuno et al.(*460*) continued to explore heat production with a better calorimeter. The method was replicated in two cases(*250, 461*) and failed in another.(*462*) A variation of the method is being commercialized by Godes.(*202, 463*) The method also produces nuclear products as described in Section 2.3.3.

As shown in Figs. 59 and 60, increased excess power results from increased applied current. However, this increase in current also causes applied power to increase. Swartz(*464, 465*) has described the relationship between generated and applied power by applying an engineering concept he calls the optimal operating point (OOP). The ratio of excess power to applied power (excess/applied) is used to show the energy efficiency of the process. As expected, maximum efficiency occurs at a particular applied power, depending on cell design and many other variables. This concept has limited value at present because efficiency is not sought during research studies. Instead, the ability to make energy well in excess of any error is the goal, which may actually occur when engineering efficiency is poor. His concept is discussed in detail in Section 4.1.0.

Baranowski et al.(*466, 467*) applied 6 kbar of D_2 to PdD and produced an estimated composition near $PdD_{0.86}$. Although these samples were held under pressure for two years at 298K, no helium was detected in the gas, as would be expected if LENR had been initiated. Apparently, this condition did not allow the NAE to form even though the D/Pd ratio was in what is considered by some theories to be in the active range.

2.6.5 Effect of bulk metal treatment

Success in making heat energy using palladium depends on its characteristics, previous treatment, and purity.(*468-470*) In other words, the process is very sensitive to the chemical and physical conditions of the initial material. Very pure palladium treated in conventional ways shows very poor success in making energy or nuclear products. Part of the reason is the inability of "normal" palladium to achieve the required high loading. Nevertheless, this condition does not seem to be the only requirement. For example, the Pd-Ag alloy has been reported(*471*) to give success even though this alloy does not acquire high compositions.

A rich literature shows how success can be improved but not guaranteed. How reproducible results might be achieved is discussed in Section 5.6.0.

2.6.6 Effect of temperature

Increased temperature causes power production to increase, all else being unchanged, as shown in Fig. 62.(*472, 473*) This effect is also seen when temperature is increased to the boiling point of the electrolyte.(*474-477*) Because the measured temperature is not that of the NAE, the activation energy is only an approximation of what is happening where the nuclear process actually takes place. Nevertheless, the temperature effect is too small to have a direct effect on the nuclear reaction itself. Perhaps of significance, the slope of the line in Fig. 62 gives activation energy consistent with a process involving diffusion – perhaps diffusion of D from its normal site to the NAE.

Temperature provides positive feedback to energy production, which creates the possibility for runaway temperature and self-destruction of the device. Controlling the effect of temperature is one of the basic engineering challenges and limitations to application of LENR, as explained in Section 5.1.0.

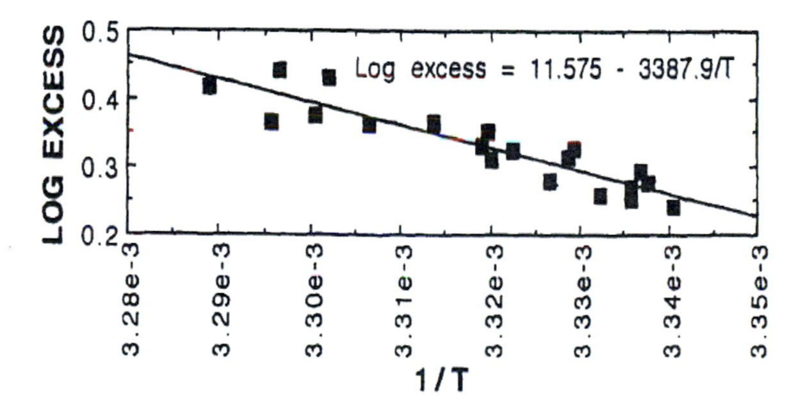

Figure 62. Effect of cell temperature on the log of excess power production.(*472, 473*)

2.6.7 Effect on power production of adding light hydrogen to D_2O

Addition of H_2O to a cell containing D_2O when excess power is produced will cause reduction in power or even its apparent elimination. When the effect of light hydrogen was first discovered, the H was thought to poison the process. A more plausible explanation involves the amount of energy obtained from H+e+H fusion being so much less than from D+e+D fusion. As a result, the heat production becomes too small to detect but the nuclear reaction perhaps does not actually stop.

Because hydrogen is favored in palladium over deuterium, the concentration of H acquired by a PdD cathode is greater than that in the electrolyte, as shown in Fig. 63. A similar increase in H content in PdD is expected when D_2 gas containing a little H_2 is used to create the hydride.

2.6.8 Energy produced using H_2O and H_2

Ordinary light hydrogen — designated here with the symbol "H" — can be used to generate excess energy in the same manner as deuterium. Initially, energy was proposed to be generated by the small deuterium impurity in all hydrogen. This explanation is no longer supported by observed behavior. The light hydrogen would appear able to also produce energy by fusion.

Ironically, ordinary water was thought to be inert and proposed as a blank to test for calorimeter error. However, this assumption proved not to be true.(*343*) As the previous discussion of radiation and transmutation testify, ordinary hydrogen in any chemical form can produce LENR and, consequently, can be expected to produce energy. This fact was first discovered by Fleischmann and Pons, much to their dismay and to the delight of skeptics who then had another false reason to reject the claims.

While H can participate in a nuclear reaction, the resulting energy will be less than that produced by D, requiring a better calorimeter to demonstrate a real effect even when the rare active conditions have been achieved. For this reason, most studies report no excess energy and pay very little attention to the possibility of energy actually being made using H_2O.

Swartz et al.(*478*) electrolyzed a nickel cathode in H_2O without addition of the usual ions to improve conductivity. Instead, the cell was open to air and claimed to be in equilibrium with CO_2. He found adding up to 7.4% D_2O caused excess power to increase. However, the nickel suffered damage that prevented further use.

2.6.9 SUMMARY OF ENERGY PRODUCTION

Anomalous energy has been produced using a variety of materials, including Pd, Pd-Ag, Pd-B, Ti, Ni, and Ni-Cu, exposed to a variety of conditions containing both hydrogen and deuterium. Bulk material in the form of wires and sheets, powders of various particle size, and thin coatings on inert material have been found successful. In a few cases, correlation between nuclear products and energy production has been obtained. This work indicates that most of the energy results in formation of ^{4}He when deuterium is used. This helium is produced within a few microns of the surface, which means the correlated energy is produced in the same location. The nuclear reaction resulting when protium is used has not been identified. Power production is increased by increased hydrogen isotope concentration and increased applied energy in any form, including increased temperature and laser radiation. Detected excess power has ranged from a few nano-watts to a few kilowatts.

2.7.0 PROPERTIES OF PALLADIUM HYDRIDE

The basic chemical and physical properties of PdH and PdD are well known. This understanding is useful even though palladium hydride is not actually present in its pure form at the site of the nuclear reactions. Nevertheless, the known properties of pure hydride help to evaluate various theories and provide an example of how similar materials might behave.

2.7.1 Phase relationship

Two phases are stable in the palladium-hydrogen system under conditions normally used during LENR studies as can be seen in the phase diagram shown in Fig. 63 for various temperatures, pressures, and compositions. The alpha phase forms at low hydrogen content shown on the left in the figure and beta phase (β-PdH) is on the right. Both phases have a face-centered-cubic (fcc) structure, as can be visualized by examining Fig. 64. The relationship between pressure and composition at 300K, a temperature close to conditions in many electrolytic cells, is shown in Fig. 65. These two phases merge at high temperature and pressure to form a continuous solid-solution of H in Pd above about 275°C.

An ordered phase forms near 50 K(*481, 482*) in which the D[H] no longer occupies a random location in the beta phase and instead forms a super-lattice with a $I4_1$/amd structure.

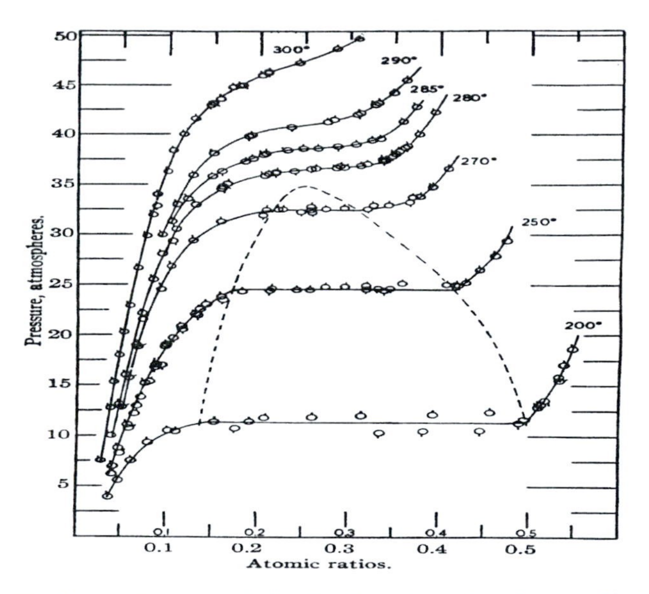

Figure 63. Phase diagram of the Pd-H system showing applied pressure for various temperatures. The two-phase region disappears above about 275°C to form a single-phase solid solution.(*479, 480*)

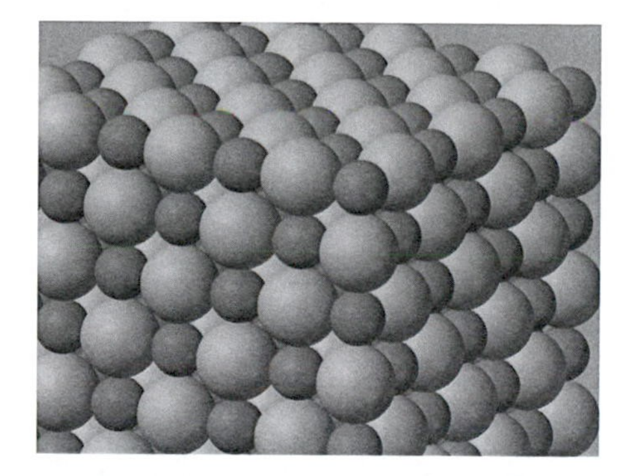

Figure 64. Example of a face-centered cubic structure (fcc) consisting of two identical interpenetrating lattices with H (small sphere) being in the octahedral sites relative to the Pd.

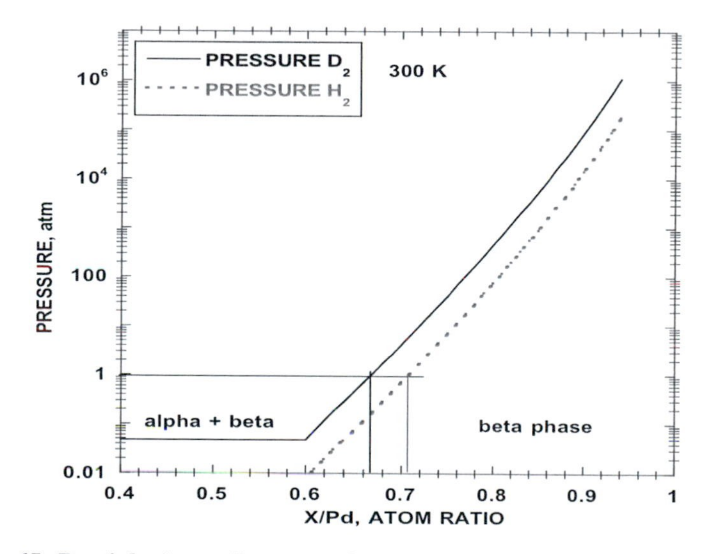

Figure 65. Partial phase diagram of Pd-H and Pd-D systems showing the relationship between applied pressure and the resulting atom ratio at 300K. The composition formed at 1 atm is within the single-phase beta region at 300 K.(*479, 483*) Pressure is uncorrected for non-ideal effects. Consequently, the values at high pressure need to be described as effective activity rather than pressure.

Impurities can reduce the lower boundary of β-PdH to near $PdH_{0.5}$, which some studies falsely attributed to the actual lower boundary of the beta phase.(*484, 485*)

Fukai(*486-488*) subjected Pd to 5 GPa ($5x10^4$ atm) of H_2 pressure and 700-800°C[18]. This condition caused formation of a second phase, in addition to β-PdH, with a proposed composition of Pd_3H_4 in which a high concentration of Pd vacancies were claimed to form. A similar study was undertaken by Miraglia et al.(*489*) with similar results. Although this phase remained stable upon cooling after pressure is removed, no evidence supports the ability of conditions present during LENR to form the Pd_3H_4 phase.

Nevertheless, extra space can form in PdH when it reacts with hydrogen. A value for the apparent free volume can be obtained by comparing the volume calculated using the lattice parameter to the measured density. Storms(*490*) did such a study using a large number of samples prepared by loading Pd with H or D using electrolysis. The deviation between the volume obtained from the lattice parameter and that from the measured volume was, in this case, called "excess volume" and attributed to cracks and voids produced in the material as a result of forming the hydride.

As shown in Fig. 66, the volume obtained from the lattice parameter is the same as the measured volume until hydrogen is added to the beta phase, whereupon excess volume forms. The amount of excess volume increased each time H is removed and replaced, indicating accumulation of excess volume in the material as a result of conversion from alpha to beta phase, not by formation of a new phase. This deviation is highly variable, as shown in Fig. 67, where the excess volume at D/Pd = 0.85 is compared to the maximum composition that could be achieved by extensive electrolysis. Apparently, the same kind of volume attributed by Fukai to metal atom vacancies works against being able to achieve the high composition required to generate energy

---

[18] These conditions are expected to produce a composition near $PdH_{0.72}$ for the fcc structure based on the data used to calculate the values plotted in Fig. 65.

from LENR, with about 2% excess volume being the limit for success in producing detectable energy from PdD.

The amount of extra volume created when H is reacted with Pd depends on how the metal is treated before being converted to the hydride. No evidence exists in the literature showing formation of Pd_3H_4 in this composition range when Pd is reacted under normal conditions.

The alpha phase forms when a small number of hydrogen atoms enter the metal structure and occupy random locations between the metal atoms, with a limit near $PdH_{0.1}$ at room temperature and 1 atm. Once this limit is reached, a change takes place in the electron energy state such that additional hydrogen can transfer its electron to the conduction band and enter the structure as ions, thereby forming the beta phase.(*492*)

Rapidly increasing force (Gibbs energy) is required to add hydrogen as the upper limit of the beta phase is approached. Indeed, compositions near $PdD_{1.0}$ can be achieved only by using electrolysis because the process requires chemical activity equivalent to over 10^7 atm at the surface of the cathode.

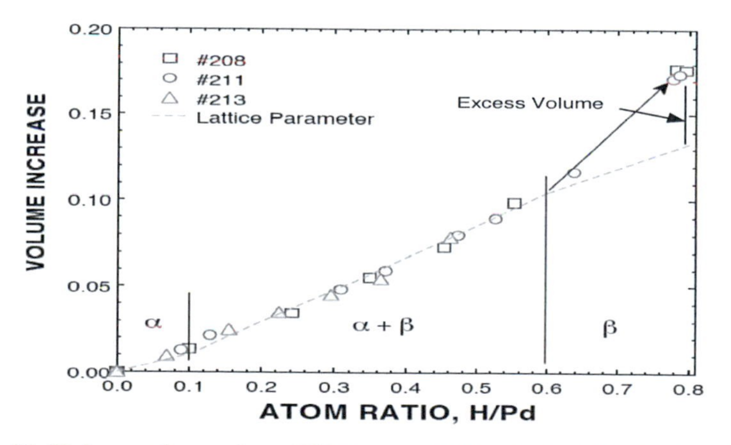

Figure 66. Volume of sample vs H/Pd content. The dashed line is the volume based on the lattice parameter and the volume for the plotted points is obtained from a physical measurement. Excess volume grows larger as additional hydrogen is added to the beta phase.(*490, 491*)

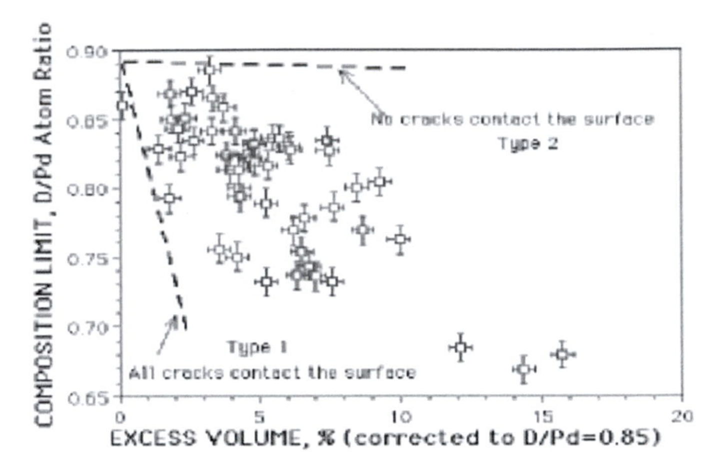

Figure 67. Excess volume in $PdD_{0.85}$ for a variety of Pd samples treated initially in different ways. The excess volume introduced when the beta phase forms is highly variable and generally reduces the ability to achieve a high composition as a result of leakage of D_2 through surface penetrating cracks.(*490, 491*)

As hydrogen is added to the beta phase, the resistance of β-PdD increases and then passes through a maximum near $PdD_{0.78}$ followed by a decrease as additional electrons populate the conduction band. This maximum apparently results when the conduction band involving the d energy level is filled, causing additional electrons to populate the s-p level.(*493*) This behavior is described as a resistance ratio (R/R_o) between the resistance of the beta phase (R) and pure palladium (R_o) and is plotted in Fig. 68 for the PdH and PdD systems.

The break in slope of R/R_o at the upper boundary has been attributed to the hydrogen entering tetrahedral sites.(*466*) Instead, formation of a second phase[19] is indicated when composition reaches the characteristic upper limit of the beta phase. A similar

[19] The sharp break in behavior is consistent with how all materials are known to behave when a second phase forms at a phase boundary. If the H simply changes location or Pd vacancies formed, a gradual change in properties would be expected as the upper phase boundary is approached. This expected behavior is not observed.

break in slope can be seen in the temperature coefficient of resistance plotted in Fig. 69. These data indicate the upper phase boundary for the beta phase is in the range H/Pd = 0.98-1.01, with a mixture of the beta phase and another unidentified phase forming in the region beyond. Consequently, material occasionally found to have a D/Pd ratio greater than D/Pd = 1 would actually consist of two phases, one being the beta phase with D/Pd near 1 and the other having a much greater composition, perhaps as high as H/Pd = 2.(*494*) Whether this phase is the Pd_3H_4 (H/Pd = 1.5) structure proposed by Fukai(*495*) is not known. In any case, the second phase can be expected to have a structure more complex than the beta phase.

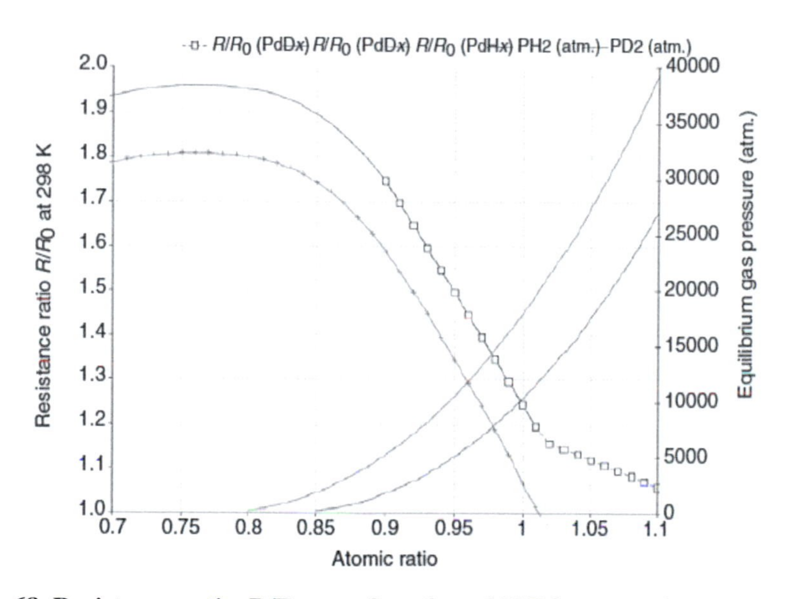

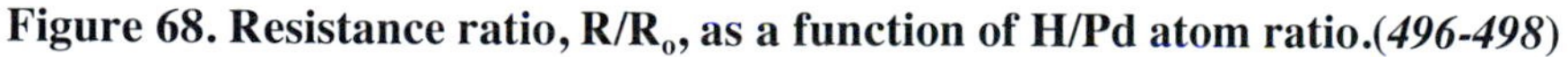
Figure 68. Resistance ratio, R/R$_0$, as a function of H/Pd atom ratio.(*496-498*)

The theoretical description of resistivity provided by Luo and Miley(*499*) is not consistent with the known phase relationship because the resistance ratio between α-$PdH_{0.05}$ and the lower phase boundary of the beta phase lies in a two-phase region. This condition requires a linear average between the resistance ratio of the alpha and beta phases be used to actually describe behavior. Because the composition of the beta phase in this region will be slightly variable and the mixture will not be uniform, the average between these two separate phases will be slightly variable. In any case, use of a smooth function, as is frequently drawn, is not consistent with the known phase relationship. To be consistent with phase behavior, a smooth fitting function only applies within the single-phase region, with a straight line being applied in the two-phase regions and with a break in slope at the lower phase boundary of the beta phase.

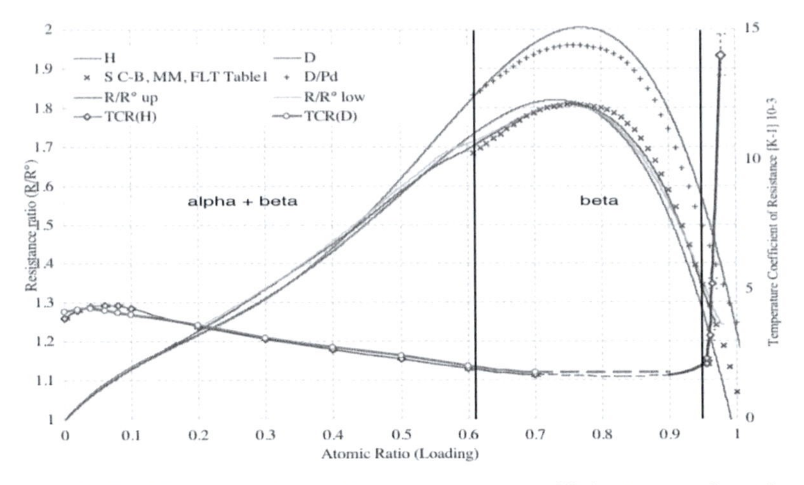

Figure 69. Resistance ratio and temperature coefficient as a function of D/Pd and H/Pd atom ratio.(*416, 500*) Lines added to clarify relationship to phase diagram.

Deuterons in PdD are located, on average, at sites that are separated by known distance(*481, 501-510*) too great to allow nuclear interaction, as shown in Section 2.7.2. Based on neutron diffraction studies,(*481, 505, 511-513*) the deuterons are located at clearly defined locations in a fcc lattice structure with each potential location being filled randomly. When a gradient is applied, diffusion shows conventional behavior(*479, 497, 514-521*), as shown in Section 2.7.3.

The deuterons appear to have an average positive charge of 0.30±0.05(*522*) at the lower composition of the fcc phase instead of a charge of unity as has been calculated to be the case for some hydrides.(*523*)

2.7.2 Lattice parameter

The lattice parameter of the beta phase, shown in Fig. 70, is well known throughout the accessible composition range with a clear relationship to the known phase behavior shown in Fig. 63. A linear relationship between composition and lattice parameter is evident (Fig. 70) with no indication of additional phases or metal atom vacancies being produced in the composition range studied.(*510*) The lattice of PdH is slightly larger than that of PdD at the same composition.

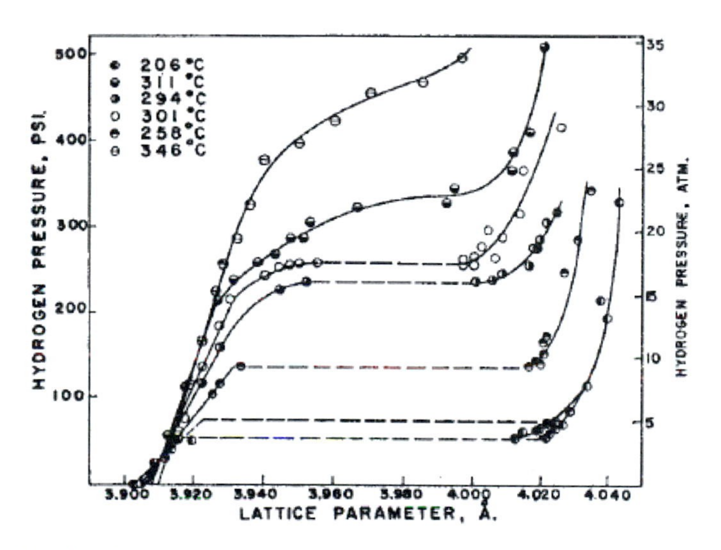

Figure 70. Lattice parameter of fcc phase for various conditions and hydrogen contents.(*524*)

Addition of H to Pd results in a lattice parameter change from 3.889Å to 3.894Å for the alpha phase.(*525*) Further addition results in abrupt increase when the beta phase forms near $PdH_{0.6}$ to give a lattice parameter of 4.025 Å. This change in size produces stress within the material resulting in cracks, dislocations, and pitting of the surface, all of which contribute to the measured excess volume as physical voids, not as vacancies in the metal lattice. When this process occurs in regions subjected to local heating by energy generated by LENR, considerable rearrangement of atoms would be expected as the alpha-beta transformation cycles during temperature change, making energy generation unstable and of limited duration as is observed.

A linear variation of lattice parameter, seen in Fig. 71, would not be expected if metal atom vacancies formed or if the hydrogen atoms shifted from octahedral to tetrahedral sites within this composition range.

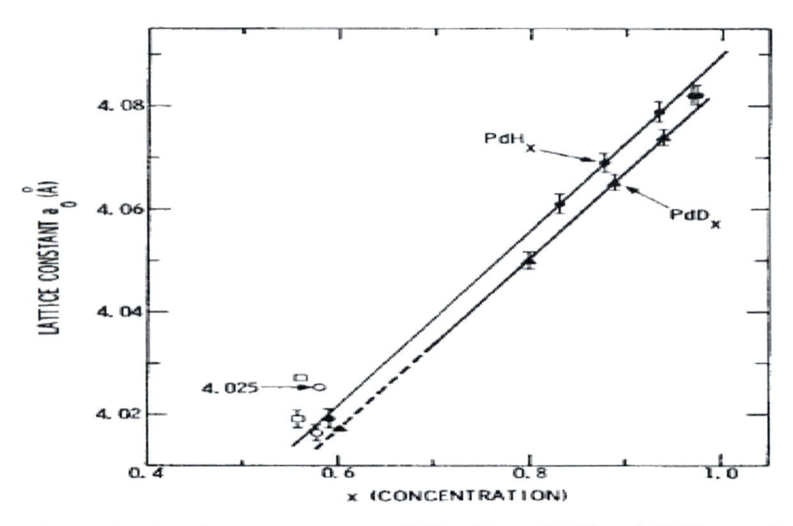

Figure 71. Lattice parameter at 77K of beta PdH and PdD as a function of atom ratio. The open points at the arrow are values for the beta phase at room temperature.(*526*)

2.7.3 Diffusion constant

The behavior of diffusion as a function of composition reveals conditions affecting hydrogen ions in the lattice. Notice in Fig. 72 how the diffusion constant decreases in a uniform manner as deuterium is added to the beta phase at temperatures normally present during LENR. This behavior is typical of what happens when octahedral jump-sites are increasingly filled, thereby reducing the places a hydrogen ion can occupy when it moves from one site to another. This behavior shows no indication of additional jump-sites being formed, such as would occur if metal atom vacancies, available tetrahedral sites, or another phase formed in this composition range. The situation at lower temperatures is more complex but not relevant to conditions existing during LENR.

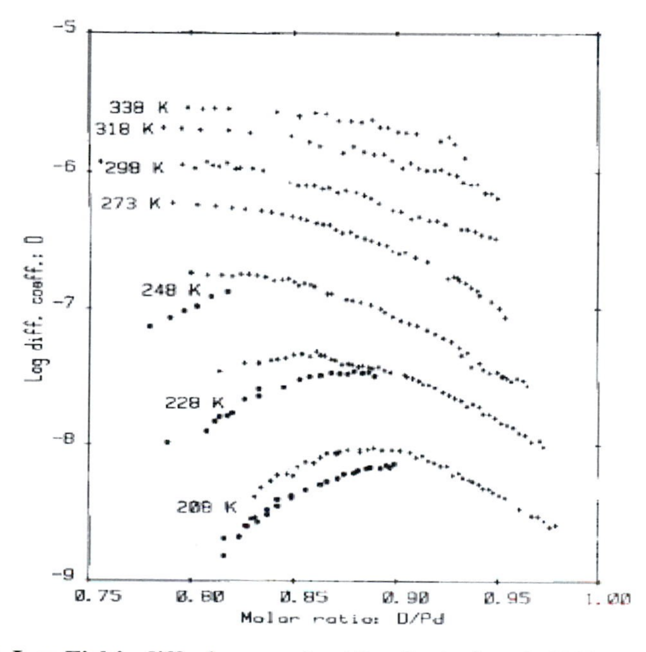

Figure 72. Log Fick's diffusion constant for deuterium in PdD as a function of atom ratio at various temperatures.(*521*)

Temperature has the expected effect on the diffusion constant, as shown in Fig. 73. Constant activation energy for diffusion exists over the temperature range, with similar values for both H and D. This behavior indicates both H and D experience the same diffusion process and this mechanism remains unchanged within this temperature range.

Using diffusion through a thin sheet of Pd subjected to electrolysis, Mengoli et al.(*527*) observed increase in the diffusion constant as the H/Pd ratio increased within the two-phase region. The effect of composition in the beta region was relative constant but scattered, with a value for the diffusion constant at H/Pd = 1 near $5x10^{-6}$ cm^2/sec at 25°C.

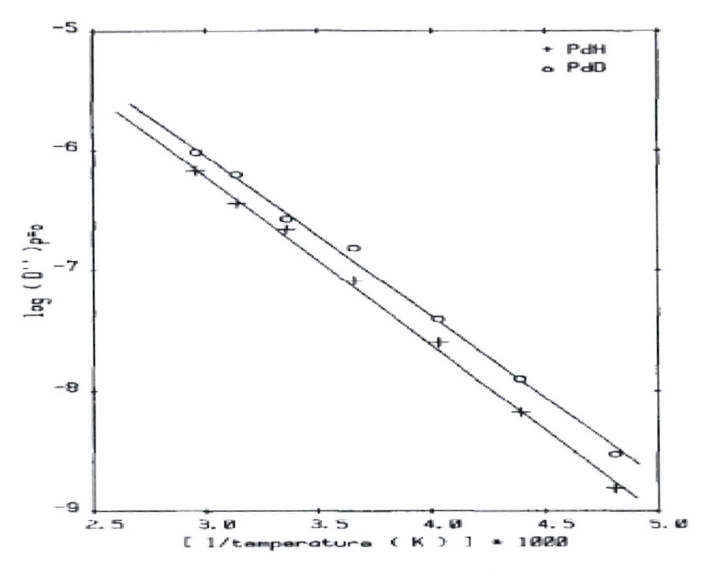

Figure 73. Effect of temperature on the log Fick's diffusion constant for β-PdD and β-PdH.(*521*)

2.7.4 Thermal conductivity

Bertolotti et al.(*528*) measured the thermal conductivity of PdH and presented the data shown in Fig. 74. The measurements apply to two-phase samples in most cases.

The relationship in Figure 65 shows that the thermal conductivity of PdH at the low-hydrogen boundary of the beta phase is about 2.5 times smaller than that of palladium metal and this trend to lower values continues as hydrogen is added to the beta phase. Since heat is carried mostly by electrons in a metal, this behavior suggests formation of PdH initially reduces the availability of electrons to carry heat as well as electron current, as shown by the resistivity. Consequently, although electrons are being added along with the hydrogen, the process initially reduces the freedom of electrons to move through the structure and vibrate in place. This freedom increases once H/Pd exceeds 0.75. This behavior indicates a change in electron bonding occurs near this composition, with electrons being freer to move as additional hydrogen is added to the structure. During this process, the crystal structure remains unchanged, with the only result being linear expansion and smooth change in the other properties.

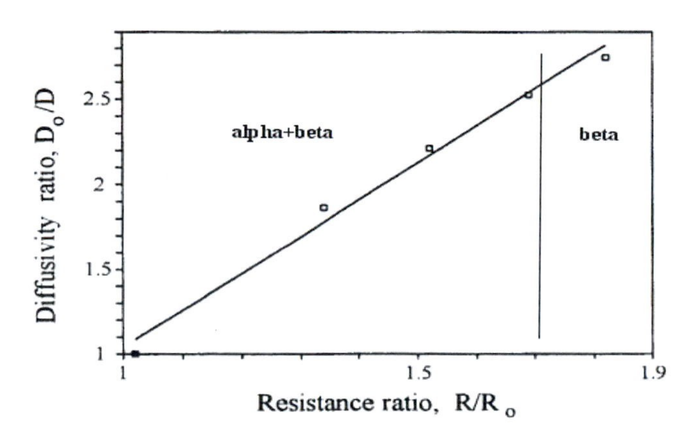

Figure 74. Relationship between thermal diffusivity ratio and electrical resistance ratio for PdH. The measurements apply to the two-phase region where a linear relationship between the properties of the two phases would be expected.(*528*)

2.7.5 Calculations describing the atom arrangement in PdH

A great deal of effort has been devoted to describing the PdD(H) lattice using a variety of models and calculations, only a few examples of which are cited here. These examinations are focused on where the hydrogen is located and how close each hydrogen ion can get to another hydrogen. Switendick(*529*) used self-consistent plane-wave total energy calculations and the Hedin-Lundquist exchange to conclude that the PdH_2 and PdH_3 phases are unstable with respect to $PdH+H_2$ for standard conditions. A large energy barrier was found as the distance between the H ions approach the distance in the H_2 molecule. This conclusion is consistent with calculations based on a molecular-dynamics simulation provided by Richards(*530*), with the calculations by Wei et al.(*531, 532*), and with local-density-functional cluster calculations by Dunlap et al.(*533*) An almost universal conclusion results from about 30 similar studies to show that the deuterons in PdD cannot get close enough to fuse at the observed rate.

Berrondo(*501*) uses the empirical interaction of charge density to conclude that some D might occupy tetrahedral sites in fcc structure. Nevertheless, even this reduced distance is considered too great to permit fusion.

Rather than use these calculations as a reason to reject LENR, as was done in the past, a different location for the process must be sought. This challenge is undertaken in Chapter 5.

2.7.6 Laws of Thermodynamics as applied to chemical structures

Because LENR occurs in a chemical structure, of which PdH[D] is an example, the Laws of Thermodynamics apply to how the deuteron or hydrogen ions behave before fusion takes place. In addition, formation of the structure in which the fusion process occurs must be consistent with the thermodynamic properties of surrounding atoms. This relationship is as important to chemistry as quantum mechanics is to physics. Consequently, neither the Laws of Thermodynamics nor behavior expected based on quantum mechanics may be violated. Because these laws are frequently ignored, a brief tutorial is useful.

The three Laws of Thermodynamics describe how energy flows in a chemical system and how atoms are arranged as a result of this energy. The First Law states that energy is conserved in a system such that energy can change form but the amount of energy must remain constant. Energy lost from the system must come from a source in the system and energy gained from the outside must go into the system with no energy lost in the process. If a spontaneous process concentrates energy in an atom or electron, an equal amount of energy must be lost from atoms or electrons located near where the energy is accumulating. This loss or gain is expected to cause local changes in temperature.

The Second Law states that energy always moves spontaneously from a higher value to a lower value. In other words, energy cannot spontaneously concentrate at a location in a material. If energy is applied from outside the system, it will attempt to dissipate in the material by moving always to lower energy. Nevertheless, individual atoms can acquire energy somewhat greater than the average. However, this extra energy has a strict limit and it requires, according to the First Law, for nearby atoms to lose an equal amount of energy. As a result, the energy difference in that region increases, thereby making further concentration increasingly unlikely.

The Third Law introduces the concept of entropy, which has baffled students since the concept was introduced. Entropy describes the degree of randomness contained in a structure. The product of entropy times temperature describes energy required to change the physical randomness of a structure. Such a process would also require chemical bonds to be changed, which involves energy called enthalpy. The combination of energy associated with entropy (S) and enthalpy (H) is called Gibbs energy (G), which is described by the formula $\Delta G = \Delta H - \Delta TS$. The delta symbol is used because these energies are only identified by a change. The magnitude of Gibbs energy is the driving force behind chemical reactions and determines the final concentration of resulting chemical products.

Enthalpy is the amount of heat obtained from a process such as measured using a calorimeter. In contrast, the Gibbs energy is related to the pressure or activity associated with the process. For example, the pressures shown in Fig. 65 are calculated from values of enthalpy observed when D_2 reacts with Pd. This measured value is combined with the entropy change as deuterons move from their arrangement in D_2 gas to form a new arrangement in fcc PdD. This reaction is spontaneous because Gibbs energy is created and is lost from the material in the process. All spontaneous changes must result in a loss of Gibbs energy from the material. For example, a cluster of deuterons cannot form spontaneously in PdD unless the deuterons in the cluster have greater Gibbs energy compared to where they are located in the fcc structure. A chemical system always tries to reach the lowest total

Gibbs energy allowed by applied conditions by creating Gibbs energy and losing it from the system as a result of a process. Presumably, such cluster formation might be forced on the system by supplying Gibbs energy to the system as a result of increased pressure or temperature, but the source must be identified rather than assumed.

The need to consider these laws makes LENR unique and unrelated to the requirements that apply when a nuclear reaction takes place in plasma, as is the case with hot fusion. This issue is not considered during normal nuclear reactions because energy is available either from an external source, such as energy of the bombarding particle, or from the reaction itself. In the case of LENR, these laws must be considered because energy sufficient to overcome the Coulomb barrier is not available to the LENR process in the normal chemical environment.

The description provided above is a simplified version of these laws and their consequences. Many books are available and careers in science are based on their study. Modern chemistry could not exist without these laws being well understood and LENR cannot be correctly explained unless these laws are applied.

2.7.7 CONCLUSIONS ABOUT THE PdD SYSTEM

The properties of the structures formed when either D or H react with Pd are very similar and show that a single phase having a fcc structure exists in the composition range formed when LENR is observed. No evidence supports the assumption that tetrahedral sites are available for occupancy by hydrogen ions or that metal atom vacancies form in which clusters of hydrogen ions might assemble.

The fcc structure terminates when almost all octahedral sites available to hydrogen are filled, at which point a second phase forms having the ability to accommodate a greater H(D)/Pd ratio. The crystal structure of this new phase is not known.

Both the lower and upper limits of the beta phase are altered by impurities introduced into the structure from deposits on the surface of cathodes. Consequently, a clear relationship between the behavior of pure β-PdD, on which most theories are based, and the material actually present when LENR is produced is difficult to establish. Nevertheless, a theory must apply to the actual condition existing at the site of LENR.

The reaction with hydrogen to create the beta phase causes cracks, gaps, holes, and dislocations to form in the material, especially on or near the surface. These defects exist outside of the fcc structure and can provide a path for loss of D_2, thereby preventing the composition from reaching a value required to fuel detectable LENR. Nevertheless, all PdD found to produce LENR have such defects on their surface and within the material with a wide range of size. Transmutation products and tritium seem to be only present where such cracks and gaps are formed on the surface. However, many cracks and pits also form without showing any evidence of LENR. Apparently, the nature and size of the cracks are important. The critical features are discussed in Chapter 5.

2.8.0 EFFECT OF ION BOMBARDMENT AND FRACTOFUSION

The ability of electrons to screen the Coulomb barrier between deuterons in Pd was explored early in the history of LENR.(*534-536*) Calculations were able to account for the small neutron flux claimed by Jones et al.(*537*) Now that LENR is found not to emit significant neutrons, the effect produced by Jones looks increasingly like hot fusion, not LENR as discovered by Fleischmann and Pons.[20]

In spite of two different types of nuclear reactions being involved, the concept of screening continues to be applied to LENR.(*538-540*) As a result, the important question of whether screening is relevant to explaining LENR needs to be addressed.

---

[20] This situation creates some irony because the claims made by Jones forced premature announcement by F-P because both claims were thought to be examples of cold fusion, differing only in their rates. Instead, apparently two entirely different phenomena were actually being studied, with both identified as cold fusion.

As early as 1934(*541*), fusion resulting from bombarding a material with deuterons was known to cause a greater fusion rate than when the reaction occurred at the same applied energy in plasma. Beuhler et al.(*542-546*) published studies of bombarding TiD with D_2O^+ and D^+ clusters (200-325 keV) which produced a greater rate for hot fusion compared to free-space, *i.e.* plasma. This encouraged the belief that hot fusion might show a transition to cold fusion when sufficiently low energy is applied to a compound containing deuterium. This idea was explored by Rabinowitz et al.(*547*) where they point out errors in a similar evaluation by Echenique et al.(*548*) This screening effect has been studied extensively, with only a few of the many studies listed here.(*540, 543, 544, 549-562*)

The effect of screening is determined by comparing the rate of hot fusion resulting from ion bombardment to the rate when the reaction occurs in free space at the same energy, *i.e.* in plasma. Fig. 75 compares the fusion rates for various combinations of hydrogen isotopes to the applied energy in plasma. Notice in the figure how rapidly the fusion rate decreases as applied energy is reduced, with a very small rate expected at 1 eV, typical of the maximum energy present during LENR.

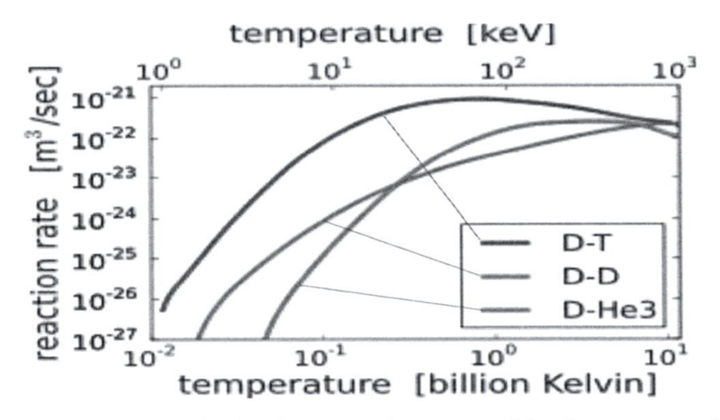

Figure 75. Rate of hot fusion between isotopes of hydrogen as a function of applied energy in plasma.[21]

[21] (http://en.wikipedia.org/wiki/Nuclear_fusion)

In addition, the screening behavior does not lead to increased fusion in the energy range present during LENR. This can be seen in Fig. 76, where the calculated "screening energy" is listed for several metals. Although the rate achieved by bombardment is increased over that in plasma, the absolute rate in the range of measurement is not increased. As can be seen in Fig. 77, the fusion rate drops rapidly as the energy applied to D^+ is reduced regardless of screening. Measurements below about 1 keV have not been possible because the rate of neutron production on which the fusion rate is based drops below the detection limit.

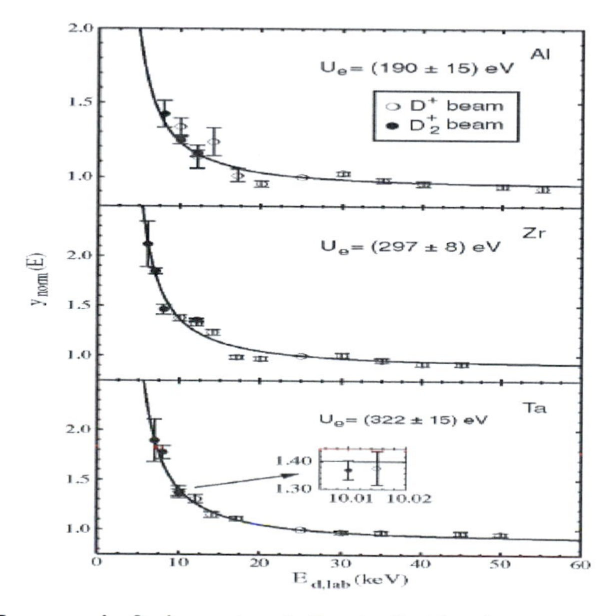

Figure 76. Increase in fusion rate relative to that in plasma as a function of energy of D+ used to bombard various targets.(*549*)

In order to apply this effect to LENR, extrapolation based on theory is required. This extrapolation would be plausible if the process causing cold fusion were the same as the process operating during hot fusion. Unfortunately for this approach, these two processes are different. Consequently, this behavior does not support the assumption that screening, based on the behavior of hot fusion, can increase the rate of cold fusion under LENR conditions.

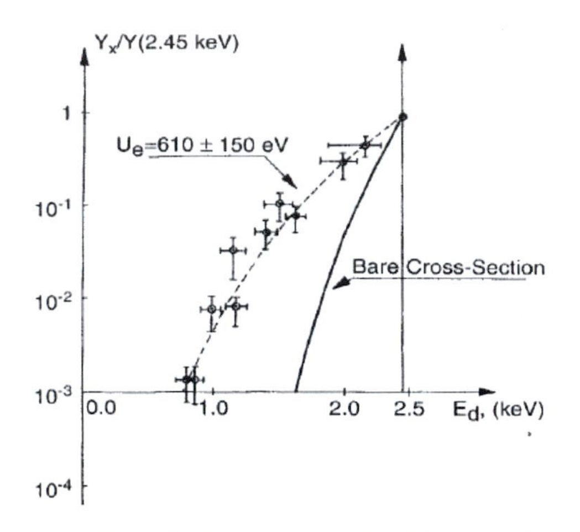

Figure 77. Increased rate of hot fusion as a result of D+ bombardment of Ti compared to the rate expected to result in plasma (Bare Cross-Section).(*313*)

This difference can be described as follows. When a material is bombarded by ions containing deuterium, the rate of the hot fusion reaction is influenced by the concentration of electrons in the material, which are proposed to partially lower the effective Coulomb barrier and allow tunneling. Once the barrier is overcome, the resulting nucleus fragments to dissipate the mass-energy and momentum. In the case of LENR, once the barrier is lowered, the resulting fusion product does not fragment. This result is only possible when the process of lowering the barrier is able to also dissipate mass-energy. Clearly, hot fusion and cold fusion are caused by two entirely different processes to which screening would have different effects.

Ion bombardment on a small scale can be made to occur in materials when cracks form either spontaneously or by applying external stress. This so-called fractofusion(*563-567*) occurs when the voltage created during crack formation accelerates D+ and causes rare and brief hot fusion events in the wall of the crack. Some low-level neutron bursts detected during LENR probably result from this process.

This phenomenon is also called piezonuclear and has been explored in detail by other people over the years resulting in at least 100 papers about the subject. The behavior is frequently observed when titanium reacts with deuterium.(*568-577*) Crack formation can apparently trigger other nuclear reactions as indicated by a brief burst of neutrons even when deuterium appears to be absent.(*578-583*)

2.9.0 SUMMARY AND CONCLUSIONS ABOUT OBSERVED BEHAVIOR

LENR generates helium as a result of D-D fusion, which produces the main amount of observed power. Fusion of D+e+H results in tritium. Reaction of target nuclei heavier than hydrogen with one or more H or D results in transmutation. Transmutation can result from two different nuclear mechanisms. Each of these reactions occurs in the surface region of the active material and is associated with cracks or pits that comprise the NAE.

All transmutation is observed to occur at a very small rate, much too small to produce detectable energy. Some transmutation products are radioactive with very short half-life. These reactions and the radioactive products produce radiation that is generally too weak to be detected outside the apparatus. Consequently, radiation needs to be measured inside the apparatus and radioactive transmuted isotopes need to be sought by placing the active material in a detector very quickly after LENR is detected.

Observed behavior and calculations show that the hydrogen (deuterium) in a chemical lattice cannot get close enough to fuse. In addition, the laws of thermodynamics and observed behavior are inconsistent with formation of clusters within the chemical structure, and energy cannot be increased enough in local regions to overcome the Coulomb barrier or to create neutrons. Consequently, LENR is not able to take place in the chemical lattice structure.

All the observed nuclear reactions appear to take place at the surface where the material is generally not pure or chemically similar to bulk material. Therefore, theory must focus on the material actually present in the NAE and not on the ideal, pure PdD or NiH.

CHAPTER 3

REQUIREMENTS OF AN EXPLANATION

3.0.0 GENERAL REQUIREMENTS

LENR requires several basic conditions to occur. The various theories of LENR have to show how these and other required conditions are accomplished. Hundreds of theories of LENR have been proposed by using different combinations of similar features for this purpose. Rather than examining each theory in detail, this chapter evaluates the general features to identify their flaws.

The process of finding flaws in the various theories can be thought of as identifying the boundaries of conceptual space in which a plausible explanation must reside. A determination only needs to be made of whether the basic features of a proposed theory are correct and therefore fall into accepted conceptual space. Search for the correct explanation can then focus on major features without being distracted by nitpicking detail. Based on this approach, use of a major feature having basic conflict with natural law would invalidate a theory no matter how correct the other assumptions might be.

The process starts by identifying the basic events of the LENR process to which explanations must apply. Three basic events can be identified regardless of how the overall theory is structured. First, two or more nuclei must come together at the same location at the same time. Second, a mechanism must take place at that location to overcome the Coulomb barrier. Third, the resulting excess mass-energy must dissipate into the surrounding atoms and appear as heat, but without producing significant energetic radiation. Each of these events must work in collaboration, be compatible with the physical and chemical conditions at the site of the process, and not violate any natural law. This combination of requirements significantly limits the nature of any proposed theory.

The first and most important feature of any theory is to identify where in the material the LENR process takes place. Some theories assume the nuclear reactions occur within the chemical lattice where they involve atoms that make up the chemical structure. In contrast, other theories propose that LENR takes place outside the chemical structure in cracks, gaps, dislocations, or on the surface. Later sections will describe limitations these possible sites impose on the LENR process.

Some theories propose a high concentration of deuterium in the normal PdD structure is the only condition required to initiate the LENR process. This high concentration is proposed to force deuteron clusters to form in vacant sites within pure PdD. A different approach places the nuclear reaction in a unique condition or structure called the nuclear active environment (NAE), as Storms first identified in 1995.(*12*) (See: Chapter 1.) These sites would be rare, widely scattered, difficult to create, and not part of the normal lattice structure. Regardless which of these two viewpoints are accepted and applied, one concept is critical. LENR cannot start until normal material is changed in some critical way regardless of the hydrogen concentration or any other condition proposed by theory. The site and nature of the active environment must be clearly described for a theory to be accepted.

Some theories propose cold fusion is just a different version of hot fusion. Evaluation of this approach requires a clear distinction be made between hot fusion and cold fusion. As can be seen in Fig. 78, hot fusion of two deuterons produces four energetic fragments containing all of the released mass-energy. This is not what happens when cold fusion occurs. Something else must happen to explain LENR. This "something else" creates severe limitations to an explanation. This point is so important, it deserves repeated emphasis, which is applied in various ways.

When evaluation is made of any theory, it is essential the process not violate natural law. The laws of conservation of energy, mass, and momentum must apply; the Laws of

Thermodynamics must be honored; and the basic chemical behavior must be acknowledged. To do this, conditions present at the actual site where cold fusion takes place must be identified and made central to the explanation. For example, assuming the ideal properties of PdD or NiH apply is not useful when such material is not present at the NAE.

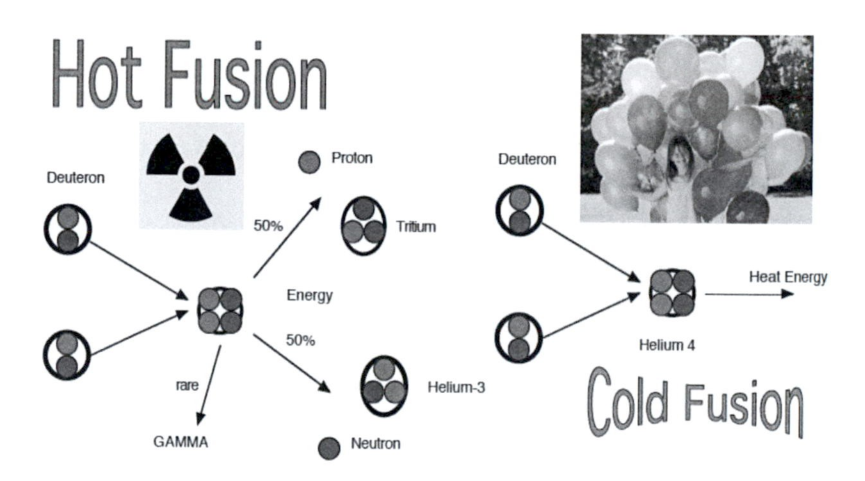

Figure 78. Comparison between hot and cold fusion.(Storms, ICCF-18)

This chemical behavior must consider the concentration and diffusion rate of deuterons in the material, both of which determine how fast fuel can reach the site of the fusion reaction with a rate sufficient to produce observed power. If a large number of nuclei are required in a cluster, the time to assemble this number is determined by how fast they can arrive at the site by diffusion. After all, a cluster does not form suddenly and spontaneously. Instead, it is constructed over time by single hydrogen nuclei arriving at the site one at a time and remaining there while the size gradually grows. This process takes time and controls the rate at which the nuclear process can occur. In addition, some unique attractive force must make this site more attractive to a hydrogen atom than any other location.

Because the effect has been so difficult to produce and so rarely detected in nature, the proposed theory must be consistent with and explain why this difficulty exists. If the process is proposed to occur in a typical chemical structure, such as PdD, something very rare and unique about that structure must be identified and this feature must be consistent with known chemical properties and behavior. In addition, the theory should not predict chemical behaviors that are known not to take place or are in conflict with known accepted natural laws.

A chemical system is not like plasma where isolated ions can acquire energy without limit. A chemical system is based on bonds between atoms. These bonds have a fixed energy and a fixed relationship as to how they connect ions and atoms together. Any change in energy, applied from either the outside or generated within the system, will change this arrangement and be quickly limited in value. This behavior is basic and must be acknowledged by any theory of LENR.

In addition to these reality-based limitations, some issues of style are worth considering. Anything can be explained if enough assumptions and variables are used. Furthermore, assumptions having limited application are also not useful. Instead, the best explanation results when only a minimum number of universal assumptions are used containing only a few variables.

Another problem results from hiding assumptions and variables in complex mathematical equations. When a mathematical description is used, which is generally based on quantum mechanics, it must result in a quantity that can be compared to physical measurements. Too often, mathematical equations are used to simply restate the assumptions without adding anything new or anything that can be tested.

Predictions are frequently made based on what is already known. A true prediction describes a physical behavior that can be detected but has not yet been observed. It is important to realize that "predicting" physical behavior that is impossible to detect is not a prediction.

3.1.0 PARTICULAR REQUIREMENTS IMPOSED ON LENR

Each of the three events occurring during LENR has a particular requirement. These requirements are examined here and applied in Chapter 4 to evaluate a carefully selected collection of typical theories. The new theory discussed in Chapter 5 is designed to meet these requirements.

A collection of atoms can be described using several different vocabularies. The structure can be assumed to have a general interaction between the parts to which a mathematical description called a Hamiltonian can be applied. Use of this approach requires many assumptions to achieve a unique relationship between the different kinds of the interaction and physical behavior, with many of these assumptions not being obvious or testable. This approach will not be described here because too many assumptions are *ad hoc*, result in visual displays that only restate the assumptions, and produce predictions that cannot be tested.

Nevertheless, a method to describe the LENR process must be developed because LENR obviously takes place somewhere in a material. Where this active material is located is difficult to discover because it is rare, difficult to produce, and not a pure material. Nevertheless, certain basic requirements apply that are discussed individually below. These requirements create limitations that can be discussed in terms of distance between atoms, ability to assemble atoms, energy available to overcome the Coulomb barrier, and the method used to dissipate energy.

3.1.1 Limits to energy concentration

Solid materials are made of atoms arranged in a choreographed relationship between the electrons associated with atoms. This arrangement produces various crystal structures having a characteristic relationship between atoms and a fixed distance between the nuclei. For this reason, internal energy is normally not available in such materials. If it were available, the structure would have to change to allow its release. Consequently, energy is not normally available to overcome the Coulomb barrier. If it is proposed to exist, its source must be clearly identified and be consistent with known sources of energy available in the structure.

Such energy would have to come from energy holding the chemical structure together and stored in its heat capacity, which is limited to a few eV. Consequently, the average energy within the structure available to cause a change in electron or atom relationships is very small. A process able to concentrate this energy at local sites is limited by very strict rules, including the Laws of Thermodynamics. These laws say, in paraphrase, that energy will not spontaneously concentrate in one location above well-understood limits[22] as described in Section 2.7.6. Although these laws apply to average behavior, a local variation must always combine with its opposite behavior to keep the average consistent with the laws. If a local region acquires energy, another region must lose an equal amount of energy.

For example, for an electron to gain 0.78 MeV, as required for a neutron to form, the process of concentration would require this energy to be drained from energy contained in surrounding atoms. Because energy has to be conserved, this process would require a large fraction of nearby material to reach temperatures near absolute zero.[23] Energy stored and identified as heat capacity is the only source because all other sources are involved in creating the chemical structure, which are not an energy source unless the structure is modified. Consequently, an explanation of LENR requiring such localization of energy, even for a brief time, violates basic experience and the Laws of Thermodynamics.

---

[22] http://en.wikipedia.org/wiki/Ludwig_Boltzmann

[23] For example, the heat capacity of Pd is 26 J/mole-deg = $3x10^{-4}$ ev/atom-deg. To concentrate 0.78 MeV in one site, about $3x10^{9}$ surrounding atoms would have to be reduced in temperature by 1 degree or about $9x10^{6}$ atoms would have to be cooled to absolute zero. What process can remove energy from so many atoms and concentrate it in one electron or one hydrogen nucleus? What event would cause such a process to start? Such a proposal is not plausible unless these questions are answered.

When energy is applied to the system, this energy also has limits because it cannot increase beyond the energy holding the atoms together or holding electrons to the structure. Even when energy is applied to an individual electron or deuteron, this limit is quickly enforced by some of that energy being transferred to surrounding atoms and electrons. This energy limit is obviously too low to initiate a nuclear reaction. Consequently, a nuclear process based on increased energy, resulting either from proposed spontaneous concentration or from energy being applied to the structure, cannot occur in an atomic structure. Even when energetic deuterons are caused to bombard PdD, hot fusion results, not cold fusion. Consequently, even if energy should concentrate by a mysterious process, only hot fusion would be expected.

To further emphasize the difficulty of proposing the chemical lattice as the location of LENR, a mechanism able to concentrate energy in a chemical lattice would influence the chemical structure long before the energy level reached a magnitude able to influence a nuclear reaction. After all, the atoms and electrons are able to rapidly share energy as can be demonstrated by applying energy in any form to a chemical system. If a mechanism were available in nature to spontaneously concentrate energy, unstable chemicals could not exist because the extra local energy would quickly initiate decomposition. For example, if such a mechanism were available in a chemical system, an explosive could not exist for long. Consequently, a LENR theory based on the concentration of a small amount of energy, even for a short time, must show why the process applies only to a nuclear process without having any effect on conventional chemical behavior.

3.1.2 Limits to cluster formation

Fusion requires two or more nuclei to assemble at the same location at the same time. This process takes time for each hydrogen nucleus to come to the active site from an increasingly distant location as local hydrogen becomes exhausted. This delay in reaching the required size must be considered because it will limit the amount of power produced. In fact, the larger the group required by a proposed theory, the less power can be produced because the rate of arrival has an upper limit determined by the rate of diffusion and local concentration. Once the cluster has reached a size claimed to initiate fusion, it will be destroyed by the fusion event and have to reassemble causing further delay until another fusion event is possible. The longer the delay between fusion events, the greater the number of active sites is required to generate observed power. The question then becomes, "How many active sites can be expected to form in a material to which the theory is applied?" The plausibility of a model depends on it being able to justify a cluster formation rate combined with a number of sites where they can form that in combination can explain the observed overall rate of energy production.

The concept can be explained using the situation in PdH. The PdH[D] crystal lattice has a limited number of unique sites where a cluster might form. Additional hydrogen might add to the normal octahedral location; the tetrahedral positions might become filled; sites where metal atoms are missing might attract additional hydrogen ions; various defects where the crystal structure is broken (cracks) or distorted (dislocations) might host hydrogen ions; or additional hydrogen ions might populate unique surface sites. Double occupancy would not be expected until all possible sites are filled by single occupancy because additional Gibbs energy is required to achieve double occupancy. Consequently, as has been proposed in several theories, double occupancy of the normal octahedral sites would only happen near the upper limit of the beta phase. This same limitation would apply to double occupancy of the tetrahedral sites and to metal atom vacancies as well because these sites would only permit double occupancy after most are filled by single occupancy, provided single occupancy were even possible. This requirement would limit LENR only to very high D/Pd ratios, which is not observed. In addition, the available D/Pd ratio in the beta phase would be much larger than is observed. For these reasons, cluster formation anywhere in the chemical lattice is very unlikely and not supported by observation.

Each site where clusters might form has different limitations but shares one basic limitation. For hydrogen nuclei to assemble in a chemical system, loss of Gibbs energy must occur as explained in Section 2.7.6. This means the Gibbs energy of each hydrogen ion in its "normal" location must be greater than its Gibbs energy in the cluster. Except for the assumed formation of the Bose-Einstein Condensate (BEC), no source of bonding energy has been identified to drive cluster formation within the chemical structure. In addition, this formation process must be inhibited in some way to prevent cluster formation except under the very special conditions identified by the theory. These requirements help explain why such clusters have never been observed within the PdD lattice and calculations(*531, 532, 584, 585*) reject their formation within a normal lattice site, as discussed in Section 2.7.5.

These various limitations make justification of cluster formation in the chemical structure very difficult. Something unusual must occur in the material after the chemical structure has formed for hydrogen nuclei to assemble. A solution to this problem is explored in Chapter 5.

3.1.3 Limits on dissipation of mass-energy

All spontaneous nuclear reactions release their energy by emission of radiations in the form of photon, electron, alpha, neutrino, neutron, positron, and/or fragments of the nucleus. These particles carry away the mass-energy while at the same time conserving momentum. In contrast, LENR creates only one nucleus when either helium or tritium is the nuclear product. Likewise, transmutation results in one nucleus when a hydrogen isotope adds to another nucleus, such as when deuterons add to Cs, as described in Section 2.3.3. The mechanism proposed to cause such nuclear reactions must dissipate energy without emission of detectable radiation or fragmentation of the product while at the same time conserve momentum.

Because tritium, helium, and transmutation production have the same problem of energy release, it is plausible to assume they each solve this problem using the same unique process. This assumption places severe limits on proposed mechanisms. These limits are discussed in Section 3.1.5.

3.1.4 Limits on production of radioactive products

The products resulting from LENR are rarely found to be radioactive, with tritium being the most obvious exception. Why radioactive products are apparently not produced must be explained. Perhaps radioactive isotopes are actually generated by LENR but they decay away before the radiation can be measured. Indeed, many of the predicted nuclear products would have short half-lives, making their detection difficult. This possibility should encourage search for such radioactivity by rapidly removing active material from the apparatus so that the full range of type and energy of the radiation could be determined without intervening absorbers.

3.1.5 Limits on production of radiation

Radiation is normally produced during and after a nuclear reaction is initiated. Why LENR fails to show this behavior must be explained. For example, if neutrons were required to initiate the reaction or were released by the reaction, a significant fraction could easily leave the apparatus and could be detected. In every case, the detected neutron flux is trivial compared to the flux required to generate the observed energy.

Before conclusions are drawn based on observed radiation, the fact that most expected radiation cannot be detected needs to be considered. In the case of energetic particles[24], their detection is limited by their range and inability to escape the apparatus. Even photon radiation will be significantly reduced in intensity by intervening material. Unless the detector is installed inside the apparatus near the source, this radiation could be invisible regardless of its intensity. For this reason, present lack of

---

[24] Neutrons, neutrinos, and very energetic photons are the only radiation able to escape a typical LENR apparatus.

correlation between radiation and power production has no meaning other than to suggest the radiation energy is too low to allow escape or to produce significant secondary radiation.

Energetic alphas (^{4}He) and energetic deuterons cannot be directly detected outside the apparatus, although they can produce secondary radiation when their energy is sufficiently large. Hagelstein examined possible secondary radiation resulting from energetic alpha and deuteron emission in several papers.(*284-286, 586*) When alpha energy is above 6.2 keV to 7.7 keV(*286*), he predicts secondary neutron emission from D-D hot fusion might be detected. This fusion reaction results when energy is transferred to a deuteron by the alpha so that it can fuse with another nearby deuteron. In other words, alpha energy above this range can be detected as secondary radiation and alpha radiation below this energy range would be invisible outside the apparatus. Of course, the alpha flux would have to be large for this secondary radiation to reach detectable levels.

If energy were released directly as energetic deuterons, such as fragments of a cluster in which fusion occurred, the energy limit of each deuteron must be less than 900 eV to avoid potentially detectable neutrons. Absence of this secondary radiation sets a lower limit of 27600 to the number of deuterons in a cluster from which detected secondary radiation would not be produced by a single fusion event. In turn, this limit sets a lower limit to the time required to form such a cluster based on the rate at which this number of deuterons can move to the same location. Finally, the assembly process creates a limit to the rate of fusion such an explanation can support. A theory proposing to use deuterons to dissipate energy much justify how these limits are consistent with the observed power, aside from explaining how cluster formation is not in conflict with the laws of thermodynamics and where in the lattice such a large structure can form. This challenge has not been met by any proposed theory.

Because radiation can result from a variety of reactions, with each having a wide range in intensity and energy, detected radiation can be difficult to assign to a single source. In addition, radiation is detected only after it has passed through all absorbing material between the source and the detector. Generally, only very energetic photons and neutrons can do this successfully, and then only after being reduced in intensity by the process. As a result, the measured intensity outside the apparatus will always be less than the intensity at the site of the nuclear reaction.

When radiation is detected, it might be used to identify the nature of the nuclear reactions under ideal conditions. Nevertheless, even when the source is unknown, detected radiation with the observed energy demonstrates the unambiguous presence of a nuclear process where none is expected based on conventional theory.

3.1.6 Limits on the number and kind of nuclear mechanisms required

Several nuclear products from LENR have been identified, including helium, tritium, and various transmutation products. Whether many separate and independent mechanisms are required to produce these results is unknown, but the simplest approach is to assume only one basic but unique mechanism is operating. If separate mechanisms are involved, each must be consistent with all the rules described here. Each mechanism must show how a Coulomb barrier can be overcome at a significant rate when the barrier ranges from 1 between two hydrogen nuclei to as high as 80 when some transmutation reactions occur. Each of these processes must also dissipate the mass-energy while being consistent with all other requirements. These requirements are hard enough to imagine being followed by a single mechanism, making a multiple mechanism approach to explain the different nuclear products even more difficult to justify.

3.2.0 COMMON FEATURES USED IN THEORY

All theories are based on a few basic mechanisms combined in creative ways. Each mechanism has certain problems requiring acknowledgement and justification. A few basic mechanisms are summarized here along with the problems their use creates because of conflicts with the requirements described above. The individual theories discussed in Chapter 4 are evaluated by referring to these sections rather than by repeating the problems and conflicts for each case.

These problems and the questions they raise can be considered deal-breakers. That is, if the questions cannot be answered in a plausible way, the model can be rejected without further consideration.

3.2.1 Resonance of deuterons in PdD lattice to achieve nuclear interaction

All atoms in a lattice vibrate as consequence of their temperature. However, this oscillation alone is obviously not sufficient to cause a nuclear reaction. Some theories use quantum mechanical arguments to demonstrate resonance vibration is sufficient to cause LENR, generally after a novel process has increased the amount of local energy. This process is frequently based on the concept of phonons[25], which is a mathematical description of energy residing in the vibration of atoms and electrons.

Problems: Resonance vibration is possible only because the atoms are attached to a structure that holds them in place. As local energy is increased by this process, bond strength holding the atoms to the structure places a limit on how much energy can be contained in this vibration. Too much energy and the bonds between atoms break, thereby limiting further local increase. For example, solid can become liquid and liquid can become gas.

A basic assumption is made when a nuclear reaction is proposed to occur between two resonating hydrogen atoms before the bonds holding them to the structure break. This assumption is based on a quantum mechanical equation (Hamiltonian) that assumes the structure can be treated as plasma with energy shared between all members. As long as this energy is kept within the

---

[25] http://en.wikipedia.org/wiki/Phonon

amount known to exist in a chemical system, the approach works well to explain chemical and physical behavior. However, this energy alone has not been found to be large enough to trigger a nuclear reaction.

Can vibration focus enough energy on two atoms well in excess of the amount associated with their chemical condition? An answer has to take into account the size of the forces and energy holding the lattice together, which is only a few eV[26]. Local energy exceeding this amount would tax the ability of atoms to remain in the lattice. In other words, such a resonance (vibration) would be expected to cause chemical effects that could be detected and have influence on local behavior. While local chemical effects might not be detected, they would drain energy from a process attempting to concentrate energy.

Vibration is proposed to reduce the distance between deuterons for a brief time, perhaps too brief for a chemical effect to occur. This assumption does not solve the problem because when the distance between deuterons is reduced by muons, hot fusion results. Clearly, simply reducing the distance does not result in cold fusion.

Some theories propose temperature in a material is not uniform, with some atoms having large energy while other atoms are relatively cold. These differences are proposed to exist for only a brief time, but long enough to initiate a nuclear reaction either by using local high energy to overcome the Coulomb barrier or by forming a Bose-Einstein-Condensate (BEC) in the local cold region. If this assumption were true, why are nuclear reactions not common in very cold materials? For example, PdD containing a high concentration of D has been cooled below 10K without producing an observed LENR reaction that would have been easy to detect.(*587-589*)

---

[26] The energy required to move H from PdH and place it in H_2 is 106 kJ/mole H = 1.1 eV/atom H. The energy required to move a Pd atom from the solid and place it in the gas is 362 kJ/mole Pd = 3.8 eV/atom Pd. These limits are far too small to provide enough local energy to affect nuclear interaction.

Wave formation is proposed in several theories to eliminate the Coulomb barrier. Even atoms in a BEC are assumed to ignore the Coulomb barrier. No plausible justification or evidence has been provided to support these assumptions.

3.2.2 Tunneling or electron screening

If the amount of energy needed to overcome the Coulomb barrier is less than a theory predicts, the process is proposed to result from tunneling through a barrier or screening of the barrier. For an explanation to be useful, it must identify how these methods of reduction can occur in a material frequently enough to support the observed rate of cold fusion and not result in hot fusion.

Problems: Ion bombardment, as described in Section 2.8.0, shows tunneling or screening can occur in a chemical lattice and the effect of screening increases as applied energy is reduced. Nevertheless, the total fusion rate decreases as applied energy is reduced to at least 1 keV, with no indication of an increase in rate as energy is further reduced (Fig. 76). Consequently, for the observed reaction rates to occur at applied energy near 0.1 eV, typical of LENR conditions, something else must be proposed besides tunneling or screening. This issue is also discussed in Section 3.1.1.

3.2.3 Formation of clusters of certain sizes and located at certain sites in the lattice

Clusters as small as two or as large as thousands of deuterons are proposed to form, sometimes as a BEC. The clusters are assumed to grow in the chemical lattice itself or on its surface. Further discussion is found in Section 3.1.2.

Problems: Cluster formation requires loss of Gibbs energy, as described in Section 2.7.6. PdD, and most similar materials, are not known to contain clusters(*480, 513, 590*) and calculations described in Section 2.7.5 argue against their presence.

Large clusters require time to assemble from the surrounding D+ ions, which limits the reaction rate and available heating power, as discussed in Section 3.1.2.

Introduction of the BEC concept raises additional questions. For example, formation of a BEC involving deuterons at room temperature would seem to be as amazing as is LENR itself. Even if the BEC formed, why would it then cause only cold fusion and at what rate? How many D must be in such a cluster before fusion could occur without detected radiation and how long would assembly of this number take? How is the energy released without causing the hot fusion process? Lack of answers to these questions impact the plausibility of the explanation.

3.2.4 Formation of neutrons or interaction with stabilized neutrons already present in the lattice

Because neutrons have no charge, they can enter a nucleus without having to deal with a Coulomb barrier. Nevertheless, some nuclei are more easily entered than others. In addition, the neutron is unstable, decaying to a proton, electron, and neutrino with a half-life of 881.5±1.5 sec.[27] Consequently, neutrons available for nuclear reaction in a material have to be made *in situ* or be held in a stabilized form to be available for LENR. Creation of a neutron from a proton and electron requires about 0.78 MeV at its birth along with a neutrino. This energy is modified somewhat by the energy of the neutrino.[28] Failure to capture the required neutrino can be expected to limit the rate of this reaction even when the required energy is supplied.

When neutrons are assumed to accumulate in a material, they must have a form that does not decay and does not interact with any other nuclei until the NAE forms. To be unnoticed, they must be present at too low a concentration for sensitive analytical or X-ray techniques to see. Just how their presence would survive purification or the various chemical reactions preceding their proposed involvement in nuclear activity is not explained. For

---

[27] http://en.wikipedia.org/wiki/Neutron

[28] http://en.wikipedia.org/wiki/Neutrino

example, PdD is made from palladium metal that was separated by a complicated purification process from the other elements in the ore and then melted. The D_2 gas is separated from water by a complex chemical process. When would stabilized neutrons enter the PdD and how would they accumulate? In other words, in order to explain LENR, such an assumption requires a condition even more unexpected, hard to justify, and novel than LENR itself. Furthermore, a circular argument is created when the presence of neutrons can only be revealed when they are proposed to initiate the very behavior they are used to explain.

Although neutrons are seldom found emitted by an operating cell, exposure to an external source has been found to significantly increase neutron emission(*591, 592*), but only by a small amount.

Problems: The amount of energy required to form a neutron is not available in a chemical lattice and cannot be concentrated in an electron or nuclei without violating the Second Law of Thermodynamics, as explained in Sections 2.7.6 and 3.1.1.

Several LENR theories propose neutrons are stabilized in special forms and are present in all materials. However, these special forms have not been directly detected. How do they manage to be present during all methods used to initiate LENR listed in Section 2.6.2 and in all the materials known to support LENR? This basic question needs an answer before a proposed model can be accepted.

Although neutrons can react with many nuclei, very few of the expected products are actually detected. If neutrons were in a material, they would be expected to produce a variety of reaction products, not just one, and some of these products would be radioactive, making them detectable. The radioactivity expected to result from this process is not detected.

In addition, the measured He/energy ratio is not consistent with a value expected when neutrons interact with nuclei to create helium as shown in Section 2.2.4.

3.2.5 Formation of special sites, such as super abundant vacancies, where fusion takes place

As described in Section 2.7.1, PdD has the fcc structure without significant metal atom vacancies and with hydrogen ions located in octahedral sites, which is typical of such structures. Sites in the lattice where LENR can occur are proposed by many theories to be vacancies where the metal or the hydrogen ion is normally missing. For reasons not made clear in the theories, groups of hydrogen atoms are proposed to grow in these sites.

Problems: All materials can contain atom vacancies, defects, and flaws of various shapes and sizes, yet they do not host nuclear reactions. For these locations to be plausible sites for LENR, the special nature of each kind of site needs be considered, starting with atom vacancies.

Atom vacancies are not special because they are part of the normal chemical lattice. Their concentration is determined by applied conditions with the goal of forming the most stable atomic arrangement. This means the Laws of Thermodynamics apply to the concentration of such vacancies, as discussed in Section 2.7.6. Consequently, the process of filling vacancies by clusters of hydrogen must be consistent with these laws.

The concept of superabundant vacancies (SAV), introduced by Fukai(*486-488, 495, 593*) applies to a structure created under conditions (5 GPa and 700°C) far removed from those present during LENR. This structure is proposed to result from hydrogen vacancies forming into clusters within the metal lattice structure. The metal lattice collapses into this void space causing the lattice parameter to decrease[29], with high pressure supplying the required Gibbs energy. As a result, this structure is no longer fcc. The SAV condition results when all the hydrogen is removed leaving behind

---

[29] Only the positions of metal atoms can be located using X-rays. The overall structure shows a simple cubic relationship based on the metal sublattice. The lattice structure created by hydrogen is unknown. Vacancies are identified relative to an assumed structure and can only be identified when the structure is fully known, which is not presently the case.

large regions where atoms are missing; in this case metal atoms are missing in addition to hydrogen atoms. Although the initial ordered structure (Cu_3Au type) can be quenched to normal ambient conditions, it has not been observed to form under these conditions nor would it be expected to form during LENR. Nevertheless, Fukai claims a structure containing SAV can form in PdH under ambient conditions.

As is discussed in Section 2.7.0, even if this claim by Fukai were true, observed behavior does not provide support. In addition, the claim for the fcc lattice being able to reach a D/Pd greater than unity as support for presence of SAV can be better explained by the presence of an additional second phase, as concluded in Section 2.7.6. Consequently, the work reported by Fukai has no relationship to LENR other than to reveal a mechanism for crack formation, as Fukai notes in Section 6 of his paper.(*495*)

Formation of pores is claimed to show evidence for SAV. As shown in Fig. 79, pores are proposed to result from the vacancies in SAV, designated as □ in the formula of the compound, all moving to one location while leaving behind vacancy-free material. If the pores result from this process, the behavior actually demonstrates such vacancies are not stable and the material containing them will eventually decompose to form vacancy-free material surrounding the pores. Unknown is whether these pores exist only on the surface or throughout the material. In any case, their presence does not support a claim for SAV. Nevertheless, such pores are frequently found in the surface of nuclear active material, as seen in Figs. 18 and 43.

In contrast to vacancies, dislocations and cracks are not limited by the Laws of Thermodynamics, with their presence being generated by opportunistic release of stress. As a result, hydrogen might accumulate and form structures with no regard for the basic properties of the lattice. Cracks especially are very common with a variety of sizes and shapes. The only rare and unique feature being the gap width, which suggests this feature might be the critical parameter for producing a rare NAE with limited concentration, discussed in detail in Chapter 5.

3.2.6 Transmutation of the metal atoms in the lattice

Transmutation can occur if one or more protons, deuterons, or neutrons react with a target nucleus to produce a new isotope or a completely new element. As described in Section 2.3.2, several variations of transmutation are observed.

Problems: Transmutation reactions normally produce radiation, either during decay of radioactive isotopes or as prompt gamma radiation from the transmutation event itself, neither of which is detected in amounts consistent with observed energy. In addition, the collection of elements and isotopes attributed to transmutation seldom can be explained by simple addition of p, d, or n to a target.

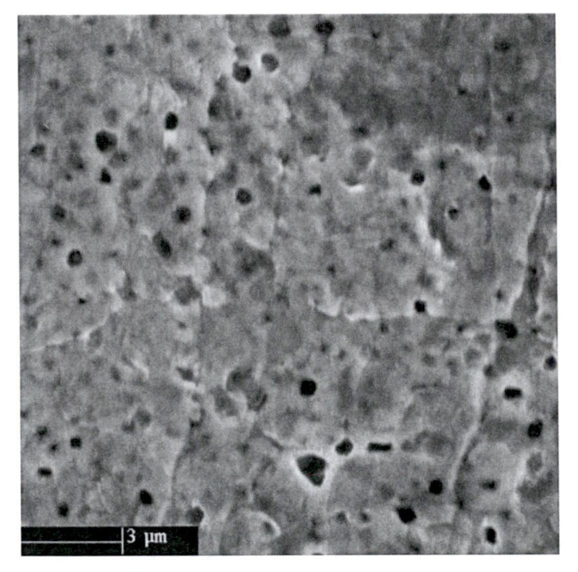

Figure 79. Pores in PdH claimed to have resulted from decomposition of $Pd_3(\square H_x)$ after annealing in hydrogen.(*489*)

If transmutation occurs at a rate sufficient to make detectable heat-energy, a large fraction of the target atoms in the material must have access to the rare NAE. This being the case, how can transmutation involve enough of the metal lattice to produce significant energy when the sites at which transmutation products are found are so rare and isolated? The theory must acknowledge this limitation and show how the amount of measured energy can be generated.

For transmutation to dissipate the resulting energy as heat, either the nuclear products must include energetic particles or the mechanism must transfer energy to the surrounding material in a manner similar to how this process takes place during fusion. Photon emission or energetic particle production is not found to occur at a rate consistent with the observed rate of transmutation, perhaps because most of this expected radiation is absorbed before it can be detected.

Overcoming the large Coulomb barrier for transmutation would require a unique mechanism to produce a detectable rate. An explanation must address this issue.

3.2.7 Shift of energy to a non-conventional state by formation of a special structure

Electrons in PdD and in similar compounds are either associated with the metal atoms or occupy energy states that allow them to move freely throughout the lattice, a condition called the conduction band. As a result of transferring its electron to the metal atoms, hydrogen in the structure appears to have a net positive charge of about +0.3(*113, 522, 594*) based on electromigration behavior (Section 2.7.0). This fractional charge can be explained by the electron having reduced probability of being near the hydrogen nucleus because it spends most of its time associated with the palladium atom.

In contrast to this conventional description, some theories propose the electron associated with hydrogen can occupy a unique energy state closer to the nucleus than provided by the Bohr orbit (the 1s level).

Problems: The electrons in PdD and in similar materials are considered to be in their lowest energy states. If new states having greater stability could form, they would be expected to populate as Gibbs energy was released until their presence became obvious.

This problem is avoided in some explanations by assuming either the state is inhibited from forming except in a few special sites or because the state is occupied only for a brief time. If this condition were fleeting, energy released by the formation process has to be retained by the electron and not be emitted as a photon as is normally the case. Otherwise, the state could not quickly revert to its original condition without having to reacquire this emitted energy, which would be a slow process. Some models assume this retained energy is used to initiate LENR. This being the case, the argument becomes circular with the energetic electron only being detectable by its role in LENR. A process made invisible by definition, except by producing the behavior being explained, is not worth considering because it cannot be tested by independent methods.

Creation of significant power as the result of these energy states requires the concentration of states be large. In other words, power is determined by the time the critical energy states are active multiplied by the total number of active states. If the lifetime of the critical energy state is small, it has a correspondingly small window of opportunity to influence fusion; hence, less power will be produced. The observed fusion rate places a limit on the time this window is open. A window of 10^{-6} sec requires 10^{17} active sites to produce 10^{11} reactions/sec, which would require about 0.0001% of all atoms in a material to be potentially nuclear-active. Is this a reasonable number given the rarity and difficulty of causing such nuclear reactions?

The reason why chemical reactions are not also triggered by this process is not given in most explanations. For example, if energy states could form with the ability to trigger a nuclear reaction, why would explosives and other unstable chemicals remain stable, as noted in Section 3.1.1?

3.2.8 Role of novel heavy particles

On occasion, evidence for unusually heavy and novel particles is reported as described in Section 2.4.6.

Problems: If such particles are required to produce LENR, they must be present in a quantity large enough to produce measured power and energy for significant time. Such a high concentration should be detected using normal analytical methods, which has not been demonstrated. In addition, if these particles were present everywhere in sufficient numbers, why is the LENR process so rare and difficult to initiate?

3.2.9 Dissipation of energy

Only two methods for energy dissipation are consistent with the fusion reaction involving D and H. These methods involve photon or phonon emission because emission of electrons or other particles would result in unobserved nuclear products and produce detectable secondary radiation. On the other hand, transmutation is expected to emit electrons, positrons, alphas, photons, and heavy fragments, as explained in Section 5.4.3.

Problems: A phonon is created by vibration of atoms, electrons, or nuclei and is identified in its gross effect as temperature. The proposed mechanism must explain how the final nucleus can shed the resulting nuclear energy by causing the surrounding material to vibrate and how this vibration is broken into sufficiently small units of energy to avoid destroying the surrounding structure. This consequence is especially challenging because energy density would be greatest immediately adjacent to the nuclear process. After all, heat production could not be sustained if the required local structure were destroyed by a single fusion reaction (Section 3.1.3).

In contrast to phonons, photons can carry a wide range of energy and this energy is gradually deposited as heat along the path of the photon, well away from the reaction site. As a result, the lattice near the nuclear event would not be destroyed even when the energy flux is high. Nevertheless, how the large nuclear energy can be broken into and carried away by small quanta of photons needs to be explained.

3.3.0 ONLY EXOTHERMIC NUCLEAR REACTIONS ARE POSSIBLE

Some theories propose impossible nuclear reactions because instead of releasing energy, energy must be consumed to make the proposed product.

Nuclear reactions generate energy by converting mass to energy. When the reaction products are less massive than the reactants, energy is released. When the products are more massive than the reactants, mass must be created for the reaction to take place. Typically, the amount of mass requires more energy for its creation than can be localized in a chemical structure. Consequently, LENR cannot involve a nuclear reaction that requires energy. Imagined reactions can be quickly evaluated to determine their energy requirements by calculating whether mass is lost in the process. If mass is not lost, the proposed reaction is impossible.

As an example of this calculation, when two deuterons fuse, the initial mass is greater than the mass of helium by 0.025601 AMU, as tabulated below. Multiplying this number by 931.4943 gives the energy in MeV for the reaction. Tables provide the mass of all the stable isotopes.[30]

d = 2.014101778 AMU
^{4}He = 4.002602 AMU
Mass change = 0.025601 AMU
MeV/AMU = 931.4943
Energy/event = 23.85 MeV = 3.8×10^{-12} J or watt-sec
He atom/watt-sec = 2.63×10^{11}

This same method can be used to calculate energy resulting from transmutation. Energy resulting from a hypothetical transmutation reaction between Ni and H is calculated in the following table.

---

[30] See: http://www.chemicalelements.com/show/mass.html,http://physics.nist.gov/cgi-bin/Compositions/stand_alone.pl

p = 1.007825 AMU
^{62}Ni = 61.928348 AMU
^{63}Cu = 62.929601 AMU
Mass change = 0.006572 AMU
Energy/event = 4.0 MeV = 6.4×10^{-13} J or watt-sec

Reactions that emit or absorb electrons require emission or absorption of a neutrino at the same time. Energy carried by the neutrino cannot be detected and would not appear as heat. Reactions that result in addition of electrons to a nucleus, such as the e + p reaction proposed to form a "cold" neutron, are rare because the required neutrino is difficult to acquire.

3.4.0 SUMMARY AND CONCLUSIONS

The various major features used in proposed explanations are examined and found to contain serious challenges. In fact, without proof for LENR actually being possible, this analysis would reasonably cause a person to reject any such claims. Nevertheless, because LENR has been demonstrated to be a real phenomenon, these flaws presently in proposed theories must be corrected and a better theory suggested.

The requirements a theory must acknowledge can be summarized as follows:

1. The LENR process does not occur in the chemical lattice or in any of its normal features, such as vacancies.
2. The LENR process does not involve formation of neutrons before LENR occurs.
3. The LENR process is not initiated by novel heavy particles.
4. The LENR process requires a special location in which small clusters of hydrogen can form without violating the Laws of Thermodynamics.
5. The LENR process has no relationship to the mechanism causing hot fusion and must not be confused with hot fusion.

When any of these requirements are violated, the explanation becomes suspect and requires considerable additional justification.

The next challenge is to locate where in the material the LENR reaction takes place and by what mechanism, all while applying these requirements. This goal is sought in Chapter 5. First, a few proposed theories are described in Chapter 4 and compared to the requirements described above.

CHAPTER 4

EVALUATION OF PROPOSED EXPLANATIONS

4.0.0 INTRODUCTION

Finding flaws in a theory is easier than discovering why the explanation is correct. Nevertheless, the fewer flaws in a theory, the more consideration it deserves. This process starts with identifying logical conflicts with observation and well-accepted natural law. If no present theory survives this process, the effort to understand LENR has to start over and take a new path in a different direction. The requirements identified in Chapter 3 are applied to a few published theories in this chapter as examples of how this approach can be applied.

Hundreds of "explanations" have been suggested and published. Many are too poorly developed or inadequately described to have value. For this reason, only a few examples of well-developed concepts are described here. The goal is to show a causal reader the basic paths taken by various theoreticians without using jargon and complex mathematics. Hopefully, this description combined with identified flaws will guide thinking in more productive directions and help avoid making the same mistakes in future explanations.

For the sake of this examination, the chosen theories are placed into categories based on their main feature. If this particular feature has a problem, each theory in that category would shares the same problem. Many theories combine several features, each of which could have additional problems.[31]

Most theories place the nuclear process within a pure chemical structure having the average properties normally attributed to the structure. In contrast, observations show that LENR occurs at the surface where the material has a complex

---

[31] The word "problem" means the conclusions and assumptions used in the theory conflict with the requirements described in Chapter 3.

morphology and uncertain chemical composition. Consequently, such theories fail by not describing the actual conditions in which LENR occurs.

4.1.0 THEORIES WITH FOCUS ON CLUSTER FORMATION

For fusion to occur, two or more hydrogen nuclei need to assemble in the same place at the same time. Consequently, cluster formation is basic to the LENR process. The only question is how and where they form in the material. The number of hydrogen nuclei proposed to assemble in excess of two depends on the model. For example, as few as four deuterons are required by Takahashi while Miley et al. propose at least as many as 157 must assemble. Just how many nuclei are needed to start the fusion process is frequently not clearly described. Formation of the BEC is sometimes used to justify cluster formation in PdD at room temperature without specifying the number of deuterons.

P. Hagelstein

Peter Hagelstein[32] is presently a professor at MIT and continues to explore theory in spite of very little general interest in the subject at the university. He has over 50 papers about LENR to his credit (*595-611*) and organized ICCF-10 held at Cambridge in 2003. His ideas have evolved over the years — many being justified by complex mathematical descriptions, only to be withdrawn after predictions were not fulfilled. In the process, Hagelstein has been unusual by being able create many "toy" models, identify their flaws, and then move on to new ideas. This approach has helped identify ideas that can be safely ignored in the future. Unfortunately, his present model is not yet consistent with all the requirements identified in Chapter 3.

---

[32] http://www.nps.edu/Academics/Institutes/Meyer/docs/CV website%20version.pdf

Hagelstein uses photon interaction described by complex mathematical equations focusing on the chemical lattice as the site for LENR. The super abundant vacancies (Section 3.2.5) are also explored as the location of cluster formation in the lattice.(*612*)

An early example of the approach can be found in *Journal of Fusion Energy* in 1990(*613*) where he justifies d+p fusion by using what he describes as a coherent process to transfer the resulting mass-energy into the lattice one phonon at a time with a "virtual" state being formed during the process. As he says, "The basic premise of the theory results from off-resonant coupling between two fusing nucleons and a macroscopic system can occur through electromagnetic interaction." The d+p fusion reaction was initially proposed to result in tritium and a positron. Radiation resulting from positron annihilation has not been detected, which Hagelstein acknowledges invalidates this approach. He now is exploring ^{3}He production from this reaction.

In 1993, Hagelstein and Kaushik(*614*) explored neutron transfer in a lattice, the assumption being that neutrons in nuclei can spontaneously transfer from one nucleus to another. The process is proposed to release energy as gamma radiation and radioactive decay should an unstable isotope be created by the process. This idea was made untenable once the amount of helium was carefully measured, as Hagelstein has acknowledged. Once again, a proposed mechanism can be effectively ignored thanks to his efforts. However, this mechanism has no relationship to the theories based on neutron creation or on stabilized neutrons as described in Sections 3.2.4 and 4.3.0.

Hagelstein(*615*) summarizes his current ideas in the *Journal Condensed Matter Nuclear Science* (2012). He has determined, based on absence of observed secondary radiation, that ^{4}He produced by d+d fusion cannot have more than 20 keV of residual kinetic energy.(*286, 586*) He proposes mass-energy is communicated to the lattice by phonons instead of by photons and that this transfer results from what he calls a lossy-spin-boson model.(*616, 617*) In order for the resulting mass-energy to be dissipated without melting the surrounding material, it must be broken into very small phonon quanta and leave the nuclear-active region fast enough to avoid local heating. This requirement severely reduces the size of the phonon quanta and the rate of energy production from the fusion process.(*611*) Limitations of this concept are examined in Section 3.2.9.

Phonons are proposed to initiate the effect when created by applied energy. This conclusion is based on the effect of using laser energy as discussed in Section 2.5.2. He proposes that the effect of temperature, another source of energy, might be related to the ability of the lattice to get rid of the resulting helium, which he assumes poisons the process. The behavior of helium is discussed in Section 2.2.4 and the effect of temperature is examined in Section 2.6.6.

He proposes a test of his model can be made based on an assumption. He assumes the released energy can be coupled to a nucleus such that a suitable isotope can be made radioactive, thereby revealing successful transfer. Use of ^{201}Hg is suggested because it has a low-energy nuclear state of 1.5648 keV and because Karabut(*618*) showed that such states might be stimulated. This proposal is justified by using a Hamiltonian(*619*) designed to show how lattice vibrations might be coupled to internal nuclear transitions.

In summary, Hagelstein assumes D-D fusion takes place in a normal PdD lattice after the lattice is fully loaded with D. This condition is assumed to cause clusters of deuterons to form in Pd atom vacancies that are assumed present.(*436, 620, 621*) Evidence against metal vacancy formation is discussed in Sections 2.7.1 and 3.2.5. Limitations to cluster formation are examined in Sections 3.1.2 and 3.2.3. He assumes these clusters fuse and the resulting energy is dissipated as very weak optical phonons created by what he calls fractionation of the released energy over time. Energy dissipation by this method is examined in Sections 3.1.3 and 3.2.9.

M. Swartz

Mitchell Swartz is a medical doctor specializing in radiography, editor of the *Cold Fusion Times* (http://world.std.com/~mica/cft.html), and associated with JET Energy Technology, Inc. Swartz has arranged several well-organized and effective symposiums at MIT, with the help of Hagelstein, and has publically demonstrated heat production from LENR. His contribution to theory has focused mainly on the engineering aspects of LENR.

Starting in 1992(*622*), he described the rate at which the Pd cathode reacted with deuterium in an electrolytic cell by a series of equations based on the known behavior of electrolytic cells, which he called a quasi-one-dimensional model (Q1D). He later(*623*) described fusion of deuterium using what he calls a Phusor model in which fusion occurs in a fully loaded lattice site and energy is dissipated by phonons (Section 3.1.3). In 1996(*624*) he proposed production of deuterium in light water cells using nickel as the cathode.

The concept of Optimal Operating Points (OOP) was introduced by Swartz in 1998(*625*) and continues to be a feature of his thinking.(*464, 465, 626-629*) He proposes all nuclear reactions have an ideal condition in which they function at maximum efficiency. When power production is described using this concept, Swartz compares the total applied power to either the excess power or the excess/applied power. The latter variable represents the efficiency of heat production, which is determined mainly by experimental design, not by a basic property of the nuclear process.

As Swartz notes, when additional current is applied to a cathode, its use in stimulating the nuclear process can become overwhelmed by energy used to decompose the electrolyte at the surface. This saturation effect is common and can be expected whenever additional energy is used to stimulate any process of any kind. Furthermore, identifying conditions that produce the highest efficiency is a challenge for engineering only after excess power can be generated reliably at high level, not the reverse as Swartz proposes.

If typical behavior shown in Fig. 59 is used to compare excess/applied to applied power (*i.e.* OOP), Fig. 80 results. Although this plot shows a behavior similar to the OOP, greater insight is obtained from Fig. 59 itself. The important conclusion is that excess power is related to applied current, not to applied power, which is not obvious from how Swartz treats the information.

Swartz(*464*) puts each type of nuclear reaction in a different location and this location is represented by a different OOP. He predicts that the ideal system involves low current and high electric field intensity. He attempts to achieve this condition by using very pure D_2O as the electrolyte and applying high voltage to achieve useful current.(*449*) This method has not achieved as much power as has been produced using D_2O-Li and high current (Section 2.6.3).

Swartz prefers to call the phenomenon Lattice Assisted Nuclear Reactions (LANR) to emphasize his belief that the nuclear reactions occur in the lattice itself.

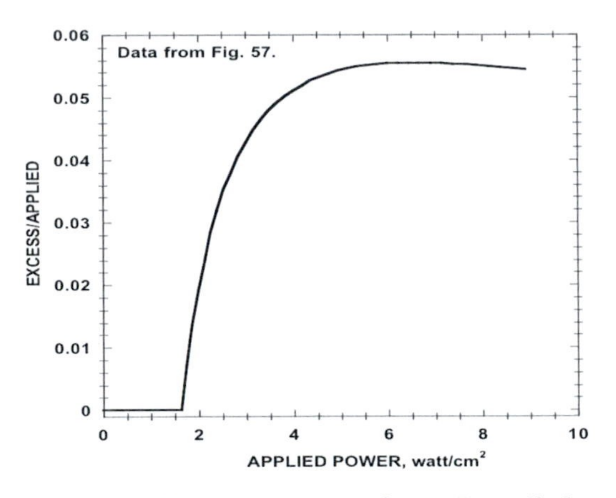

Figure 80. Excess power/applied power vs estimated applied power using data shown in Fig. 59.

G. Miley and H. Hora

George Miley is retired from the University of Illinois and has continued to publish papers about LENR.(*630*) His initial explanation proposed neutron transfer or neutron stripping from the deuteron followed by its addition to a palladium nucleus, thereby producing energy.(*631, 632*) However, the expected radioactive isotopes have not been detected.

Heinrich Hora is retired from the University of New South Wales in Australia. Together, he and Miley proposed a theory in 1990(*633*) based on earlier work of Hora et al.(*634, 635*) They assume electrons exist in PdD as plasma and can spontaneously accumulate on a surface in high concentration called a "swimming electron layer." This layer is proposed to help overcome the Coulomb barrier and bring deuterons closer together. Multilayer electrodes are suggested as a way to amplify the effect.(*636-638*) This method is claimed to produce brief bursts of energy before the site is destroyed by electrolytic action.

As noted in Section 3.2.1, simply getting the deuterons closer together is not sufficient to explain LENR. Nevertheless, Hora et al.(*639*) attempt to predict how distance between deuterons would affect the hot fusion rate. His later papers apply this concept to cold fusion.(*640*) The concept of swimming electrons on a surface also can be applied, with modification, to describe a process taking place on the surface of cracks as proposed in Chapter 5.

Miley et al.(*169, 641*) measured transmutation products (Figs. 20 and 21) found in beads provide by Patterson.(*642, 643*) The distribution of transmuted elements is proposed to be related to assumed "magic numbers"(*179, 644, 645*) used to identify especially stable nuclei. This idea is based on the natural distribution thought to exist in the universe, as shown in Fig 81. This idea is used to predict formation of a stable nucleus identified as $^{306}Xe_{126}$, based on calculations provided by Rutz et al.(*164, 647*) This unknown element is proposed to be stable enough to actually form in PdD yet be unstable by fragmenting to produce the observed transmutation products and energy without detectable radiation. Formation of such a heavy nucleus would be an endothermic process, hence impossible under LENR conditions, as described in Section 3.3.0. Significant experimental justification would be expected for claiming formation of $^{306}Xe_{126}$, but does not appear in any of the many papers by the authors.

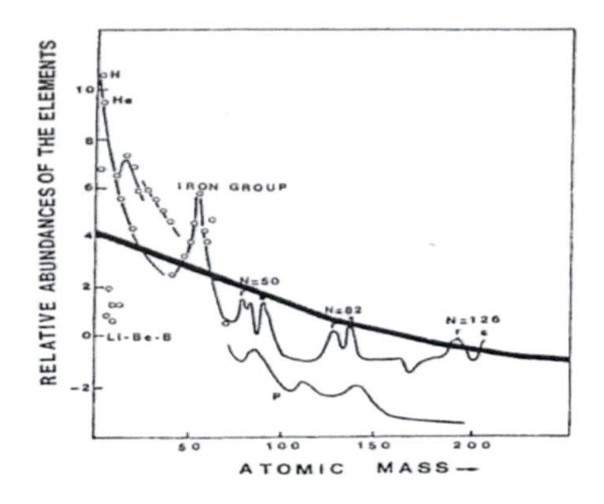

Figure 81. Proposed elemental distribution in the universe.(*646*) The heavy line describes an exponential function having the same coefficient as the line in Fig. 21.

Instead, justification is based on the following assumed process. The swimming electrons are proposed to allow the deuterons to assemble on a surface, and reduce the Coulomb barrier, after which a large number of deuterons enter the nucleus of a nearby metal atom. This process is proposed to rapidly produce a super-heavy nucleus that eventually fragments, favoring certain nuclei having special stability when the atomic number matches a proposed magic number.(*640, 648*) This mechanism violates the requirements discussed in Sections 3.1.2, 3.2.4, and 3.2.6.

A comparison between the power expected to result from formation of the measured elements and power measured using similar conditions was made by Miley and Shrestha(*649*) and used to support the model. Unfortunately, the values of power on which this comparison is based, as reported by Luo et al.(*650, 651*), are not correct because the wrong neutral potential is used to correct for gas lost from the open cell[33]. Even if the correct value had been

---

[33] A value of 2.2 V or 2.01 V was used for the neutral potential, depending on which paper is cited. The correct values are 1.482 V for H_2O and 1.527 V for D_2O [1], based on the enthalpy of formation of liquid water from the elements.

used, such a comparison is not useful. Measured power depends on applied conditions, as shown in Figs. 59 and 60, while power calculated from the total inventory of transmutation products depends on the sum of rates of each reaction during some unknown time. For this reason, agreement between the power measured and that calculated using the observed transmutation rate has no significance because these two quantities are not expected to agree. A comparison between the total energy observed and that expected might be useful provided the comparison is made using the same sample, but Miley et al. did not make such a comparison.

Miley and Shrestha(*177*) provided a review of their efforts up to 2003 at ICCF-10, which is amplified at ICCF-11 by Hora et al.(*652*) They note the similarity between their distribution of transmutation products and the distribution reported by other people as support for their concept of transmutation products having "magic" mass numbers.(*179, 644, 645, 648, 653*) This conclusion is difficult to justify when a comparison is made between the results shown in Figs. 19, 20, and 21. The (A) and (B) regions in Fig. 20 are better explained by a combination of fusion and fission of elements in the (C) and (D) regions.

In addition, the reported rate of transmutation gives implausible values for the expected rate of energy production. For example, if silver (Ag) were made by adding a deuteron to Pd at the rate shown in Fig. 20, a 1 cm^3 sample would make 15000[34] watts, which is not observed. Presumably, a much smaller but unknown volume is typically active. A comparison between generated excess energy and rate of formation of transmutation products cannot be made unless this active volume is known.

The authors' present theory involves forming a "virtual" neutron as D diffuse through the lattice causing what they call "orbital mixing." The "virtual" neutrons are then assumed to form

---

Use of this incorrect value creates an apparent excess energy that does not actually exist.

[34] $^{105}Pd + d = {}^{107}Ag$ gives 13.4 MeV/event = 2.15×10^{-12} J/event. A rate of 7×10^{15} atoms/sec-cc, as shown in Fig. 20, would produce 15000 watts/cc if this transmutation reaction occurred throughout the sample.

a "large mass compound nucleus" by a process resulting from "compressed flow during diffusion." However, this process must allow some protons to enter the final nucleus along with the "virtual" neutrons in order to create a nucleus having the required number of neutrons and protons. How this distribution is accomplished is not described in the model. How diffusion can produce neutron-like conditions is hard to reconcile with the behavior shown in Figs. 72 and 73.

A cluster of deuterons(*164*) is proposed to form in a manner shown in Fig. 82. Apparent local melting and hot spots seen using an IR camera (Fig. 57) are offered as evidence for such a high concentration of deuterons. How such a large collection of like charges can form in PdD and how this is possible in view of calculations that limit such clusters to no more than 6 deuterons are not explained (Section 2.7.5). The role of swimming electrons or how diffusion gets involved is not obvious from the drawing. In addition, the laws of thermodynamics (Section 2.7.6) require a significant release of Gibbs energy for such a cluster to form (Sections 3.1.2 and 3.2.3). The source of this Gibbs energy is not identified. Nevertheless, this model is being used to justify commercial development.(*654*)

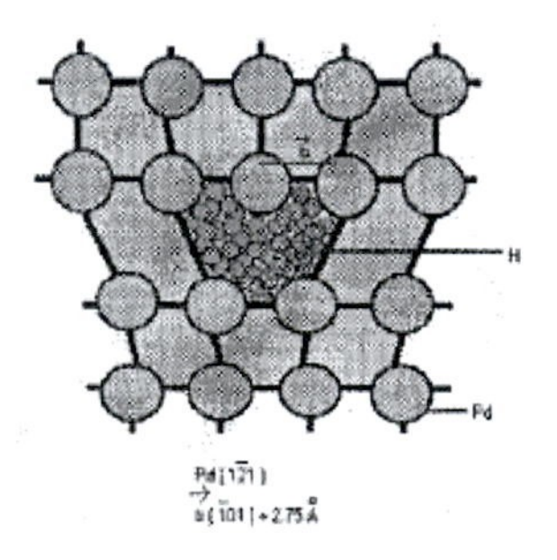

Figure 82. Cartoon provided by Miley et al. of a lattice site in PdD where a super nucleus is proposed form.(*164*)

A. Takahashi

Akito Takahashi works for Technova Inc., in Japan where basic studies of LENR are underway. His interest in the subject began in 1989 when he was able to generate excess energy by electrolyzing Pd in D_2O-Li while measuring the energy of neutron emission.(*655-658*) His general model has grown in complexity but has remained focused on multibody fusion in clusters.

Initially, the observed energy spectrum was used to support a theory based on a three-body process involving clusters of deuterons. An elaborate mathematical theory was developed using the screening effect of electrons to cause a three-body cascade within the chemical lattice. Nuclear products typical of hot fusion resulted, which at the time were thought to result from cold fusion.(*659*) The proposed process starts with D ions in the PdD lattice vibrating in place until the magnitude allows adjacent D nuclei to fuse (Section 3.2.1). This process was proposed to involve three or more D in order to be consistent with the measured neutron energy in the 3-7 MeV range. Repeated addition and removal of D from the PdD cathode was proposed to provide additional vibration energy. This process seemed to have the expected effect, although the detected neutron flux was too small to account for the energy being observed.

Over the years, the basic concept of multibody fusion has been expanded and further justified. Ohta and Takahashi(*660*) described at ICCF-8 a process to cause fusion of 4D at a common site in the PdD lattice. Four D are proposed to come together at a single site in the lattice, where coherent motion combined with enhanced screening produced by "heavy electrons" leads to fusion. Later papers(*661, 662*) describe the process as four deuterons converging on a single lattice site to form what is identified as a BEC(*663*) for which the fusion rate is calculated. The Coulomb barrier is proposed to be screened by a process called the "two times Green function method."(*664*) Thus is born what Takahashi calls the 4D/TSC (tetrahedral symmetric condensate) model, which addresses the problem of how energy is released while momentum is conserved. The so-called 4D/TSC is proposed to form on a surface, obviously by a different mechanism than used to justify the BEC(*665*) (Sections 3.1.1 and 3.1.2).

The process is proposed to form ^{8}Be, which is assumed to have a lifetime sufficiently long for energy to be emitted by gamma emission before the nucleus finally fragments into two alphas without exceeding the Hagelstein energy limit (Section 3.1.5). This decay process is based on ^{8}Be forming a proposed halo nucleus with the ability to dissipate energy before eventual fragmentation into two alpha particles takes place. Unfortunately, several additional problems are created by this model. Even though a mechanism for overcoming the Coulomb barrier can be imagined using several plausible assumptions, the problem of dissipating the resulting mass-energy without obvious and perhaps harmful radiation is an acknowledged problem with no easy solution using this approach (Section 3.1.3).

The model has been described in increasing detail over the years with the sequence shown in Fig. 83 provided recently to show the overall process. Summaries of the process can be found in the proceedings of the ASTI-5[35] conference in 2004(*667*), and more recently in the proceedings of ICCF-17 in 2012.(*668*)

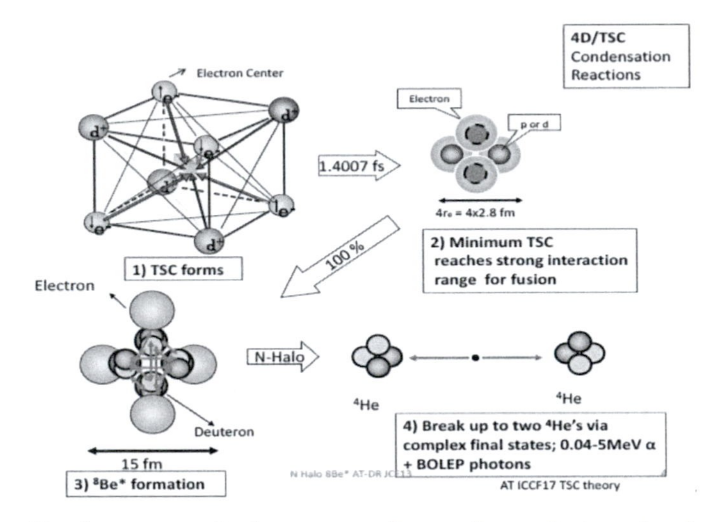

Figure 83. Sequence of the proposed reactions that make heat and helium.(*666*)

---

[35] See: http://www.iscmns.org/meetings/Asti/asti.htm

The most recent version can be summarized as follows. The TSC structure consists of four deuterons and the accompanying electrons that condense in a site within the PdD lattice to form a very small cluster. This cluster, once formed, immediately experiences fusion to produce ^{8}Be. The total fusion rate in a material is determined by the product of the formation rate times the fusion rate once TSC forms. The TSC is said to form within 1.4 fs[36] in active sites described as "fractal surface with many sub-nano-holes (SNH) and inner finite lattice (Bloch) structure." Once formed, all the TSC immediately fuse. This means the amount of observed power depends only on the rate at which TSC forms while ignoring the concentration of rare special sites in which a TSC must form. The number of such sites is important because a variable and limited number of active sites are possible since the entire chemical lattice is obviously not available. Furthermore, such sites will take a variable time to form before they can be occupied by the TSC (Section 3.1.2), a delay not considered.

The description provided by Takahashi goes on: "D_2 molecule is adsorbed and dissociated at deep trapping potential, due to many dangling chemical electron bonds there at SNH on surface and then diffuse into inner finite lattice sites by quantum-mechanical (QM) 'tunneling.'" In other words, a chemical structure forms at chemically active sites (SNH) on the surface and the resulting TSC structure diffuse into the PdD where it immediately reacts to form ^{8}Be at a rate determined by "Fermi's golden rule under the Born-Oppenheimer approximation."

When light hydrogen is present instead of deuterium, the TSC is thought to spontaneously collapse and form a virtual neutron by reaction between p and e. This virtual neutron interacts with other members of the TSC to create ^{4}Li that decomposes to give ^{3}He + p and energy (7.45 MeV/^{3}He). The process is proposed to be the source of the energy when H_2+Ni is used. It remains to be determined whether ^{3}He is actually produced under these

---

[36] To which part of the fusion process this time interval applies is not clear.

conditions or whether a neutron having a proposed virtual existence can actually interact with other atoms like a real neutron.

Transmutation(*665*) is explained as being the consequence of the hydrogen fusion reaction. The very small TSC is proposed to penetrate the Coulomb barrier of Ni or Pd before it has had time to fuse, thereby producing the observed transmutation products by the process Ni+4p or Pd+4d followed in some cases by fragmentation of the final nucleus. In other parts of the description, energy released by the fusion process is proposed to force p or d into the metal nucleus, thereby overcoming the difficulty created by the large Coulomb barrier. In other words, the dominant fusion reaction and occasional transmutation are both explained by the same initial process (Section 3.2.6).

Another path to transmutation(*166, 669*) is also proposed by Takahashi to involve coherent photon radiation providing energy near the giant resonance found to exist in some nuclei near 15 MeV. A nucleus can be caused to fission when energy at this level is applied. This radiation is proposed to result from the fusion process and be absorbed by the lattice before it can exit and be detected. In this way, the process avoids the high Coulomb barrier and uses energy generated by fusion to cause transmutation.

Justification of the model was sought by bombarding(*552*) TiD with 300 keV D^+. Energetic charged particles were produced as expected, some of which resulted from enhanced hot fusion (Section 2.8.0). Evidence for D+D+D fusion was sought and claimed based on a few emitted particles in addition to those produced by the much larger D-D fusion reaction. Whether this result supports the proposed mechanism for D-D-D fusion during LENR depends on several assumptions being accepted. For this claim to be related to LENR, the cluster must actually form in the lattice by a spontaneous process, which cannot be justified except by complex calculations (Sections 3.1.2 and 3.2.3). In addition, the formation of such a structure would violate the laws of thermodynamics (Section 2.7.6 and 3.1.2) and the known behavior of PdD (Section 2.7.0).

An alternative explanation might involve chance encounter between the bombarding D and two D in the lattice where hot fusion products attributed to D-D-D fusion might form during intense ion bombardment.

Y. Kim

Yeong Kim[37] is a professor of physics at Purdue University. Kim started his search for an explanation of LENR by proposing a variety of mechanisms in a series of reports published by Purdue University.(*670-677*)

In 1998, Kim et al.(*678*) focused on the Bose-Einstein-Condensate (BEC) of deuterons being a fusible cluster. Deuterons are considered bosons because they can occupy the same quantum space, have an even number of fermions (e, p, and n), and have integer spin. Atoms having these characteristics have been found to form very weak structures only at temperatures near absolute zero in free space. For deuterons to collectively form a BEC in a chemical lattice near room temperature, while at the same time eliminating the Coulomb barrier, is an extraordinary claim. Nevertheless, Kim et al. have explored this idea for the last 16 years in numerous papers.(*679-691*)

The theory(*688*) rests on ion traps being present into which the D can fall. In chemical terms, these traps are considered lattice sites where deuterium ions would normally be located. These sites would obey the Laws of Thermodynamics and occupy stable positions in the lattice with known separations (Section 2.7.5). The fusion rate for deuterons occupying such a trap is calculated by the authors using the number of D in the trap, the trap diameter, and the BEC ground state occupancy. The equation is made to fit the observations by adjusting the value for ground state occupancy. The released mass-energy is proposed to be shared between all the deuterons in the BEC after a single fusion event in the BEC. The remaining deuterons scatter and dissipate the released mass-energy in the surrounding material. A BEC containing at least 4450

---

[37] http://www.physics.purdue.edu/people/faculty/yekim.shtml

deuterons is assumed to dissipate the energy. Hagelstein(*284*) places this number as high as 26,000 deuterons to avoid each deuterium having enough energy to produce detectable hot fusion (Section 3.2.9). Where such a large BEC could form in a fcc structure and why fusion is delayed until the cluster has reached this limit are questions not answered. Low temperature is predicted to increase the rate of fusion despite studies of superconductivity at temperatures as low as 2K showing no evidence of extra energy being produced at low temperatures (Sections 3.1.2 and 3.2.3).(*588, 692*)

Kim and Ward(*691*) now acknowledge that the BEC cannot form in a lattice. Instead, they propose a small metal particle is required where "surrounding grain barriers" provide a site. The cross-section for the reaction is calculated using the cross-section for hot fusion. They claim their equations do not involve the Gamow factor, which they claim only applies in free space because the Coulomb field disappears between the deuterons located in the center of the sphere comprising the BEC. This conclusion assumes the charge on each deuteron is somehow combined with every other nuclear charge and its accompanying electron to create a uniform field at the center of the BEC. Once a sufficient number of deuterons have assembled, fusion takes place at the center. This results in what is called monopole vibration of the resulting ^4He nucleus, with mass-energy gradually communicated to the surrounding Pd atoms as phonons or e^+e^- pair production. In this manner, having to explain why the expected secondary radiation produced by energetic deuterons is not detected can be avoided by proposing an entirely new and novel process.

4.2.0 ROLE OF RESONANCE

For deuterons to get close enough to fuse without forming a cluster, vibration of individual deuterons in their normal lattice sites must bring them close enough to fuse on occasion. This process is also assumed by several theories of LENR to create a wave structure able to eliminate the Coulomb barrier, with each author using different mathematical constructs to justify the assumptions. Why these vibrations and wave structures do not affect the chemical behavior in an detectable way is explained by assuming this process happens very seldom and at very low level, with only enough rate to produce observed nuclear products but not often enough to produce detectable chemical interaction. This mechanism is described in more detail in Section 3.2.1.

G. Preparata

Giuliano Preparata[38] (1942-2000) worked at the Dipartimento di Fisica dell'Universita in Milan until his death in 2000. He with two other authors, Bressani and Del Giudice(*693*), published the first paper in 1989 describing an approach that Preparata(*535, 694-700*) went on to develop with enthusiastic support from Fleischmann and Pons.(*701-703*) For this reason, the model had important influence on how cold fusion was understood at the time and even today. His work is regarded so highly that his name is associated with an award given to people in the field whose work deserves recognition.[39]

He proposed that the atoms of deuterium form plasma in the PdD crystal structure where coherent oscillation (Section 3.2.1) could occur around the classical positions in the face-centered-cubic (fcc) lattice. This coherent process is assumed to lower the Coulomb barrier so that barrier penetration (tunneling) (Section 3.2.2) could occur with a significant rate between two deuterons that would normally be located on average about four Angstrom apart. This idea was challenged by Leggett and Baym(*704, 705*) based on the normal separation between deuterons in the lattice. Preparata(*696*) rejected this critique because he assumed the nuclei could get closer as a result of the proposed coherent oscillation.

No special condition other than a fully occupied lattice of deuterium (D/Pd = 1) was proposed necessary. The surrounding electrons are proposed to carry away the resulting mass-energy before fragmentation (hot fusion) could occur. The authors

---

[38] http://en.wikipedia.org/wiki/Giuliano_Preparata
[39] http://www.iscmns.org/prizes.htm

predicted, as did Schwinger(*706*), that p-d fusion would be much faster than d-d fusion, which has not been found to be the case.

In 1991, Preparata(*695, 696*) reviewed several theories competing with his own. At the time his review was written, the importance of having a high deuterium concentration in the PdD structure was known(*707*), production of tritium was observed and confirmed(*36, 346, 708*), and the difficulty in producing the effect was being used as a reason for rejection.(*709*) In addition, some methods(*537, 573*) produced products and radiation typical of hot fusion, which confused explanation. His review found the models suggested by Schwinger(*710*) and Hagelstein(*711*) to be inadequate compared to his own. Preparata(*563*) notes that the occasional burst of neutrons can result from fractofusion,[40] but this cannot explain the observed sustained production of energy (Section 2.8.0).

According to Preparata, fusion in a lattice requires either increase in ambient energy, for example by coherent oscillation of the deuterons, or lowering of the barrier by electron screening. Preparata rejects the electron screening approach. Instead, he emphasizes need for coherence in the lattice between the deuterons to initiate fusion and coherence between electrons to dissipate mass-energy. The nature of this coherence process and how it might be initiated is not identified, although he suggests use of quantum mechanics and quantum field theory to find the answer. Preparata quite rightly concludes that cold fusion does not violate any known law, yet he proposes the fusion process can occur in a "normal" lattice in spite of such reactions not being consistent with known nuclear and chemical behavior (Section 3.0.0).

In 1993, Rabinowitz et al.(*712*) wrote a critique of the Preparata theory as requested by Preparata. A detailed critique of various other concepts followed in 1994.(*9*) Their main criticisms are:

---

[40] A brief voltage gradient occurs between the walls of a crack when it first forms that can accelerate deuterons to sufficient energy to fuse and produce hot fusion products for a very brief time.

1. Preparata failed to give proper credit to others who first developed the mathematical models he used. Since the previous authors had not applied their ideas to the cold fusion problem, this complaint is not as valid as the other criticisms.
2. Preparata falsely assumed that the electrons surrounding and associated with the Pd atoms also surrounded the d nuclei. This assumption reduced the Coulomb barrier between deuterons by too great an extent. As stated in the critique, "Solids would collapse if such close equilibrium screening were possible." This conclusion is important because it applies to all other models that assume the lattice structure can be modified enough to cause a nuclear reaction simply by addition of deuterium.
3. Preparata used incorrect values for some of the constants in his equations. For example, he used 20 fm for the d-d nuclear attraction radius while other authors(*713*) suggest values between 1.1 fm and 2.8 fm. An assumed large number reduces the amount of screening required to cause fusion. In addition, the authors suggest Preparata made a numeric error in his equation that results in the wrong value for the tunneling rate based on his assumptions.

As is common throughout the history of this field, agreement between theoreticians about even basic conclusions has yet to be achieved. Nevertheless, the concepts vigorously defended by Preparata had a significant effect on LENR theory development, even to the present time, in spite of the obvious problems.

J. Schwinger

Julian Schwinger[41] (1918-1994), before his death in 1994, was a theoretical physicist who took an interest in cold fusion after having been awarded the Nobel Prize for his work on quantum electrodynamics (QED). Even this award did not protect him from

[41] http://en.wikipedia.org/wiki/Julian_Schwinger

the scorn of cold fusion skeptics, resulting in his resignation from the American Physical Society. His experience is one of many showing just how irrational the behavior of scientists became when this subject was discussed or attempts were made to publish papers in some journals.

Schwinger(*710*) first suggested his explanation in 1990 at a conference that eventually became a regular series called the International Conference on Cold Fusion (ICCF)[42]. The ideas were published in more detail in *Z. Phys. D*(*714*) along with an unprecedented disclaimer by the editor. Schwinger accepts the claims of F-P, but suggests the actual nuclear reaction producing heat is the fusion reaction, d+p = ^{3}He. He assumes lattice coupling reduces the barrier and that liberated energy is "transferred to the multi-phonon degrees of freedom of the lattice." A complex series of equations describing phonon modes is used to justify this assumption. Later studies of the PdD system show that the main heat producing reaction is formation of ^{4}He, not ^{3}He (Section 2.2.0).

As is common among physicists, Schwinger viewed PdD as an assembly of energy states focused mainly on the d nucleus. Consequently, he uses equations describing phonon modes in a Hamiltonian applied to the lattice. These phonons are described as existing in "vacuum" and they are given a purely mathematical character by being called "virtual." The lattice itself is considered to play an undefined role in how the phonons behave. Consequently, how this model relates to the common description of the phonon[43] is not clear. He assumes a p or d can move to the immediate vicinity of another hydrogen nucleus and form an "excited state" of helium. How or why a hydrogen ion in PdD would move in this manner was not explained.

---

[42] This conference series has been held in nine countries with the most recent being ICCF-18, which was held at the University of Missouri in July, 2013. The next conference will be held in Italy in 2015.

[43] http://en.wikipedia.org/wiki/Phonon

S. Chubb and T. Chubb

Scott[44] (1953-2011) and Talbot[45] (1923-2011) Chubb (nephew and uncle) worked together at the Naval Research Laboratory in Washington DC for many years before their deaths and co-authored many papers together, although they did not always agree on all details of their model. The basic idea involves formation of ion bands in the normal fcc fully-loaded PdD ($PdD_{1.0}$) lattice that are assumed to interact like waves, thereby reducing the Coulomb barrier between deuterons and dissipating the resulting fusion energy. They call the resulting interaction a Bose-Bloch Condensate (BBC).(*715-735*) They identify wave-function overlap as the requirement for fusion to occur, which only happens according to their model very near $PdD_{1.0}$. Over the years since 1990, Scott and Talbot worked to justify this concept with increasingly complex arguments.

The most recent paper(*736*), written by Talbot and Daehler, summarizes their final thinking. They maintained their belief that the entire lattice is involved in the fusion process by comparing their approach to what Swartz(*628*) describes as lattice-assisted nuclear reactions rather than to the Storms description of nuclear-active environment (NAE).(*12*) This distinction is used even though they identify a special condition they call "distinct, ordered reaction-site volumes" as the special required condition, *i.e.* a NAE. They identify another NAE when they quote Bass as locating the nuclear reaction in "an extremely thin slice of metal next to the ionic crystal. Its volume is the reaction site volume where fusion reaction takes place." As a result, they failed to correctly apply the NAE[46] concept.

They propose local melting results from each fusion reaction even though this would destroy conditions required to initiate the effect again at the active location.

The Chubbs believe that the world of the nucleus and the chemical world coexist together and interact, although this

---

[44] http://www.infinite-energy.com/images/pdfs/ChubbMemorial.pdf
[45] http://www.infinite-energy.com/iemagazine/issue101/talbotchubb.html
[46] The concept of NAE is explained in detail in Chapter 5.

connection is not observed and not accepted as occurring in conventional science. They propose a standing wave exists between the deuterium nuclei in PdD such that deuterons can fuse upon applying stimulation by using what they call momentum shocks. They assume once wave function overlap occurs the Coulomb barrier disappears. In other words, in their theory, the force repelling two like charges ceases to exist once the particles exhibit wave behavior. Why wave behavior does not always result in loss of charge separation in PdD and occur in other materials is not made clear.

R. Bush and R. Eagleton

Robert Bush and Robert Eagleton worked together at California State Polytechnic University, with Bush providing the theoretical understanding. The theory was first proposed in 1989(*737*) and published in 1990.(*738*) The basic assumption rests on the deuteron ion in the PdD structure being a boson and therefore having a tendency to "clump" together into what they identify as the Bose-Einstein-Condensate (BEC)[47], with these authors being first to use this concept. They imagine the PdD to have what they call phase-space where this clumping is increased by a greater number of bosons being placed in the space. They propose this clumping occurs at an interstitial site or vacancy in the lattice structure. Of course, once the first atom of a clump enters a vacancy, the site is no longer vacant. The second member of the growing clump has no way of distinguishing this site from any other occupied site. Consequently, they are proposing the same multiple-occupancy process that has been studied and rejected by other people (Sections 3.1.2 and 3.2.3). They provide a series of assumptions to explain why the BEC can form at 300K, why the size of the BEC is limited, how much expected power is produced, and the influence of d concentration.

The authors justify the observed production of tritium by using the Oppenheimer-Phillips process where a neutron is

---

[47] http://en.wikipedia.org/wiki/Bose%E2%80%93Einstein_condensate

stripped of one deuteron to leave behind a proton. This neutron is then added to other nearby deuteron to produce tritium and release 4.0 MeV, similar to the neutron exchange process rejected by Hagelstein.

They recognize that local heating would drive out deuterium, thereby starving the reaction briefly, which they note would cause pulsed heat production, the average of which is the measured power. The authors attempt to show the relationship between calculated power production from each fusion reaction and the average measure power. Such a comparison is impossible because the number of active sites making power at any one time is variable and unknown.

In 1994, Bush(*739*) described his approach as "a unifying model for cold fusion," from which he claims 11 successful predictions, partly summarized as follows.

The fusion reaction is proposed to result from "oscillatory collision of two nearest-neighbor d" in the PdD structure, in other words, resonance (Section 3.2.1). This process is described by a series of equations from which a fusion rate is calculated and related to the concentration of d. Using a fitting factor, the equations were found to fit the data provided by McKubre et al.(*437*)(Fig. 58) and Kunimatsu et. al.(*740*) even though the composition reported is the average in the material, not the composition at the site of the fusion reaction as the model requires. In addition, the fusion reaction is now believed to occur in a very small fraction of the material located near the surface, not in the entire sample as these authors assume. Although the model appears to fit the data, this does not justify the claimed support for a prediction because the model is not applied to conditions actually present where fusion takes place. The role of lithium provides another failure to predict because although it might occasionally play a role, the LENR effect has now been produced in the absence of lithium. Tritium production is proposed to peak near $PdD_{0.8}$, but no evidence supports this prediction. The last contribution to this approach was published in 1997.(*170, 741*)

Here we see a common problem revealed when theories are tested. Such a test is faced with two questions. Does the theory really apply to the conditions being tested? How much of the observed behavior is caused only by chance combination of behaviors having no relationship to the model? For example, this model introduced assumptions many other theories have used, such as the BEC and the need for a high concentration of D. Are any of these assumptions related to actual conditions present in the NAE? If not, we must find the assumptions and predictions to be invalid.

R. Bass

Robert Bass founded a company called Innoventek Inc.[48] His theory describes what he calls the quantum resonance triggering principle (QRT) that proposes a resonance exists in PdD between the deuterons. He extends the process proposed by Turner(*742-744*) based on how a beam of light behaves when in a resonance cavity between two semitransparent barriers. In this case, the idea is applied to deuterons in PdD that are proposed to fuse because an assumed resonance is proposed to reduce the barrier between them. In addition, the model further assumes the lattice consists of "multiple slits" through which the deuterons diffuse, thereby producing an interference pattern that allows two deuteron waves to superimpose and fuse. Such an effect has never been observed. The resonance concept is evaluated in Section 3.2.1.

The ideas proposed by Parmenter and Lamb(*745-747*) are held in high regard by Bass and are worth noting here. Based on screening by conduction electrons, these authors estimate a fusion rate of 10^{-18}/sec. The Oppenheimer-Phillips process is proposed to increase the rate by a factor of 1.77. In 2002(*748*), Parmenter suggested a proton and deuteron mixture will increase the fusion rate between deuterons. The process starts by fusion of d+p to create energetic ^{3}He that reacts with a nearby deuteron to release a proton and create ^{4}He. This two-stage process is proposed to occur

---

[48] His papers are available at www.innoventek.com.

more rapidly than direct fusion of two deuterons. The presence of lithium is also suggested to increase formation of ^{4}He. Formation of ^{3}He, except as the decay product of tritium, has not been observed, lithium does not appear to be important to formation of ^{4}He, and added protons actually reduce heat production in the PdD system. Consequently, this model has no experimental support.

These concepts allow Bass to propose what he calls the Schwinger Ratio, which is $s = L/V$, where V is the mean-square oscillation amplitude of the resonating deuteron and L is lattice parameter of the lattice. He goes on to assume the fusion rate[49] is related to $\exp[-(1/2)\ s^2]$, which is given in units of W/cm^3. Finding justification for the observed behavior of power production based on this concept is difficult because only the lattice parameter is a known variable in the equation and units of power are not given by the equation. Maximum power is proposed to occur "if s/p is closer to an odd rather than an even integer." Based on the equations, he proposes the triggering energy required to initiate fusion is 5.1 eV, which is proposed to result in maximum power if applied to the deuterons. Using these concepts, Bass suggests future work should focus on creating material having the required lattice parameter and applying the critical energy to deuterons.

Bass(*749, 750*) proposed applying the Kalman-Leverrier algorithm to data resulting from heat measurements. Swartz provided data from his measurements for this treatment, which resulted in a joint paper(*751*) showing how this method of fitting noisy data with a smooth function might work.

This method is simply a way to fit data to an arbitrary function. The process does not give insight about the mechanism, role of the variables, or allow extrapolations outside of the data range. Nevertheless, it is useful to smooth scattered results and provide uncertainty for the values when the correct algorithm is used. The so-called prediction made using this method is similar to but perhaps more accurate than obtained using the more common least-squares method of data averaging. The function is useful in

---

[49] This equation is made more complex in later arguments by Bass but the basic concept still applies.

engineering where it can be used to improve control and help identify the sensitivity of a process to variables that affect the output of a complex machine or chemical process. Its role in explaining LENR is doubtful because it does not apply to individual nuclear events.

Bass summarizes his theory by describing it as, "The only conventionally viable cold nuclear fusion theory."(*752*) He agrees that deuterons combined with Pd or Ni will fuse to produce helium and heat, and predicts that protons in nickel will fuse to make deuterium and heat, with the mass-energy being released as phonons. The fusion process in his model involves a linear assembly of hydrogen and electrons as shown in Fig. 84, where a deuteron surrounded by an electron cloud is free to resonate and fuse with its neighbors. The condition is described by applying what he calls the Schwinger Ratio that describes the separation between deuterons required to produce fusion, with only certain separations within the chemical structure being active.

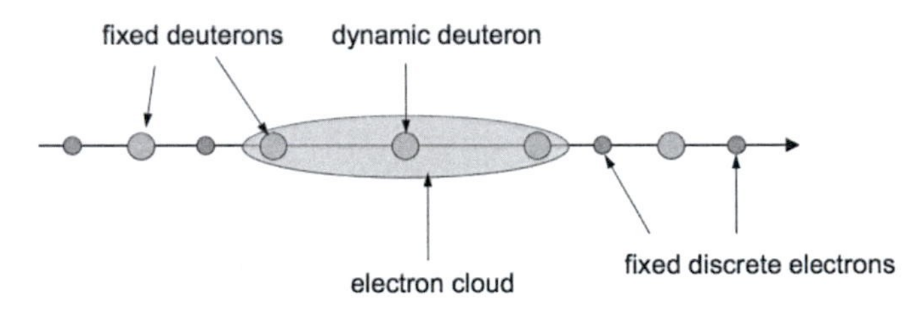

Figure 84. A proposed assembly of hydrogen that initiates fusion in a PdD lattice.(*749, 750*)

V. Violante

Vittorio Violante works at Associazione Euratom ENEA in Italy where LENR studies have been underway since 1989.(*753*) The models described by Violante et al.(*754*) assume the fcc structure of PdD can accommodate D in excess of the ideal limit at $PdD_{1.0}$. The extra D is proposed to occupy tetrahedral sites where it can experience harmonic oscillation initiated by an oscillating electron cloud.(*755, 756*) The oscillation is proposed to bring the deuterons close enough for fusion to occur.(*757-761*) Limitations of such a resonance process are described in Section 3.2.1.

Violante et al.(*762*) propose that a charged particle in a lattice can acquire energy as high as several thousand electron volts based on a Monte Carlo calculation and on the behavior of plasmons and polaritons on a surface.(*763, 764*) While this energy is too small to affect the Coulomb barrier, it is more than sufficient to cause local melting, but this consequence is ignored by the authors.

This high concentration of energy might result by applying external energy, but the model confuses how electrons spontaneously behave within a lattice with how they are observed to respond to applied energy on the surface. In addition, for energy to spontaneously accumulate in a lattice site, an equal amount of energy has to be removed from the surrounding sites, which violates the Second Law of Thermodynamics (Section 2.7.6). The accumulated energy and the close approach between two deuterons caused by the resulting vibration are proposed to initiate fusion, with released energy coupled to the lattice by the plasmons. This process is proposed to take place within the chemical lattice near a surface.

The role of stress in PdD created by reaction with hydrogen is explored. This stress is proposed to limit reaction with hydrogen, even though stress is caused by this reaction(*144, 765*), thus creating negative feedback. Annealing conditions and a procedure for current loading are proposed to increase the final deuterium content. Further complicating the process are the deposited impurities that change Young's modulus and the diffusion paths for hydrogen. Some of this stress is relieved by formation of cracks in the surface, a few of which are large enough to allow loss of H_2 (D_2) from the surface and further limit the maximum composition. These cracks become visible by being decorated by bubbles, both large and small.(*766*) Crack formation would limit loading but at the same time would perhaps create the NAE. This result of stress is ignored.

4.3.0 ROLE OF NEUTRONS

The limitations of neutrons in a model are explained in Section 3.2.4.

A. Widom and L. Larsen

Lewis Larsen[50] is the CEO of Lattice Energy, LLC and Alan Widom is a professor of physics at Northeastern University, Boston, MA. Together they published a theory, first in 2005 in arXiv,(*767*) with a version available in a journal in 2007.(*768*) This theory has achieved an unexpected level of acceptance with interest being expressed by NASA(*769, 770*) and by other theoreticians. Although the theory has been often criticized(*1, 771-773*), no effort has been made to answer most of these complaints, neither in public nor in private, the exception(*774*) being a response to Ciuchi et al.(*771*) who challenged the mathematical analysis of their wave functions used to support neutron formation. Larsen and Krivit[51](*775*) have repeatedly claimed fusion is not the source of energy based on the assumptions used in the Widom-Larsen (W-L) theory. The authors justify their approach using complex assumptions as summarized below.

1. Electrons can acquire sufficient energy as a result of "electromagnetic field fluctuations" to allow them to combine with a proton to make "ultra low momentum" neutrons. This energy is proposed to appear as extra mass identified as producing "heavy electrons." Furthermore, the process is further described(*64*) as being radiation induced and "theoretically described within the standard field theoretical model of electroweak interactions." Hagelstein(*773*) goes to great lengths to show the many conflicts with basic conventional behavior this claim contains (Section 3.2.4).

---

[50] The author worked with Larsen for several years and has units of ownership in his company.

[51] http://www.newenergytimes.com/

2. The resulting neutron is proposed to target only lithium in the surface region of an electrolyzing cathode to produce ^{8}Li from ^{7}Li, which decays by beta emission to ^{8}Be that then fragments into two ^{4}He. The energy obtained from this reaction, according to their calculations, is 13.4 MeV/He. This energy is not consistent with the measured value when helium forms (Section 2.2.4). Formation of ^{3}He is also proposed to result from the reaction with lithium, which is not detected. The authors' objection to the term "cold fusion" is based on the proposed reaction between deuterium and lithium not being acknowledged as fusion.
3. The neutrons are proposed to react with other nuclei, such as Ni or Pd, to produce observed transmutation products.(*776*) Experimentally observed transmutation products do not match predictions made by the model (see: Figs. 19, 20, and 21).
4. The heavy electrons are proposed to convert any gamma radiation generated in a material into photons of lower frequency, thereby hiding this kind of radioactive decay from detection and providing a potential gamma-ray shield.(*777*) No evidence exists to support this claim.

Chang(*778*) also proposed formation of neutrons from reaction between an electron and proton or dineutrons when deuterium is involved. This structure then finds another deuteron with which to fuse to make ^{4}H, which decays by release of an electron to form helium.

R. Godes

Robert Godes formed Brillouin Energy, Corp. and Profusion Energy, Inc. to commercialize his claims. He describes the mechanism in a patent, at ICCF-17(*202, 463*), and in *Infinite Energy*.(*779*)

Apparently, phonons are proposed to accumulate until energy sufficient to create a neutron from p+e is acquired. The common understanding of the "phonon" is distorted by assuming they can accumulate by what he calls phonon activity and involvement of what he calls "Heisenberg confinement energy." This description is equivalent to assuming temperature can locally increase by 0.78 MeV, an amount required to form a neutron. Limits on spontaneous accumulation of energy are discussed in Sections 2.7.6, 3.1.1, and 3.2.4.

The process starts with an energetic electron being captured by a proton, similar to the mechanism proposed by Widom-Larsen. If deuterium is present, a dineutron is created that reacts with another deuteron to make ^4H. This unstable isotope then decays by beta emission to create the observed helium. If only protium is present, the generated neutrons are proposed to keep adding to the nuclear product to make first ^2H, then ^3H, and finally ^4H. No effort is made to evaluate the implausibility of this process based on the low probability of a neutron finding the small amount of ^2H or the even more rare ^3H before the neutron decays.

J. Fisher

John Fisher(*325-328, 384, 780-788*) is a retired scientist who has been explaining his concept of polyneutrons since 1992. Fisher describes a very complex structure of neutrons based on the "liquid drop" model of the nucleus. These polyneutrons are proposed to react with various nuclei and add additional neutrons using the rules Fisher has identified. His model requires ^6H to be a stable isotope of hydrogen but this isotope has never been detected.

The theory is based on ^6H being the source of polyneutrons as a result of its interaction with other isotopes. This process adds neutrons to already formed polyneutrons, thereby producing energy and additional ^6H (Section 3.2.4). Even though none is found in H_2 gas, he suggests ^6H is concentrated in D_2O where it causes the fusion process by formation of polyneutrons rather than by direct fusion of deuterium.

Evidence supporting the process is based on finding nuclear reactions to occur in the space surrounding an electrolytic cell(*325, 327, 384, 784*) where the generated polyneutrons being emitted are assumed to have interacted with a nucleus in the air to produce radiation detected using CR-39 (Section 2.4.3).

H. Kozima

Hideo Kozima[52](*789-791*) is a scientist working in Japan who has focused on a structure of neutrons described as a "neutron drop" or a cluster(*792*) he calls the TNCF model.(*793*) Since 1990, he has described his model in over 50 papers by using "11 premises." He assumes bands of neutrons can become stable in a periodic lattice, especially in a small thickness (1 μm) near the surface of a cathode. The initial "ambient" number of neutrons is assumed to increase by breeding, but the initial source and why they are not released for detection when the lattice is destroyed are questions not answered (Section 3.2.4).

The observed complex collection of transmutation products is proposed to result from different numbers of neutrons being added to target nuclei followed by accelerated emission of electrons (beta decay) to form new stable elements. The neutrons were also thought to accelerate the hot fusion reaction between deuterons even though the required energetic products and expected secondary radiation are not detected during LENR. Reaction of the neutrons with lithium is also proposed to produce the detected energy and account for observed helium. Energy created by this reaction of 9 MeV/He is significantly less than the measured value (Section 2.2.4).

4.4.0 ROLE OF SPECIAL ELECTRON STRUCTURES

Electrons normally can get no closer to the nucleus on average than allowed by the Bohr orbit or the so-called ground state, as accepted QM theory requires. Nevertheless, arguments both conceptual and mathematical have been used to justify how the electron can get closer to the nucleus as a way to explain unusually high-energy production and unexplained spectral lines. (See Mills section.) In addition, if the electron could get closer to

---

[52] http://newenergytimes.com/v2/sr/Theories/KozimaTheory.shtml

the nucleus, it might shield or hide the Coulomb barrier enough for fusion and other nuclear reactions to occur. However, when a muon brings two deuterons close enough for fusion to occur, hot fusion products are the result — not cold fusion. Consequently, the theories must find a way to avoid this expected result. In other words, the collapsed state must do more than simply bring the two nuclei closer together. Dissipation of the energy without the energetic radiation typical of hot fusion is also required. This problem is discussed in Section 3.2.7.

A. Meulenberg and K. P. Sinha

Andrew Meulenberg is a physics professor at Universiti Sains Malaysia and Krityunjai Sinha is a professor at the Indian Institute of Science. They base their theory on the Dirac description of quantum states in which the electron in a hydrogen atom is proposed to enter an energy state well below the ground state. This condition is said to look like a "fat neutron" and is called a Deep Dirac Level (DDL), following the ideas of Maly and Va'vra.(*794, 795*) The condition is similar to what Mills calls the hydrino(*796*) but the details of the energy state and the mathematical justification are different. Rice et al.(*797-801*), among others, reject this description of electron behavior.

According to Meulenberg and Sinha, the energy level an electron can reach nearest the nuclei is called the naught orbit (n = 0), with the conventional ground state being defined as n = 1).(*802*) Energy loss or gain between these two states is proposed to be slow and can be described as leakage of energy rather than a single quantized transfer as is characteristic of conventional transitions. Presumably, the electron can get no closer to the nucleus unless enough energy is added to form a neutron. The n = 0 condition is said to bring two deuterons close together, similar to the effect of a muon acting in the same role. Nevertheless, the authors argue that the close approach produced by the muon is different from the effect of the DDL because the energy levels are different. Consequently, they claim hot fusion resulting in fragmentation can be avoided.(*802*)

According to Dirac, this transition to the DDL state happens all the time in all atoms but only for a brief time. The authors assume during this brief time, a nuclear reaction would occur at a useful rate if the collapsed hydrogen atom could be stabilized. This stabilizing condition is assumed to involve electron pairs called lochon (local charged bosons).(*803-805*) These electron pairs are assumed to form in D_2 molecules using the normally present "s" electrons when D_2 molecules become trapped in cracks, grain boundaries, or similar linear channels. These special sites have now been extended to include interfaces and vacancies.(*802, 806*) Resonance exchange between the resulting D^- and D^+ results as the electrons change location and is proposed to reduce the Coulomb barrier(*805*) by a process described as reducing the fragmentation energy of the resulting ^4He.(*806*)

However, if fragmentation does not occur, how is the mass-energy expected to escape and appear as heat? In this model, most mass-energy is assumed carried away by the lochon interacting with the surrounding lattice to produce phonons. However, under some undefined conditions, one of the electrons in the lochon can be drawn into the fusion process while the other one is emitted. This process would predict deuterium formation when protons fuse with emission of an energetic electron and a neutrino.(*538*) Because the neutrino would be expected to carry most of the energy, reduced observed heat energy would be expected to result from such a reaction.

Capture of naught electrons by a nucleus is proposed to result in transmutation products as protons contained in the target nucleus are converted to neutrons.(*802*)

In some sentences Meulenberg and Sinha(*803, 807*) propose the nuclear reactions cannot occur in the lattice itself because the structure restrains how close the nuclei can get to each other. Consequently, they also focus on defects of various kinds in which two deuterons combine where more freedom of motion is possible to produce a lochon. In another paper they describe the process as taking place in the lattice and being influenced by forces in the lattice that can bring the nuclei closer and increase the mass of electrons.(*807*) In each case, they claim energy is released as phonons.

The DDL formed by the electrons bonding two deuterons is assumed to have a very large energy (500 keV), thereby changing the nature of the fusion process. How or why such a stable state would have a short lifetime is not explained. The problems are discussed in Section 3.2.7.

Sinha and Meulenberg(*805*) propose as an explanation of how applied laser energy affects LENR, "the optical potential can enhance both the probability of localized D^- formation and its stability along with the probability of D^- and D^+ to come closer and fuse." This description would mean that laser radiation can create additional DDL and accelerate fusion of the lochon, although how or why this happens is not made clear.

J. Dufour

Jacques Dufour, while working at Shell Research in France, was able to initiate heat production by creating a spark in D_2 between Pd and stainless steel using pulsed DC. He explained the behavior using the ideas of Vigier(*808*), who proposed the reactions p+p+e = d + neutrino, d+p+e = tritium + neutrino, and d+d+e = ^{4}H + neutrino might be made to occur in capillaries as a result of Ampere forces upon passage of high current. Since the neutrino would be expected to carry away most of the energy, the measured heat would be much less than the expected Q value for the reaction. Vigier also suggested some heat results from collapse of the electron to a position closer to the nucleus than permitted by the Bohr orbit but not close enough to form a neutron. This process is expected to reduce the magnitude of the Coulomb barrier and allow fusion to start(*809, 810*), similar to the model of Meulenberg and Sinha. In contrast, as explained below, Mills proposes most energy results from the change in electron energy, not from subsequent nuclear reactions.

Dufour observed blackening of nearby photographic film, suggesting ^{4}H decays by emitting 50 keV electrons with a half-life of 20 min. This energy is observed in spite of most energy being expected to reside in the neutrino where it would be undetected. This observation is important in showing beta decay of ^{4}H that results in formation of ^{4}He.

Dufour carried the idea of a shrunken orbit further and named the process Pico-chemistry, which generates energy without a nuclear process being involved.(*811, 812*) He calls the shrunken atoms hydrex or deutex(*813-816*), similar to the notation of hydrino used by Mills.(*817, 818*) The difference between the two theories seems to be in the proposed amount of shrinkage and the mechanism used to achieve the condition. In any case, a skeptical person might expect the concentration of this shrunken hydrogen would eventually grow large enough to be detectable after production of significant energy, which Mills has claimed to be the case. Evidence for the hydrino supports the basic idea of a collapsed state in hydrogen being possible.(*819, 820*)

R. Mills

Randell Mills is a medical doctor and CEO of Blacklight Power Inc.[53] He and J. J. Farrell, his physics professor at the time, first described the concept in 1990. The hydrino concept is described in a small book published in 1992(*821*) and later in a large book published in 2006.(*822*) He proposes that the electron in a hydrogen atom can move closer to the nucleus below the Bohr level and occupy fractional quantum states.(*796*) This process causes the atom to lose energy, which is communicated to a surrounding atom, called a catalyst. This transfer of energy is possible when the catalyst has an energy level that exactly matches the loss in energy experienced by the electron in the hydrogen atom when reaching one of the fractional quantum levels. This energy is transferred without a photon being emitted. Instead, phonon interaction is proposed followed by the conversion of this energy to heat.

A great deal of work(*820, 823-830*) has been published describing the properties of the hydrino. According to Mills, the

---

[53] http://www.blacklightpower.com/ and http://en.wikipedia.org/wiki/BlackLight_Power

process can make useful heat(*831*) and the hydrinos can form unusual chemical compounds(*820*), all without involving nuclear reactions. The process is justified by complex mathematical arguments that have resulted in conclusions beyond this process. For example, dark matter is explained and various energy transitions in atoms and molecules are calculated with impressive accuracy. Nevertheless, the mathematical justification for these predictions has been questioned.(*832*)

Although claims for nuclear reactions being initiated by hydrino formation are not made, tritium was found on one occasion after heat production using electrolysis of Ni in $H_2O+K_2CO_3$.(*833*)

4.5.0 ROLE OF TRANSMUTATION

Some theories focus on energy being produced by transmutation rather than by fusion. However, because the process is greatly influenced by the Coulomb barrier, such claims are difficult to justify. The problems are described in Section 3.2.6.

F. Piantelli

Francesco Piantelli is a physics professor at Universita di Siena, Siena, Italy who, with several collaborators, has explored excess energy, transmutation products, and radiation produced while exposing nickel to hydrogen gas.(*199, 332-338, 834*) This process has been described in several patent applications.(*330, 340, 341*)

The explanation can be summarized as follows. "Micro/nanometric" crystals are said to form on the surface of the nickel (Ni), or on other elements, in which hydrogen is absorbed as H^- ion clusters when the material is heated. The H^- ions are captured by the Ni nuclei to form copper. The resulting excess mass-energy is transferred to some H^- ions and these are eventually ejected as energetic particles. The process is enhanced by thermal, mechanical, magnetic, or electrical shock. The description makes no effort to show how this process can occur while remaining consistent with known behavior of hydrogen in nickel or known limitations experienced by all nuclear reactions. Nevertheless, excess energy appears after the nickel surface has been made nuclear active after a special propriety treatment is applied to the surface.

A. Rossi

Andrea Rossi, an Italian businessman and engineer, has found a way to cause Ni-H_2 to generate significant power with the help of Prof. Focardi at the University of Bologna. Although this claim has not been described in scientific papers, the method is the basis for two patent applications(*201, 203*), is discussed on many internet sites(*201, 835, 836*), and is the subject of two books.(*837, 838*) A US patent application filed in 2008 was recently rejected.[54] The claimed process can be summarized as follows.

Powdered nickel is subjected to an undisclosed treatment that makes it able to react with H_2 and initiate a nuclear reaction. When this material is heated in H_2, protons are proposed to enter the nucleus of the nickel, thereby converting it to a copper isotope that makes heat energy. This explanation follows the work of Piantelli who reported finding transmutation products that were assumed the source of all power. Fusion of protons is not considered. A similar claim is made by Defkalion Green Technology Corp.[55](*839*)

Power is removed either by flowing water or by radiation when high temperatures are achieved. A ratio of [(output power)/(input power)] as high as 415 is claimed. However, his design gives stable control only when the ratio is near six.

4.6.0 ROLE OF TUNNELING

The flaws in tunneling as an explanation for LENR are discussed in Section 3.2.2. Although this concept is popular and

---

[54] http://coldfusionnow.org/us-examiner-addresses-andrea-rossi-us-patent-application/

[55] http://www.lenr-canr.org/acrobat/Defkaliondefkaliong.pdf, http://coldfusioninformation.com/companies/defkalion-green-technologies/

frequently combined with other mechanisms, only one theory is discussed below as a general example.

X. Z. Li

Xing Zhong Li is a professor in the department of physics at Tsinghua University in Beijing, China. Li along with several co-authors developed a theory based on tunneling described first at ICCF-3 (1992).(*840*) The model assumes the Coulomb barrier can be penetrated by resonance tunneling when "the incident nucleus is in resonance with the virtual energy level of the potential well between the barriers." The concept was expanded in 1994(*841*) when the idea of "centrifugal potential barrier" was introduced. A concept of double barrier fusion as a method to increase the calculated penetration factor is proposed. Presumably, several barriers rather than the usual single barrier allow a better match between resonance conditions involving deuterons and the source of energy causing resonance.

The life-after-death behavior is used to propose a fusion lifetime of 10^4 seconds.(*842*)

Three years later in 1999(*843, 844*), the mechanism was further developed by assuming a long lifetime (~3h) nuclear energy state could form in the PdD lattice. In this case, the deuterons are assumed to be much closer than in a molecule, but not close enough to form ^{4}He. This condition is then made nuclear-active by resonance (Section 3.2.1). Justification for the mechanism is based on the behavior of nuclear reactions when initiated by high applied energy (Section 2.8.0).(*845*)

The proposed explanation is demonstrated by passing deuterium through Pd metal and observing a small amount of excess power. The power was found to correlate with the flux of deuterium diffusing through the metal.(*846*) Reports are noted of finding increased mass-6 on the surface. However, this is very unlikely to result from T_2 as claimed. If tritium were present, it would be contained in the DT (m/e = 5) molecule because the dominant isotope in the environment is deuterium.

Production of tritium is explained(*847*) to result from the reaction p + d = t + e^+ + v or p + d = k-capture. The annihilation radiation resulting from positron emission has not been detected. How the resulting excess mass-energy can be released by k-capture is not explained.

No matter how this process is justified, it addresses only a small part of the problem. The process is described as occurring in a chemical lattice without any mechanism being proposed to start or sustain resonance and without any method to dissipate the excess mass-energy short of the hot fusion reaction.

Furthermore, the life-after-death behavior used to support the theory can be explained in a different way. The fusion reaction continues after electrolytic power is turned off because fuel is supplied to the NAE from the large supply of deuterium contained in the bulk PdD rather than being supplied to the surface by electrolytic action when power is applied. As a result, the duration of life-after-death has no relationship to the lifetime of the nuclear process. The duration only depends on how long deuterium contained in bulk material can last.

4.7.0 ROLE OF CRACKS AND SPECIAL STRUCTURES

To avoid conflict with chemical conditions in the normal lattice structure, the role of cracks or other deviations from the organized structure have been suggested as the site of the nuclear process, the so-called NAE. Cracks are very common and many authors have explored their roles, both as location of the NAE as well as providing a path for deuterium to leave the cathode during electrolysis.

In general, cracks are harmful when the electrolytic method is used, because they allow the D_2 to leak out. In contrast, they are useful when gas loading is used because they allow greater access to the surrounding gas.

Cracks are produced by stress relief and populate the surface in complex and random ways with a variety of gap widths. Hydrogen embrittlement(*848-853*) encourages the process.

F. Frisone

Fulvio Frisone is associated with the department of physics, University of Catania, Italy. Frisone started his study of LENR in 1996 by examining the mathematical relationship between how plasmons might lower the Coulomb barrier and the effect of impurities on this process within the lattice.(*854-857*) In 2000, he shifted his attention to microcracks.(*858-860*) The number of active cracks is proposed to grow as the nuclear reactions produce micro-explosions. However, the emphasis on tunneling to achieve fusion without a mechanism to dissipate the resulting energy makes his mathematical analysis less useful.

E. Storms

Edmund Storms is an independent researcher who is retired from the Los Alamos National Laboratory where he was successful in making tritium(*708*) and excess energy(*861, 862*) using the electrolytic method. He has been successful in making energy and radiation using electrolysis(*863*), gas discharge(*314, 864*), and gas loading(*258, 344*) in his private laboratory.

A new and unique concept has been recently suggested by Storms(*63, 865-871*) based on the LENR reaction taking place outside the chemical lattice and releasing excess mass-energy in small units before the fusion process has been completed. In this case, two or more hydrogen nuclei assemble in a nanogap to form a special chemical structure called the Hydroton. This structure allows mass-energy to be slowly released as photons. In the process, the assembly shrinks and becomes the final nuclear product only after most mass-energy has been released from each nuclei, whereupon the individual nuclei fuse together with an electron. This book and especially Chapter 5 explore this idea in detail.

4.8.0 SUMMARY OF PROPOSED THEORIES

A survey of current theories of LENR has revealed several flaws. They describe only part of the process, are proposed to take place in the chemical lattice, introduce assumptions that are impossible to test, and conflict with many observations. Most models are described by mathematical arguments based on quantum mechanics with very little agreement about how QM should be applied to the process. These flaws are described in Chapter 3 and repeated here for emphasis.

Five major mechanisms are identified, with several frequently being combined to create the proposed model. The flaws are identified by the sections in which they are discussed in detail. This list is only provided as a general summary of the potential flaws.

1. Neutrons are made by fusion of electrons with protons (deuterons) or released from a stable structure in the lattice. These neutrons interact with the surrounding nuclei to make heat and helium (Section 3.2.4).
2. The protons or deuterons react with the surrounding metal atoms to cause transmutation that produces observed heat (Section 3.2.6).
3. Two or more D come together to form a cluster. Two deuterons in this cluster eventually fuse (Sections 3.2.3 and 3.1.2).
4. The Coulomb barrier is overcome by screening, tunneling, or being forced closer by a resonance process. This resonance process is frequently not applied in a manner that would achieve the required result (Sections 3.2.2 and 3.1.3).
5. The resulting excess mass-energy is dissipated by fragmentation of the structure or by being communicated to the surrounding atoms by phonons (Sections 3.2.9 and 3.1.3).

All of these features have serious conflicts with either observed behavior or with basic laws of nature as described in Chapter 3.

CHAPTER 5

DESCRIPTION OF A NEW EXPLANATION

The major observations have been described and a few examples of published efforts to explain LENR have been evaluated. Based on these arguments, a person might reasonably conclude that LENR is impossible. This conclusion would be reasonable if the observations supporting LENR were not so many and so well done. As a result, a plausible explanation is necessary and possible no matter how difficult the search.

Apparent failure of present models requires the search to take a different path. First, the mechanism causing LENR needs to be removed from the chemical lattice and placed in nano-cracks located outside the chemical system. Second, the process of overcoming the Coulomb barrier must be combined into one mechanism along with a process that dissipates the resulting mass-energy. The following discussion describes in detail how these two basic conditions are combined into a mechanism proposed to cause both fusion of hydrogen nuclei and transmutation.

First, the assumptions on which this explanation is based need to be described.

5.0.0 BASIC ASSUMPTIONS

All theory starts with basic assumptions that guide the logic. These assumptions cannot be proven individually, but must be initially accepted on faith until they are proven correct by the entire theory being consistent with what is observed, both in the past and in the future. Although stated previously, these assumptions are gathered together here for clarity and again used to provide a general description of the theory.

1. The LENR process does not conflict with any known law of chemistry or physics, but instead reveals a new phenomenon.
2. All isotopes of hydrogen can fuse and this process occurs only in a rare and unique NAE that forms in the surface region of solid material.
3. A universal process causes all observed nuclear reactions in the same kind of NAE regardless of the material, the methods used to form the NAE, or how the nuclear process is initiated.

In summary, this theory identifies a new phenomenon, shows how it functions, and where in the material it occurs, without creating conflict with what is known about chemical and nuclear processes. Very few additional assumptions are needed.

5.0.1 General description

Energy production using the LENR effect can be broken into three separate parts, with each having a separate function while retaining a logical connection between each part.

The first item involves the fuel for the nuclear reaction, which are the isotopes of hydrogen. This fuel functions exactly like any other fuel, with nothing special being required just because LENR is caused. The behavior of hydrogen in the lattice can be described by using conventional chemical and engineering understanding. This fuel is like gasoline waiting patiently in the tank for it to be pumped into the engine, with the pump rate determining the amount of power produced.

The second item identifies the NAE and how it might be created outside of the chemical structure. This can be identified as the engine that uses the gasoline. Learning how to make the NAE in large and controlled amounts is the basic problem limiting power production and commercial success. A material without the NAE is as useless as an automobile without an engine, even when the gas tank is full.

The third item describes what happens in the NAE and the nature of the nuclear process. This can be imagined as the process that actually takes place in the cylinder to convert gasoline to forward motion. This part of the process contains the great mystery cold fusion has forced science to confront. However, this knowledge provides no ability to make the effect occur at useful and predictable rates. That goal lies solely with understanding and creating the NAE because once the NAE is created the nuclear process takes place immediately without additional help.

Using the automobile analogy once again, once the gas is in the tank and the engine is in the car, no further knowledge is required to start the car moving. A person does not have to know what happens in the engine to produce energy. Nevertheless, most theories of LENR attempt to understand what happens in the engine before they have learned where to find an engine and how to place fuel in the tank.

No part of this description is in conflict with conventional understanding. Nevertheless, the third item contains new and novel ideas requiring creative thinking. This is where the great mystery of LENR lies and from which future Nobel prizes will emerge. Because this item does not have to be understood to generate useful energy, items #1 and #2 are discussed first.

In summary, LENR has three basic parts:

1. Availability of fuel to the nuclear-active sites.
2. Formation of nuclear-active sites.
3. Conversion of mass-energy into heat energy by a nuclear process.

Item #1 has major influence on the rate of energy production, understanding Item #2 is essential to make LENR functional, and Item #3 has scientific interest for the future.

5.0.2 Item #1: Variables affecting the process outside the NAE

The source of fuel for LENR is hydrogen that is distributed in the chemical structure of the active material or contained in the surrounding gas in molecular form. Sites able to convert mass-energy in the hydrogen to heat-energy are small, generally few in number, and present only in special locations, generally in the near-surface region. For energy to be released, a source of fuel (hydrogen isotopes) must be present near a generating site and the hydrogen nuclei must move from their normal location in the chemical structure to the rare sites where the nuclear reaction can take place. Consequently, the rate of energy production is determined by how fast fuel is delivered to the isolated generating sites, in this case by diffusion within a solid structure. This structure is not pure hydride but instead a mixture of components having complex morphology. Nevertheless, access to fuel is determined mainly by diffusion and hydrogen concentration in the material surrounding the nuclear active sites.

Energy density of delivered fuel is also important. Because deuterium has a greater energy density than normal light hydrogen, it will produce more power, all else being unchanged.[56] For this reason, adding H to D can reduce power production so that detectable power can disappear without the nuclear process actually being stopped. As summarized in Table 10, ^{4}He is the final product when D alone is present. When H and D both are present in the material, tritium forms. If only H is present, deuterium results. As is apparent, presence of H will eventually produce first deuterium (D) and then tritium (T), thereby producing a variable amount of total power with increasing amounts of tritium as the nuclear products further react in the Hydroton. Tritium production can be avoided only by using pure D. In this model, the ability to cause LENR has no preference for the metal used or the hydrogen isotope. While success has been achieved mostly by using D with Pd and Ni with H, the reverse is proposed to work just as well once the universal NAE can be created without fail.

---

[56] For consistency and simplicity, H refers to a proton (p) and D refers to a deuteron (d). The designation p and d are used only in equations describing nuclear processes. The designation "H" refers to normal hydrogen while ignoring the small amount of D. The word "hydrogen" is used to identify both isotopes. Normal hydrogen contains 156 D/1,000,000 H.

With this description in mind, power production can be described by the following equation:

$$Power = K*[X*A*C*exp(-B/RT)] \qquad [1]$$

X = a value determined by which hydrogen isotope is reacting. If mostly D is present, this number will be large. If mostly H is used, the number will be small and variable. An unknown contribution by transmutation can add energy, as described in Section 5.4.0.

A = number of NAE in the sample. The greater the number of active sites in which the fusion reaction can occur, the faster energy can be generated by a sample of active material.

C = concentration of hydrogen isotope in the material surrounding the NAE. This value depends on several variables including the chemical characteristics of material surrounding active sites, temperature, applied hydrogen activity, and rate at which the hydrogen can enter the material through the surface. Surface activation, ion bombardment, and high pressure (activity) can be used to increase this concentration.

B = energy required to move hydrogen within the material. Several different conditions can be used to move hydrogen ions. This equation is based on diffusion being enhanced by temperature. Concentration gradients or application of an electric field also can be used to increase rate of movement in addition to the effect of temperature. If these methods are used, this term becomes more complex. Nevertheless, this term shows how these conditions would affect power production.

A flux of hydrogen has no role in initiating the nuclear reaction. This flux only allows D^+ to move more rapidly through the material, which increases the opportunity for D^+ to encounter the isolated and rare NAE sites. Without this gradient, the deuterons would encounter active sites by random diffusion, which is a much slower process.

T = temperature (K) of material surrounding the NAE. This temperature has several effects, including changing the composition, changing the diffusion rate, and changing the rate at which energy can be lost from the NAE. Although these effects make the equation more complex, conventional understanding alone is required to describe the behavior. Temperature is not expected to have a direct effect on the nuclear reaction taking place in the NAE.

K = constant used to make the units for the different variables consistent.

R = gas constant.

As this simplified equation shows, power production increases with increased temperature, as observed, which means a runaway condition can destroy the device. This event is prevented only when energy can be dissipated at a rate greater than the rate of production. Most LENR devices are stable only because this condition is easily met at low temperature where they are normally studied.

Loss of energy occurs several different ways, including by radiation, conduction, and convection. For this discussion, the controlling loss will be assumed to result from conduction through the material surrounding the NAE, as described by the following equation.

$$\text{Power Loss} = \Delta T * \lambda \qquad [2]$$

ΔT is the average temperature difference across the barrier having a thermal conductivity of λ. Because energy generation increases rapidly with increased temperature while loss is nearly a linear function of temperature under most conditions, temperature is the most important variable in controlling power production. This behavior is different from most energy sources and creates problems for successful control of energy production under variable loads.

These two equations identify conditions having influence on energy production, how the system will behave under various conditions, and how power can be controlled. Values for the variables can be easily determined once a generator design is accepted.

Because power production is increased by increased temperature, the process can go out of control under certain circumstances. This runaway process can be understood by examining a plausible relationship between power and temperature shown in Fig. 85, where a runaway condition can be seen to occur once power produced exceeds the ability of this power to leave the material, as identified by the dashed line starting where power generation exceeds power loss. If temperature is too close to the runaway temperature, unanticipated changes might push the system into the runaway state.

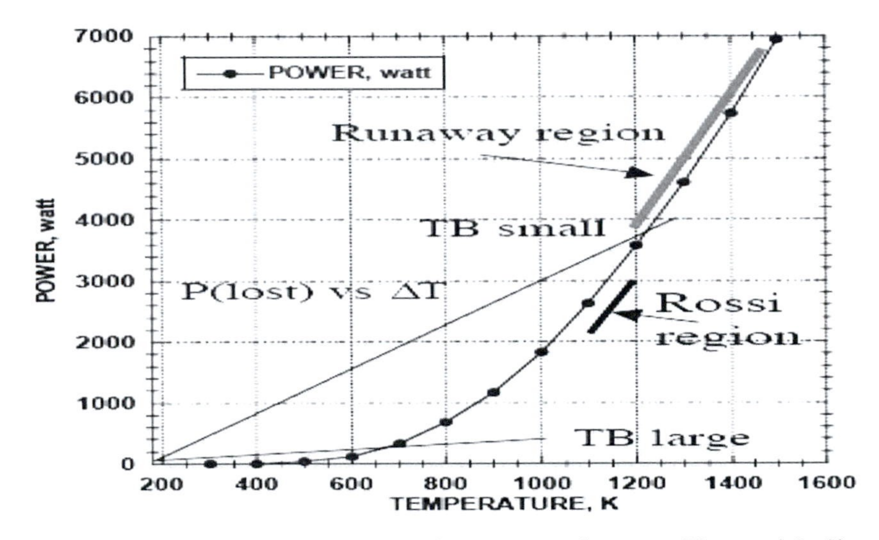

Figure 85. Effect of temperature on the amount of power. The straight lines show the relationship between power loss and delta T for two different amounts of thermal resistance (TB). Runaway occurs when power produced exceeds power loss, as indicated by the dashed line. The "Rossi region" indicates where the E-Cat HT is operated below the runaway region.(*866*)

To avoid this possibility yet achieve a significant amount of power, Rossi has chosen to control the E-Cat by using variable applied power supplied by electric heaters. These heaters increase temperature until the runaway region is approached, whereupon applied power is turned off so that the material can briefly cool, causing reduction in nuclear power. Use of temperature cycling close to the upper limit maximizes power production while maintaining control. Because power must be applied, efficiency is reduced. If this power were not applied, the temperature would either remain too low to produce useful power or control might be lost if the temperature were increased too close to the runaway condition. This method is one of several ways control can be achieved.

5.0.3 SUMMARY OF ITEM #1

The concentration of hydrogen isotopes in the material in which the NAE is located[57] is determined by thermodynamic properties of the material in which temperature and pressure of hydrogen play a role. Rate of delivery to the generating sites is determined by local temperature, concentration of hydrogen, and concentration gradients, each of which is present in the structure surrounding the NAE.

Most variables affecting heat production can be identified and related to conventional chemical and physical behavior without understanding how power is generated. An equation is proposed describing how power production can be increased and controlled in a working generator. The effect of a few other variables that can increase the fusion rate, such as laser light and RF radiation, are discussed separately in other sections.

This source is unlike all others because power production is very sensitive to temperature, especially at high temperature. This behavior creates a problem for engineering design to extract heat energy while keeping all of the active material hot enough to be productive yet still below the runaway temperature. Should any significant part of the active material reach the runaway temperature at any time, its local ability to host LENR will be destroyed. This problem does not apply on a very small scale

---

[57] The properties of the bulk material are not relevant. Only conditions in the material in which the NAE forms are important. Frequently, this active material is falsely assumed to have the same properties as bulk material.

because as temperature increases, the hydrogen concentration in the small region will decrease, thereby denying fuel to the local Hydrotons.

This process can be used to explain why regions are observed to flash to high temperatures as shown in Fig. 57, Section 2.6.4. In this case, local temperature goes up, fuel is lost from a site, the site cools, acquires more fuel, and reheats. Average power is the sum of power produced by all individual sites going through this sequence. The material remains stable as long as all regions can lose fuel rapidly enough upon heating to limit their maximum temperature. Once fuel availability and concentration of active sites are sufficient to produce a critical local temperature, the local NAE will be destroyed as the local region melts. Controlling this process becomes a major engineering challenge.

5.0.4 Item #2: Production of nuclear active sites

Clearly, a condition normally absent and difficult to form must be created before the process can occur. In addition, as argued in Chapter 3, conditions in a normal chemical structure are incompatible with conditions required to cause a nuclear process. Since the nuclear process obviously occurs, suitable conditions outside the chemical structure must be sought.

A number of people have suggested cracks might be a suitable location(*805, 808, 858-860, 872-882*), but without providing a full and plausible description of how cracks were uniquely able to host the nuclear process or how the gap dimension played a role. Previous descriptions by Storms of how cracks might function (*865, 868-870*) are expanded here.

If the process occurs in a crack, conditions producing an active crack structure must be rare and unique because cracks are found everywhere, yet they do not host nuclear reactions. The gap width is proposed to be the critical parameter, with a critically small width being able to support the process. This small width prevents loss of hydrogen as the molecule while creating conditions needed to form the unique Hydroton structure.

Since cracks have a complicated structure, a typical crack can be expected to contain a few regions having a nuclear-active width while most of the crack is inactive. In addition, a typical surface will contain cracks having a wide range of gap widths, with only the larger gaps being visible using conventional magnification. Such large cracks have been found to reduce power production because they prevent the concentration of hydrogen from increasing as gas leaves the material through such paths. Obviously, the NAE is not large cracks.

Because cracks of the critical width would be largely invisible, except to the best scanning electron microscopes, these sites can be easily overlooked. For this reason, identifying the gap width at which a crack becomes active is still a matter of speculation. Nevertheless, the crack is the only condition in a chemical structure where a nuclear process might take place without violating various natural laws, as explained in Chapter 3.

Cracks form in materials when bond strength at local regions is exceeded as result of stresses generated by various chemical reactions. These local regions are normally crystal grain boundaries or where other flaws weaken the structure. As an example, stress can be generated by reaction between hydrogen and the metal as the lattice expands unevenly. This stress would produce cracks where strength of the surface region has been reduced by reaction with other elements, such as would be the case during extended electrolysis or glow discharge. The required long electrolysis before heat production starts, as noted by Fleischmann and others, might be necessary because this delay allows enough impurity to deposit to make the surface brittle and inhomogeneous enough to form active nano-cracks.

Indeed, any material deposited on the surface appears to improve performance as long as it does not prevent reaction with deuterium. Unfortunately, many surface deposits also inhibit uptake of hydrogen, thereby denying necessary fuel to the NAE. This competition between two independent and opposite effects complicates interpretation of behavior.

Crack production can be further understood when behavior of a ductile material is considered. Under stress, a ductile material such as pure palladium will first stretch and then tear when local strength is exceeded. The resulting gap would rapidly increase in width as material around the gap moves away from the gap and returns to its pre-stress position, much like parts of a rubber band returning to their original length once it breaks. This rebound process would leave behind a gap too large to support the nuclear process. In contrast, the frequently successful Pd-B alloy(*134, 135, 883, 884*) would be sufficiently brittle to form the necessary cracks without pretreatment and the material, not being ductile, would not rebound away from the gap. Once this overall requirement is accepted, many useful alloys can be identified.

Use of pulsed current, such as the super-wave(*450, 452-456*), would also subject the surface to stress and encourage crack formation. In fact, all methods claimed to trigger the effect would be expected to cause stress cracks, which are frequently detected in the surface. While this observation is not proof that nano-cracks are important, this consistent pattern gives encouragement. This prediction can be tested and, if correct, would provide a clear path to creating the NAE either by controlled stress relief or by nano-machining a gap in suitable material. At present, these active gaps are produced by chance in random locations with highly variable concentration. A better method needs to be found.

5.0.5 SUMMARY OF ITEM #2

Of all the features present in a material, only cracks have the potential to be the site of the nuclear process. The rare nature of the process suggests the width of the site is critical to its success. Narrow cracks, in fact, are actually observed during LENR. Only the proposed critical dimension is unknown. However, the proposed active width is too small to be detected using normal magnification.

Cracks are not related to or have properties similar to dislocations or vacancies. They are regions outside of the chemical structure having an environment different from that in the structure where dislocations and vacancies reside. This makes them uniquely attractive as a way to avoid the problems other sites create.

Once a site having the ability to support LENR is created, the nuclear reaction will be initiated in the site without additional effort being required. However, the rate can be changed as described above in Section 5.1.0.

5.0.6 Item #3: A process for converting mass-energy into heat-energy without radiation

The great mystery behind LENR resides in the NAE where two "miracles" need to take place at the same time. For fusion to occur, the Coulomb barrier must be overcome while energy is dissipated in small units and deposited in surrounding material as heat without significant radiation being produced. Because fusion and transmutation have the same release-of-energy requirement and are found to occur in similar locations, a similar mechanism can be assumed to apply to both. These requirements place a limit on the possible mechanism. First, the chosen mechanism must start by creating an assembly of hydrogen atoms.

A structure of hydrogen atoms called the Hydroton is proposed to form in the NAE. This is proposed to be a linear molecule bonded by covalent electrons having a novel energy-state that follows normal and well-understood chemical rules, as explained in Section 2.7.6. Any number of hydrogen atoms can be present in a Hydroton, with the number being determined by how fast hydrogen nuclei can assemble before fusion starts. Many Hydroton molecules are expected to form in each gap where the critical dimension exists, with each Hydroton being independent of all others.

This novel structure is proposed to form as result of the high negative charge present in and on the walls of a narrow gap, which is assumed to force the shared bonding electrons into an energy level normally not occupied. The gap achieves this ability because a hydrogen atom is equally attracted to both walls and caused to hover midway between the walls. This novel situation combined with a larger than normal negative charge in the environment forces the structure into a more compact form, which bonds the nuclei closer than normal. This structure can be imagined to have the proposed properties of metallic hydrogen and act like a superconductor on a small scale. Its size is determined by how fast the hydrogen atoms can assemble in the gap before resonance starts and causes the first photons to be emitted. The size would be determined by the concentration of hydrogen in the NAE, the temperature, and applied energy, with higher composition and higher temperature resulting in increased size.

Once formed, this novel structure is able to cause the LENR reaction when coherent resonance along the line between the nuclei is initiated. This resonance starts after random vibrations, normal to all atoms in materials, gradually become coherent between the nuclei because coherent vibration along the axis of the structure lowers Gibbs energy. This process focuses additional energy within the structure to bring the nuclei closer than normally would be the case.

Resonance takes time to start, during which the molecule grows in length. Once coherent vibration starts, mass-energy emission starts and no additional hydrogen can be added to the molecule. This vibration causes two nuclei to get closer for a brief time while the opposite nuclei get further away, as shown in Fig. 86. As the structure resonates, two photons of equal energy and opposite spin are ejected in opposite directions, with each photon being ejected from each of two nuclei as they approach each other. Each approach removes a small amount of mass-energy from each hydrogen nuclei while spin and momentum are conserved. Photon emission is directed along the axis to avoid changing the entropy by creating spin between the emitting nuclei. If enough gaps are all aligned in the same direction, a substantial source of coherent photons might result. This might account for the laser-like photon emission[58] observed by Karabut et al.(*349*) (*885*)

---

[58] Although it is common to call these photons X-rays, this is not accurate unless the photons result from changes in electron energy. According to the model described here, the photons emitted from the Hydroton can be called gamma

This loss of mass-energy is novel to LENR and involves a process normally not seen to operate because it is normally overwhelmed by applied energy, such as occurs during hot fusion.

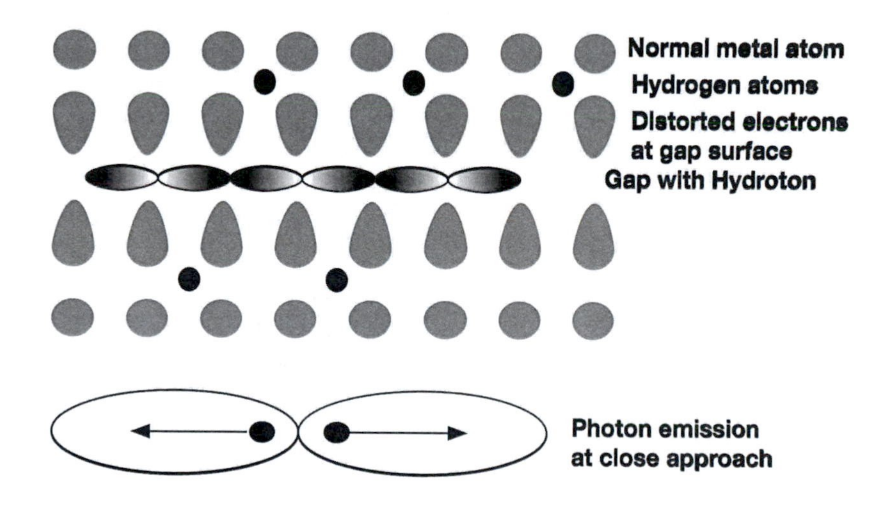

Figure 86. A cartoon of the Hydroton. The Hydroton structure is shown in the active gap with the electron density distribution shown for one part of the resonance cycle. The arrows show emission of photons in opposite directions from adjacent nuclei as they are briefly forced closer by the resonance process. The book cover shows a Hydroton at the final stage before fusion takes place.

In contrast, the nuclei move together slowly[59] and gradually approach a distance at which the two deuterons find themselves to be helium-like for a brief time with too much mass-energy. For this temporary excess mass-energy to be released, the process requires a mechanism for the two nuclei to communicate at this distance, determine that they have too much mass, and coordinate the necessary mass loss by each emitting a photon. The unique features shown by LENR have forced consideration of such a

---

rays because they are emitted from a nucleus as result of an energy transition in the nucleus.

[59] "Slowly" and "gradually" describe rates relative to how fast nuclear interaction normally takes place. Nevertheless, the interaction is fast by chemical standards.

mechanism even though such a process enters unexplored territory. Nevertheless, the mechanism has fewer limitations and conflicts than do the other suggested mechanisms (Sections 3.1.3 ad 3.2.9).

The energy of each emitted photon is determined by how long this close approach lasts[60], which is determined by the frequency of resonance. Eventually, the resonance cycle moves the two nuclei apart again before additional mass-energy can be lost. This process is repeated as each deuteron approaches the opposite nuclei. Again, a photon is emitted from each of the closest nuclei with opposite direction and spin. This process continues until most mass-energy resulting from helium formation has been lost by emission of many low-energy photons. Because all emission is along the axis of the Hydroton, the radiation from each Hydroton is coherent and in a particular direction with respect to the orientation of the gap. If a sufficient number of Hydrotons line up in the same direction, the average radiation can acquire coherent characteristics. In other words, the material might become an X-ray laser.

Once excess mass-energy has been drained away from each nucleus, pairs of hydrogen nuclei fuse and capture the bonding electron in the process. Because only a small amount of mass-energy remains when complete fusion takes place, the emitted neutrino resulting from electron capture at that time carries very little energy.

Although many details are obviously missing from this general description, the basic mechanism satisfies all requirements and allows predictions be made, as discussed below and in other sections. The atomic collection existing after the first two photons have been emitted can be identified as a "strange state of matter" having a very brief existence, terminated by the final fusion process. Whether this resonance process can be stopped, allowing the intermediate structure to be analyzed, remains to be determined.

---

[60] Close approach caused by muons and applied energy has no limit and hence results in emission of the mass-energy as a single event called hot fusion.

The nuclear product depends on which hydrogen isotope is present. The possible nuclear reactions are summarized in Table 11 along with the expected energy each would produce. When all nuclei in a Hydroton are D, fusion produces nuclei of ^{4}H, which decays rapidly into ^{4}He by electron and neutrino emission. Neutrino emission at this time carries only a small amount of energy because very little energy remains to be released by this beta emission. Nevertheless, the detectable heat energy would be slightly less than calculated from mass loss when ^{4}He forms, as noted in the table. Consequently, the lower than expected measured value for this energy (Section 2.2.4) might not only result from helium retained by palladium.

Table 11. Proposed fusion reactions and resulting energy. The open circles represent neutrons and the closed circles represent protons.

	d-e-d _ ^{4}H _ ^{4}He + e	<23.8 MeV
	d-e-p _ ^{3}H	<4.9 MeV
	p-e-p _ d	1.9 MeV
	d-e-^{3}H _ ^{4}He + n + e	<19.2 MeV
	p-e-^{3}H _ ^{4}He + e	<21.3 MeV

If the Hydroton were a mixture of D and H, tritium (^{3}H) results, which decays slowly by electron and neutrino emission, as observed. When the Hydroton contains only H (p), the result is stable deuterium (^{2}H), some of which would fuse with p to produce tritium. This prediction can be tested when H is used by observing increase in the amount of deuterium and eventual production of tritium. The simplicity and symmetry of the process provides comfort for believing the proposed process is correct.

Photon emission continues until each nucleus in the Hydroton has lost enough mass-energy to allow fusion to take place without significant mass-energy remaining in the nuclear product. This means each d must lose about 12 MeV before ^{4}H can form, which would require emission of perhaps a million low-energy photons for each helium created. Tritium formation requires each d and p to lose about 2.5 MeV with fewer photons required to do the job.

Deuterium is formed after only about 1 MeV is lost from each proton requiring ever fewer emitted photons. As a result, a greater radiation flux would be required to produce the same amount of nuclear product during ^{4}He production compared to formation of D from proton fusion. This is a testable prediction.

When tritium is made in an active cell, it can be lost by the following process. Tritium has its highest concentration at the site of its formation where it can be incorporated into the growing Hydrotons and experience fusion with p or d, depending on the relative concentrations of these isotopes. The resulting fusion removes tritium and eventually creates ^{4}He when fusion with p occurs. If tritium fuses with d, a neutron would be emitted. This neutron is proposed to be the source of the highly variable but small neutron flux detected when tritium is made (Section 2.1.5).

5.0.7 SUMMARY OF ITEM #3

The following series of steps is proposed to take place leading to fusion of all hydrogen isotopes.

1. Stress is created by reaction with hydrogen or by other chemical reactions.
2. Cracks are formed having various sizes as a result of stress relief.
3. A few cracks having a critical gap size allow Hydrotons to form in them. The Hydroton is a structure containing multiple hydrogen nuclei having the properties of metallic hydrogen.
4. The Hydroton starts to resonate. This process causes emission of photon pairs, thereby carrying away mass-energy in small quanta. Most photons are absorbed in the apparatus and converted to heat. If the concentration

of NAE is too high, the local region can produce sufficient energy to cause local melting, thereby stopping all fusion in that region. If the local concentration of suitable cracks is low, a steady rate of heat production occurs that is determined mainly by d concentration and local temperature.

5. If the Hydroton contains only D, ^{4}He results after rapid beta decay of ^{4}H, which rapidly diffuses from the site and is replaced by deuterons.
6. If the Hydroton contains a mixture of H and D, tritium results along with some ^{4}He, deuterium, and a very small neutron flux.
7. If the Hydroton contains only H, stable deuterium is produced followed by an increasing amount of tritium up to a limit.

Obviously, this process is complex and has important implications to understanding nuclear and electron interactions. Somehow, the two nuclei, when they get closer for a brief time, have to "know" they have too much mass-energy for this separation and to "know" when to eject it as photons. This knowledge represents a new kind of nuclear interaction different from the strong and weak forces normally associated with the nucleus.

Once this mass energy is released, the hydrogen nuclei are no longer like typical nuclei. They have acquired a new and unusual condition that is terminated only after they complete the fusion process. In addition, each cycle in this dance brings the adjacent nuclei closer until very little mass-energy needs to be released when the final nucleus forms as two particles and the intervening electron combine into a single nucleus. Each Hydroton is assumed always to contain an even number of hydrogen nuclei. The next challenge is to explain transmutation using this model.

5.1.0 CAUSE OF TRANSMUTATION

Transmutation is difficult to evaluate because contamination frequently confuses the apparent isotopic ratio and such small amounts of transmuted material are difficult to measure. Many nuclear products predicted by this new model are missing because suitable detection methods were not used. As a result, a rich field for speculation is available. Nevertheless, certain requirements not based on speculation must be acknowledged, starting with the Coulomb barrier. No matter how the process is proposed to function, this barrier is real and effective. After all, the universe exists only because this barrier is not ignored by nature. Nature has to go to great lengths in stars and supernova to achieve transmutation while mankind has to use special accelerators and nuclear reactors to achieve the required high energy. This source of energy must be identified for transmutation by LENR to be properly explained.

Three kinds of transmutation are observed and need to be explained (Sections 2.3.5 and 3.2.6). The two associated with LENR and involving addition of hydrogen to a target nucleus are summarized below. The third type requires a different book.

1. Transmutation #1: Addition of two or more hydrogen nuclei to the target produces a single nucleus. This process has no obvious way for the resulting energy to be dissipated while conserving momentum.
2. Transmutation #2: Addition of one or more hydrogen nuclei to the target nuclei followed by fission into two nuclei, with the number of protons and neutrons being conserved. This process conserves momentum and dissipates energy in the conventional way.

Obviously, these two kinds of transmutation require different processes to dissipate the energy. Transmutation #1 requires the energy be dissipated in small packets by a mechanism similar to that operating during hydrogen fusion while transmutation #2 can dissipate energy in a normal way.

Nevertheless, a unique process must be available in both cases to overcome the large Coulomb barrier while being compatible with the required dissipation process. The required energy to surmount the barrier is generated by the hydrogen fusion reaction. Which type of transmutation occurs depends mostly on whether d or p are involved and how the target is made available to the Hydroton.

The transmutation process is proposed to start when a target element becomes attached to a Hydroton. More than one Hydroton attached to each target atom can be visualized by looking at Fig. 87, where three of many possible situations are shown. Each Hydroton starts to resonate, as described in Section 5.3.0, and releases energy until hydrogen nuclei combine along with the intervening electron to produce fusion products listed in Table 11. As the final fusion event takes place, the fusion product nearest the target enters the target nuclei. When two Hydrotons are attached to each target, two fusion products enter the target as independent events. An even number of additions, as observed by Iwamura et al., is proposed to result because both hydrogen nuclei in the Hydroton are bound together and move as a unit into the target.

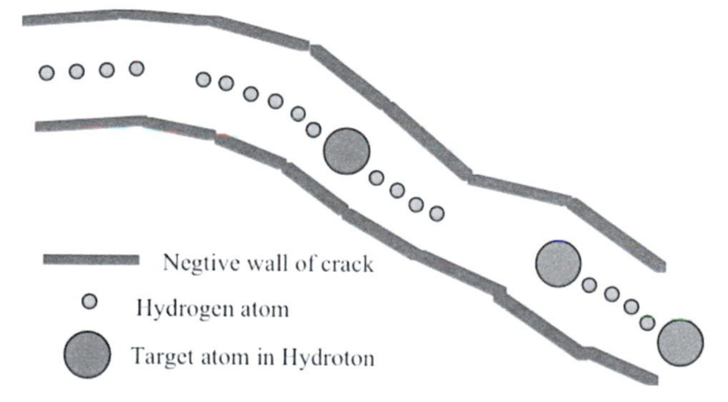

Figure 87. Examples of Hydrotons in which target atoms are captured and of one example of a Hydroton without attached target.

As Iwamura et al. (Section 2.3.3) find when deuterium is involved, from two to six deuterons or three equivalent helium can add to a target, which suggests a target can become attached to at least one, two, or three Hydrotons. This number might be determined by the chemical valence of the target and the number of Hydrotons relative to the number of targets in the NAE. Both Ni and Pd apparently react most often with only two Hydrotons. This description is assumed to apply regardless of which isotope of hydrogen is in the Hydroton.

However, the product of this transmutation reaction will depend on which isotope of hydrogen is involved. Hydrotons containing only deuterium produce Transmutation type #1 and Hydrotons containing some or only protium produce Transmutation type #2. This difference results because the resulting nucleus must not have excess mass-energy after the hydrogen nuclei are added to avoid fragmentation. If too much mass-energy remains, it can be dissipated only by fragmentation, which leads to Transmutation type #2. The role of excess mass-energy can be explained by first describing Transmutation type #1.

5.1.1 Transmutation type #1

When transmutation takes place as a part of fusion with deuterium, the mass-energy of the deuteron is mostly dissipated by photon emission before transmutation takes place, leaving very little energy to be dissipated once transmutation occurs. For example, each d has to lose about 12 MeV by action of the Hydroton for fusion to produce ^4H without much excess mass-energy needing to be dissipated at the final event. Addition of this ^4H to the target nucleus would generate about 2.3 MeV additional energy based on the final mass change. Fragmentation would not be needed to dissipate energy if subsequent emission of the beta could do the job, thereby adding the equivalent of ^4He for each ^4H added.[61] This beta radiation would not exit the apparatus and a short half-life would make its detection difficult. Nevertheless, this prediction is worth exploring by removing the transmuted material quickly and testing for beta emission.

Once the excess mass-energy created by the transmutation reaction with ^4H is released, the resulting nucleus might still be

---

[61] This process might be a source of double or triple beta emission.

radioactive and decay with its own characteristic half-life and decay mechanism. When this process is applied to Pd as the target, isotopes of cadmium (Cd) would result when one dd unit is added, with one isotope being radioactive, as listed in Table 12. Decay of ^{109}Cd by internal conversion (IC) would produce stable silver (^{109}Ag). If two dd units are added, tin (Sn) results. Decay of the tin isotopes would eventually produce stable isotopes of Cd and indium (In). The resulting initial isotopic distribution of the tin isotopes is expected to match that of palladium.

Table 12. Result of adding ^{4}H to Pd isotopes followed by beta emission. The half-life and decay mode are shown for the isotopes of Pd. IC identifies the decay as internal conversion, which produces only weak photon radiation.

^{46}Pd isotope	One dd unit added to give ^{48}Cd	Two dd units added to give ^{50}Sn
102	106 – stable	110 – 4.1h - IC
104	108 – stable	112 - stable
105	109 – 462d -IC	113 – 21.4m - IC
106	110 - stable	114 - stable
108	112 - stable	116 - stable
110	114 - stable	118 - stable

Accumulation of ^{109}Ag would slowly take place over time after the LENR process stopped, with very little radiation being emitted. Unfortunately, silver is difficult to detect at low concentration in the presence of palladium because the peaks produced by EDX significantly overlap. In addition, palladium can contain a small silver impurity that can become concentrated as a result of electrolytic action. Consequently, claims for silver on palladium using EDX need to be evaluated very carefully. Search for Cd or Sn and the resulting radiation would provide better evidence for this explanation.

This mechanism for transmutation is proposed to happen after elements are deposited on an active surface, such as during the Iwamura et al. studies (Section 2.3.3). In their case, stress created between the CaO and Pd layers is proposed to produce active cracks that radiate to the surface. Deposited material diffuses into these gaps and bonds with the deuterium-based Hydrotons. Only transmutation products resulting from deposited material are detected because they remain close enough to the surface to be detected. Other transmuted elements might be present in the material, but located too far from the surface to be detected.

If all material deposited on a nuclear-active surface is assumed to have the same experience, lithium (Li) and other elements deposited during electrolysis using D_2O would be expected to result in new radioactive and non-radioactive elements by addition of d. For example, adding 2d to Li results in isotopes of boron (B), being ^{10}B and ^{11}B, with more ^{11}B being produced than ^{10}B because more ^{7}Li is present than ^{6}Li in natural lithium. This prediction is consistent with the results published by Passell,*(68)* who found the $^{10}B/^{11}B$ ratio in Pd to decrease*(160)* after electrolysis in $LiOD+D_2O$.

5.1.2 Transmutation type #2

A Hydroton containing H produces an entirely different transmutation product than does one containing deuterium. Unlike deuterium, protium cannot dissipate enough energy during the fusion process to avoid extra mass-energy when transmutation takes place at the end of the fusion process. For example, when D resulting from p-e-p fusion is added to a target, as much as 13 MeV must be released. Because this amount of energy cannot be released by electron emission, it can only be dissipated by fragmentation, *i.e.* fission.

The following rules are assumed to apply.

1. Because the electron used to form deuterium produces a stable element, it cannot be emitted to lose excess-mass energy during transmutation.

2. As a result, all elements and isotopes experience fusion-fission when transmutation involves protons.
3. The total number of protons and neutrons are conserved in the process.
4. Stable isotopes are produced if possible.
5. When several stable isotope combinations are possible, they are assumed to form with equal probability.

These rules may not be the only ones followed, but for the present, they alone are applied consistently to show where to look for confirmation or rejection and to discover whether other behavior might be operating.

This process is described using palladium as an example. The goal is to examine all possible combination of fragment products to determine which are more likely to form as stable products. Some of the Pd isotopes have a greater chance of being in the Hydroton and producing more fragment combinations because of their greater concentration. This condition is taken into account by multiplying the number of different isotopic combinations of fragments resulting from each Pd isotope by its percent in the total amount of Pd. These numbers are summed for each fragment combination to give the relative amount expected of that combination.

An example is shown in Table 13 using ^{46}Pd to which 2 d are added followed by fragmentation into two possible fragments. The atomic weight of each Pd isotope is listed with its percent abundance shown in Column 2. If all element combinations are considered, the total number of ways each isotope can produce a stable combination is shown in Column 3. Column 4 shows the number of ways each isotope of palladium can fragment to produce different isotopes of Zr+O, as one of many possible fragment combinations. The total relative abundance in column 5 is obtained by multiplying values in column 4 by the percent abundance.

The calculation shows that all isotopes of Pd can fragment to make stable zirconium (Zr) and oxygen (O). A similar calculation is made for each of the 23 element combinations and the average total amount (sum of Column 5) resulting from each similar calculation is plotted as a function of atomic number in Figures 88-91.

This distribution can be compared to observation as a test of various assumptions. A start of this testing process is shown in Fig. 88 where one p-e-p is assumed added. Clearly, this distribution does not match the observed distributions shown in Figs. 19, 20, and 21. A better match can be seen in Fig. 89 where two (p-e-p) are added. A similar distribution but with fewer combinations results if the reaction Pd + 2(d-e-p) should happen on occasion when a mixed Hydroton is formed. Because the D concentration would slowly increase as a result of fusion, the fragment distribution would be expected to change slightly over time.

Platinum, which is a common impurity on the surface of a cathode, shows the distribution in Fig. 90 for Pt+2(p-e-p), which can be expected to add fragmentation elements when H is present during electrolysis. In this case, energetic alpha emission is a common method of energy release, which may account for the occasional detection of energetic alpha emission.

Table 13. Information used to calculate relative amount of each fragment combination for $^{46}Pd + 2(p+e+p) = {}^{40}Zr + {}^{8}O$

Atomic Weight	% total	Number of ways a stable fragment can form	Number of ways Zr+O can form	Total relative amount of Zr+O
102	1.02	32	1	1.02
104	11.14	33	3	33.42
105	22.33	19	2	44.66
106	27.33	19	2	54.66
108	26.46	7	2	54.92
110	11.72	2	1	11.71
			sum	**198.4**

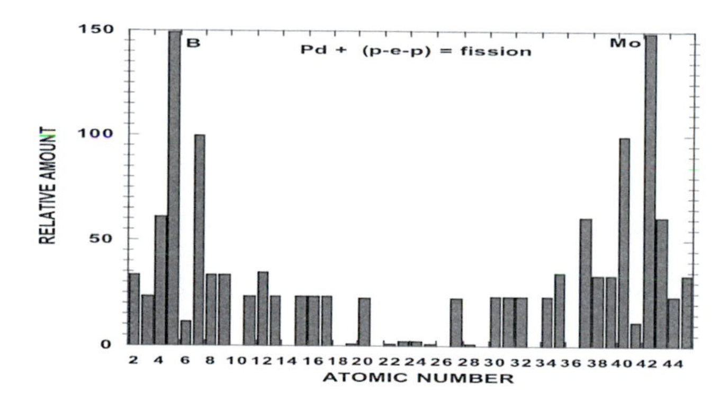

Figure 88. Fragment distribution when a single p-e-p enters Pd followed by fission. Noble gases are not shown.

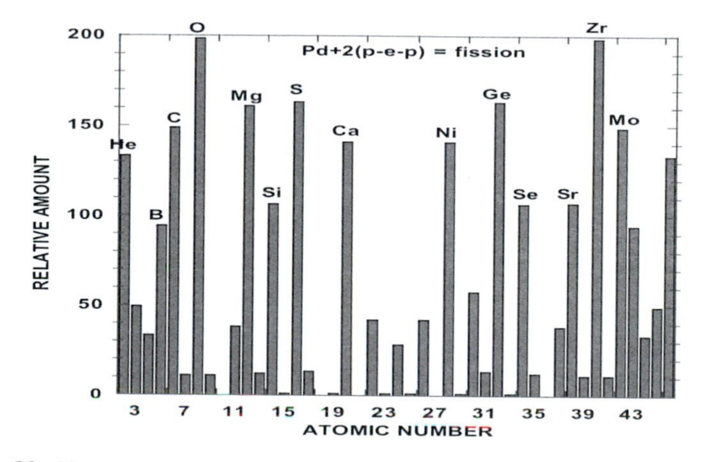

Figure 89. Fragment distribution when two (p-e-p) enter Pd followed by fission. Noble gases are not shown.

Transmutation of other elements would result in a different product profile. The profile produced by nickel to which 2(p-e-p) is added is shown in Fig. 91. Alpha emission is the most likely method for energy release. In other words, Ni in a Hydroton can convert p+e+p to helium with release of about 23.8 MeV/event. However, many of the rare Ni atoms that happen to be in the NAE would be slowly converted to other elements, such as Si and S, with less ability to release energy by alpha emission. As a result, alpha emission would be rare. Some energetic ^{3}He emission is also expected along with formation of some radioactive elements. In addition, the observed distribution would contain fragments from all impurity elements as well as from further transmutation of the accumulating fragments. This process will produce a rich and complex assortment of elements and radiation that so far has not been fully explored and has defied understanding.

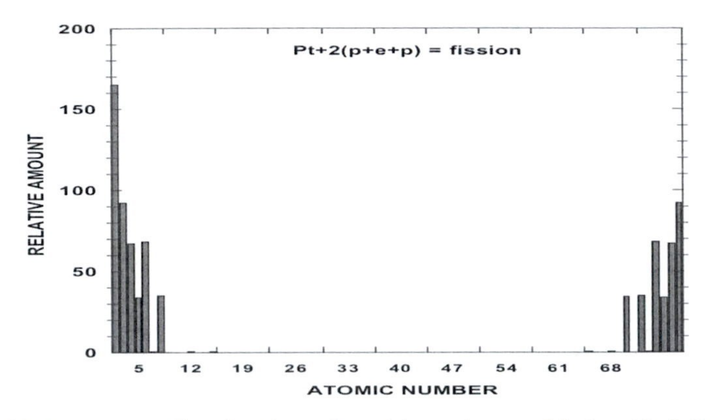

Figure 90. Fragment distribution when 2(p-e-p) are added to Pt followed by fission.

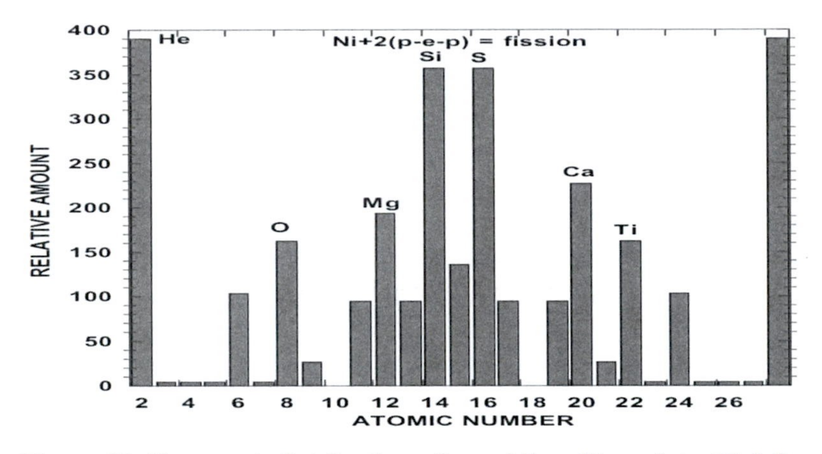

Figure 91. Fragment distribution after adding 2(p-e-p) to Ni followed by fission. Noble gases are omitted.

As shown in Table 14, all isotopes of Ni can contribute some energy by fusion-fission, with similar average energy/event.

The various fragments produce a distribution of average energy shown in Fig. 92. Some Ni isotopes contribute less energy because they have a low concentration and have fewer ways they can fragment. As the table shows, ^{58}Ni contributes the most energy resulting from fusion-fission. The fraction of total power contributed by transmutation depends on how fast Ni can move into the NAE. The much slower diffusion rate of Ni compared to H would be expected to allow fusion of H to occur at a much faster rate than transmutation and produce more overall power even though transmutation produces more energy/event.

Table 14. The number of ways the isotopes of Ni can fragment and the resulting average energy for each event is listed for each isotope. The % of total energy results from multiplying columns 2, 3, and 4.

Atomic weight of Ni	% abundance	# ways to fragment	Average MeV/event	% total energy
58	68.08	12	13.59	69.3
60	26.22	17	8.79	25.7
61	1.14	15	11.49	1.3
62	3.63	14	10.85	3.5
64	0.93	4	10.72	0.3

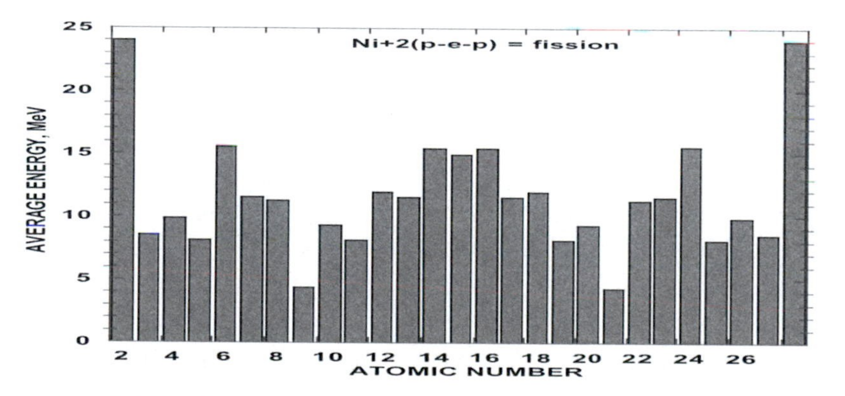

Figure 92. Distribution of energy resulting from fragmentation of Ni stable element combinations. The value of 23.8 MeV for helium is for ^{4}He. Emission of ^{3}He produces about 12 MeV/event.

Once the metal atom has transmuted, the fission products might react again as new Hydrotons form to sustain heat production. Because alpha emission leaves Ni nuclei behind, this target could remain in place, perhaps to accept two more d, and again release 24 MeV by alpha decay until the Ni is no longer available because it has transmuted into another fragment combination. In effect, the process in the Hydroton might use Ni to convert p-e-p into ^{4}He and on occasion to ^{3}He. This helium would be promptly emitted as an energetic particle identified as alpha emission.

The relative contribution to the total power favors fusion over transmutation. Nevertheless, the model shows how and why so much energy and energetic alpha emission can result when ordinary hydrogen is used.

5.1.3 SUMMARY OF TRANSMUTATION REACTIONS

Transmutation takes place in the same NAE and involves the same mechanism operating to produce hydrogen fusion. The target nuclei are mostly those deposited on a chemical structure in which the NAE is present on the surface, with the most effective target being determined by which element has easiest access to the NAE. Because a limited number of atoms can enter the gap and become attached to the Hydroton, transmutation has limited duration and rate. The process produces two different kinds of transmutation depending on which isotope of hydrogen is present in the Hydroton.

When p-e-p is added to a target as deuterium, the resulting nucleus fragments into two parts. The products are stable as long as stable isotopes can result from the fragment combination. The concentration of elements produced is determined by the abundance of the isotope from which they originate and the ability to form stable isotopes of the product elements. Alpha emission with energy near 22 MeV is expected on occasion as a possible fragment.

When ^{4}H is added as a result of d-e-d fusion, the final nucleus does not fragment, but remains intact. The resulting nucleus can be stable or radioactive, with stable isotopes being favored. Energetic beta emission is expected along with weak photon radiation.

The process does not depend on which target isotope is involved in the event — only the hydrogen isotope determines the type of transmutation. As a result, the two isotopes of hydrogen produce different results while using the same basic process to overcome the barrier. The final element and isotopic distribution is determined by the target nucleus combined with the reacting hydrogen isotope.

This kind of analysis shows several interesting conclusions. First, the lighter isotopes of Pd will be depleted by transmutation. Second, the isotopic composition of boron would show an increase in the amount of ^{11}B relative to ^{10}B, which is consistent with what Passell reported.(*160*) The process would also increase the amount of ^{89}Sr over ^{86}Sr, as observed by Bush.(*886*) The heavier isotopes of iron would be increased by the process as noted by Ohmori and Enyo.(*887*) Third, ^{4}He will result with energy near 22 MeV, perhaps accounting for the 20+ MeV alpha detected using CR-39. Emission of ^{3}He might explain the pits assigned to alpha with energy near 10 MeV (Section 2.4.3).

Iwamura et al. found H_2 produced no apparent reaction perhaps because they were not looking for the expected fragments, not because LENR had stopped. Other behavior might be misinterpreted as well. For example, use of K_2CO_3 in H_2O is proposed to generate calcium from the reaction K + p = Ca.(*260, 888*) According to the model described here, such a reaction would not occur. Instead, the observed calcium might have resulted as a fragmentation product from Pd and stripped from the cathode by electrolytic action. Other fragments would have been produced but were not detected because they were not sought.

This model predicts copper would not result from adding p to Ni. Instead, the rare Ni attached to the Hydroton would be expected to fragment into many elements in order to dissipate excess mass-energy.

This explanation shows why two different types of transmutation occur, how the Coulomb barrier is overcome, why radioactive products are rare, and accounts for the published distribution of transmutation products. This model is unique in being able to provide so many testable predictions and can account for so much observed behavior.

5.2.0 DETAILED EXPLORATION OF THE HYDROTON STATE

The next step is to dig deeper into the mechanism operating in the Hydroton. It is essential to realize that the details of how the Hydroton operates, or in fact how any explanation of the nuclear process functions, is irrelevant for making LENR useful or for explaining many behaviors. Once the critical conditions are created, the nuclear reaction will start and function without any need to understand how it works. The emphasis on explaining the nuclear process in most theories is not useful even though it gives intellectual satisfaction. Nevertheless, this satisfaction is sought here by showing how the Hydroton is related to other proposed structures of hydrogen.

5.2.1 Metallic hydrogen

During formation, the Hydroton structure is subjected to effective pressure by the walls of the gap. Because these walls have a high concentration of negative change, the electrons shared by the hydrogen nuclei would be forced into a more compact state with an average smaller distance between nuclei. This state might be related to the DDL proposed by Meulenberg and Sinha or the hydrino proposed by Miles (Section 4.4.0).

Also, this structure might be described as a version of metallic hydrogen (MH).(*889*) So far, this predicted metallic state has apparently not been created by applying extreme pressure.(*890*) It is interesting to speculate about the consequences if attempts to make MH actually create Hydrotons.

Ashcroft(*891*) proposed metallic hydrogen might be a high-temperature superconductor and Liboff(*892*) proposed fusion of D might result after its formation using deuterium, especially in the extreme condition on Jupiter. According to Horowitz(*893*), a fusion reaction made possible by this state of deuterium might explain the energy being generated by this planet. This naturally raises the question about what would happen if MH should actually form in the laboratory. If high pressure produced the metallic state that then experienced LENR, considerable energy and radiation would be produced on a small scale in the high-pressure cell. This nuclear reaction might confuse interpretation of observed behavior and make metallic hydrogen impossible to detect other than as a consequence of a nuclear fusion reaction. This speculation suggests a search for the intense but low energy radiation be made when MH formation is attempted.

5.2.2 Resonance of negative and positive charges

The concept of the Hydroton is based on a structure that resonates. This action allows mass-energy to be released over time by an unknown process. The process requires interaction between the positive charged hydrogen nuclei and the electrons bonding the structure, a process that has been described mathematically by many authors using various assumptions. However, although the basic process has been explored, many assumptions conflict with the requirements listed in Chapter 3. The resonance mechanism needs to be applied without this conflict.

Kalma and Keszthelyi(*894*) proposed interaction between a positive charged particle and an electron can provide enough energy and momentum to cause fusion with another D. The location and conditions in which this reaction can occur were not identified. Although the approach was applied to the hot fusion reaction in a metal, its application to the Hydroton might be more realistic and plausible. The Hydroton provides a unique structure not found elsewhere in nature with many features inviting mathematical analysis.

5.2.3 Role of the surface

This theory focuses on the surface as the site of the NAE and the nanogaps contained therein. This conclusion is in direct conflict with theories that focus on the properties of pure palladium in the chemical lattice. The chemical and physical properties of bulk material are important only when they affect the surface.

For example, the observed effect of average deuterium content of the bulk material (Fig. 58) on power production results because composition at the surface is determined to some extent by this average. In fact, the actual composition at the surface is somewhat greater than the average while electrolysis is underway. This increased composition combined with increased flux probably causes power to increase when applied current is increased (Fig. 59). In addition, the surface will have a greater H/D ratio compared to the bulk material or the surrounding source of hydrogen. Consequently, both energy production and distribution of nuclear products might show unexpected behavior as this ratio changes from what is thought to be present.

Because the surface layer is very thin, very little material needs to be deposited to have a large effect, both good and bad. For example, heating in vacuum containing vapor from silicon vacuum pump oil may deposit enough silicon to make the surface active even though the effect is wrongly attributed to annealing time. Other treatments can apply intentional and unknown local deposits as a result of polishing with SiC or Al_2O_3, both of which might activate local regions were the particles remain imbedded in the surface. As SEM examinations show, the surface is not uniform in morphology and composition. Where in this complex mess LENR occurs has not been identified. Until the location and local conditions are discovered, no theory applied to the pure material has any plausible relationship to experimental behavior.

Formation of nanogaps requires stress and weak sites where the lattice can break. Stress is created when any chemical reaction takes place at the surface as a result of non-uniform expansion. The weak sites are usually grain boundaries where a crystal orientation change takes place, but not always. Certain impurities in the grains can weaken the structure within the grain. As a result, the morphology and purity of the surface is very important and very complex. This complexity must be acknowledged in the proposed theories. Because so little impurity is required to affect the surface morphology, these variables are difficult to control and to large extent can make success unreliable until this control is achieved. The situation is not unlike the problems that plagued early use of transistors.

Production of maximum power is expected to result from powder of a particular particle size. Particles too big can generate enough stress to produce large cracks, which are inactive. On the other hand, very small particles cannot generate enough stress to cause gap formation. Depending on ductility, an ideal particle size will be influenced by its structural weakness and its reactivity with hydrogen. Because each particle in a sample can have a range of size and properties, a mixture will normally contain only a fraction of active material. Being able to increase this active fraction is essential to success.

5.2.4 Role of the laser and magnetic fields

What role does application of laser energy have? At the very least, the coherent waves can couple their energy to electrons, thereby causing them to vibrate with the same frequency. This process is said to produce plasmarons, plasmons, polarons, or polaritons, depending on the behavior and the author's name preference. If energy density is sufficient, ionization of atoms can result. At very high energy density, nuclear reactions might occur between atoms. These reactions are caused by high energy and would show all the behavior expected when high energy is used, which results in hot fusion. The challenge is to explain how laser radiation can stimulate LENR.

Laser radiation is dissipated within a few nanometers of the surface. Just how far the coherent vibrations can penetrate into the material before they become random and part of the normal temperature is a matter of speculation. Nevertheless, the depth is not expected to be more than a few microns, depending on laser frequency and the nature of the material. As a result, such radiation would be expected to affect material only at the surface where LENR is found to occur. The question then becomes, "What effect does laser radiation have on the NAE and the Hydroton contained in the surface?" As emphasized many times, the properties of material located away from the surface have no relationship to this question and are best ignored.

Both polarization of laser radiation and orientation of a magnetic field have been found to affect heat production(*414, 895*) by the power being sensitive to their orientation relative to the surface. This behavior is possible only if some feature of the LENR process has a special relationship to the surface. All features on a typical surface appear to have a random relationship to the overall orientation of the material except cracks. Many cracks appear to be lined up in a similar direction perpendicular to the surface. How the orientation of the laser wave front and the magnetic field can affect what happens in the crack is a matter of speculation. Coherent photon radiation entering the nanogap might cause the Hydrotons to resonate in unison, thereby creating faster release of energy. Crack orientation might allow greater absorption of laser energy resulting in greater local temperature. In this case, the large inactive cracks would be important as well to help absorb this source of local energy and convert it to increased temperature. Application of an external magnetic field could either support the resonance process or conflict with it, thereby changing the rate of the fusion process, especially if the Hydroton were to act like a superconductor. These possibilities need to be explored.

For the laser radiation to have a direct effect in the nuclear process, it must couple to the Hydroton within the nanogap. This coupling would be sensitive to gap size and the frequency of resonance within the gap. Because a variety of gap sizes will exist, and the Hydrotons in each gap will have different frequencies depending on the number of hydrogen nuclei and the isotopic composition, (with certain frequencies being able to couple to the Hydroton better than others), the fusion rate is predicted to increase as this energy amplifies the resonance process.

Because fusion is predicted to cause transmutation, transmutation is predicted to also increase when laser light is applied, as observed. Local stress produced by the local energy from the laser itself or from the initiated nuclear reactions could create additional nanogaps, hence more NAE could become available to support the process. This complicated interaction is expected to make the true process difficult to investigate and understand.

5.3.0 HOW CAN LENR BE MADE REPRODUCIBLE?

The difficulty in achieving reproducibility has been a frustration and source of rejection. This failure results because important variables are not controlled. The explanation provided here can reduce this problem and suggest a method to make LENR occur with greater reliability.

The nature of the cathode material is the most important feature determining whether LENR can be initiated at all. Once active material is made, detecting the effect becomes the next problem. Because the rate can sometimes be too small to detect by heat production, a person might overlook successful initiation of LENR that might be made visible instead by detection of tritium or radiation. For this reason, studies should be designed to monitor all known products of LENR, not just heat production. Correlation, in real time, between productions of these different products greatly reduces uncertainty about the claim.

The first step in creating an active material is to apply a small layer of impurity to its surface in order to create small regions where the surface is modified at small local sites. The goal is to produce stress as a result of this reaction as well as create local stress when hydrogen or deuterium reacts with some regions of the surface and not with others. This uneven expansion of local regions is designed to produce many local nanogaps at the boundaries between these different regions. Use of a surface made brittle by reaction with various nonmetallic elements or one covered by many small crystallites works best.

The basic requirements apply to all methods, but the electrolytic method is the easiest to use as an example. During electrolysis, this surface impurity might come from material in the cell applied by long electrolytic action. The most common impurities are Li from the electrolyte, Si from glass, oxygen from water, and Pt from the anode,(*447, 896-898*) with lithium apparently having the greater ability to modify the surface and increase heat production.

The effect of lithium involves its ability to dissolve in palladium by substituting for Pd and by forming many metallic compounds,(*899-901*) some of which will be brittle. Some of these Pd-Li compounds will dissolve in the electrolyte and the resulting Pd ions redeposit back on the cathode surface as Pd metal with modified morphology and impurity content. Some of this Pd might diffuse into surface-opening active nanogaps and be the source of observed transmuted elements.

If an inert substrate is used, such as platinum metal, the palladium can be applied as a thin coating by co-deposition(*902*) or sputtering. If co-deposition is used, $PdCl_2$ and any other sources of the chloride[62] ion are best avoided. Although such coatings have been found to produce LENR, they are unreliable for reasons too complex to address here.

Use of bulk palladium has been more reliable than co-deposition and achieved better understanding. Nevertheless, the bulk palladium must not form large cracks through which hydrogen can be lost too rapidly and where stress relief can occur without forming nanogaps. This tendency to crack can be tested in order to avoid using flawed material.

Stress will create different sized gaps and the resulting vacant space can be measured. This can be done by first loading a sample with hydrogen using electrolysis to reach a composition above about H/Pd = 0.75. The physical volume change is measured and compared to the volume change expected based on the lattice

---

[62] Chlorine dissolved in or deposited on palladium metal in low concentration inhibits reaction with hydrogen. (Storms)

parameter change using the known lattice parameter for the initial metal and the final composition of PdH.(*490, 861*) This process is described in Section 2.7.0 and Figs. 66 and 67 for palladium.

A potentially active material can be expected if the physical volume change is less than 2% of the volume change calculated using the lattice parameter. This low concentration of cracks or gaps will not always result in an active sample because other factors may make the sample inert. For this reason, the test only removes samples that cannot work at all. Nevertheless, this is a good way to avoid wasting time trying to encourage an impossible result.

Some palladium containing alloys are more likely to be active than others. For example, an alloy with silver(*471*) shows less expansion, hence less stress, for the same H/Pd ratio compared to pure PdH. This alloy was apparently used with success by F-P and identified as the alloy used in hydrogen purifiers, which is a Pd-Ag alloy.

A Pd-B alloy(*134, 135, 903*) has been used with success. This material is a two-phase mixture that is much more brittle than pure palladium and does not acquire a high concentration of hydrogen. As a result, the cracks are expected to remain too small to permit significant loss of hydrogen. Nevertheless, a complex surface must form in which the NAE can grow. The role of boron in this process is not clear but it is expected to make the surface more brittle and more likely to form many small gaps.

The surface of pure palladium can be modified to cause small gaps to form. This can be done by heating the metal in gas containing a low concentration of Si, B, or any element that forms a brittle compound with the metal. For example, palladium can be heated in air above 900°C for time sufficient to make the surface rich in oxygen, carbon, and nitrogen. This results in formation of very active blue-colored palladium oxide.(*904*)

Nickel will require the same type of treatment, however the response to the treatment will be different from palladium because nickel is less ductile, more chemically reactive, and reacts with hydrogen less easily. Consequently, each material used to initiate

LENR needs to be treated to different conditions for active-sized gaps to form on the surface. A high bulk concentration of hydrogen is not required because the active gaps are present on the surface where they have easy access to hydrogen in the gas. In fact, active material can be even obtained by depositing a thin layer of active material on an inert substrate. To achieve this access to hydrogen, the resulting surface must be able to split the molecules of hydrogen into ions, which limits the chemical nature of the surface. Ionizing the gas using electric discharge to make ions more available can reduce this problem.

Once potentially active material has been obtained, it needs to be reacted slowly with the hydrogen isotope to allow stress to be released at as many sites as possible. Causing the cathode to lose and then gain hydrogen repeatedly will add stress and can be used to produce additional gaps. However, eventually the gaps will get too wide to be active and these large cracks will prevent smaller gaps from forming. This inevitable result will terminate energy production. The physical volume compared to the volume based on the lattice parameter can be used to identify this end-of-life condition in palladium.

Growth of gap size creates a natural and inevitable end-of-life to all methods used to initiate LENR. Fine powder has a longer life compared to bulk material because each particle has a limited ability to produce stress and release this stress by gap formation. As a result, the number and size of the active gaps are limited on each particle while many active particles can be present. According to this model, these are the only important features provided by powder.

Once active gaps are formed, the amount of power can be increased by increasing the concentration of hydrogen in the surface, by causing a hydrogen flux to flow through the surface, or by applying energy to the surface using laser light or energetic ion bombardment. Because local temperature has a major influence on power production, all treatments found to increase power production must take this variable into account when analyzing the study.

Several mistakes are frequently made. First, chlorine in any form should not be applied to palladium because chlorine is known to inhibit reaction between palladium and hydrogen. For example, Aqua Regia should not be used to clean the surface and $PdCl_2$ should not be used to deposit palladium. Palladium can be effectively cleaned using fuming nitric acid. Second, copper wires should not enter an electrolytic cell even when they are covered by shrink tubing, epoxy, or gold plate. None of these protections is effective. The only metal in the cell other than the cathode should be platinum. Wires should not be attached using spot welding with copper electrodes because small amounts of copper can be transferred to the welded structure. When spot welding is used, the electrodes should be tungsten. The heavy-water and electrolyte must be kept out of contact with H_2O and CO_2 in air. Great care must be used to keep all materials free from impurity including pre-cleaning the electrolyte using a dummy cathode that is removed before the study is made.

5.3.1 SUMMARY OF NEW THEORY

The LENR process can be divided into three stages, each of which works in collaboration with the others. The first stage delivers hydrogen fuel to the site of the nuclear reaction, the second stage creates the nuclear active sites, and the third stage involves the nuclear process itself. The first two stages obey all rules required of normal chemical processes. Only the third stage is unique and requires new understanding.

Stage 1 is influenced by the concentration of hydrogen at the sites of the nuclear reaction and by the flux of hydrogen through this region, both of which are influenced by temperature and other kinds of applied energy.

Stage 2 describes the NAE generated by stress relief. The NAE appears as nanogaps or very small cracks in which the nuclear reactions take place. The nuclear process does not occur in the lattice structure itself, neither in vacancies nor in dislocations. Instead, these nanogaps occur mostly at the surface and their concentration influences the maximum amount of generated power. Powders are more active than bulk material because they have more surface area where the NAE can form and allow control of stress generation and relief.

Stage 3 involves formation of a molecular structure in the nanogap called a Hydroton, which has the ability to overcome the Coulomb barrier in stages while releasing excess mass-energy as photon radiation. All isotopes of hydrogen can fuse in this structure. When only deuterium is present in the Hydroton, ^{4}H results and promptly decays by beta emission to ^{4}He. A Hydroton made from only protium results in deuterium. When a mixture of deuterium and protium are present, the result is tritium.

Transmutation occurs as a result of these fusion reactions when an element besides hydrogen is attached to the Hydroton. If this Hydroton contains only deuterium, the target element can acquire multiple ^{4}H structures and emit the excess mass-energy as beta emission. If the Hydroton contains some protium, the resulting target nucleus must fragment in order to dissipate excess mass-energy.

The Hydroton structure contains the entire mystery of the LENR process. In order to understand the mechanism, new concepts of nuclear interaction need to be discovered and applied.

The explanation provided here shows how the special conditions required to initiate LENR can be created, how power production can be controlled, and where attention needs to be focused to understand the nuclear process. The model explains most if not all observations without introducing *ad hoc* assumptions and makes many testable predictions, all without using mathematics or quantum mechanics. The model has opened a window to explaining LENR from which many more conclusions can result than explored in this book.

CHAPTER 6

FUTURE OF LENR

"So what?" you may ask. A source of energy is proposed without an explanation while being strongly rejected by many scientists. It may even be an economic threat to conventional sources of energy. Why bother?

Energy is essential for civilization to survive at the present level. Energy sources now in use are poisoning the environment at many levels. If global warming were not important enough to get your attention, think about the radioactive material being accumulated by nuclear reactors. Perhaps the local areas being destroyed by coal mining, extraction of natural gas, or strip mining for oil are not important to you, but consider what happens when the transport of these toxic chemicals goes wrong. Consider the failure of the electric grid caused by weather or a computer virus expected in the future. Modern society cannot function without a reliable source of electric power, which is becoming increasing vulnerable and unreliable in many countries.

Even if global warming cannot be stopped, its effects will have to be countered using more energy. Dikes will have to be constructed to protect cities or cities will have to be moved, all of which will require huge amounts of energy. Droughts are becoming more common where food is grown, requiring irrigation water be pumped from the sea. For this use, seawater requires purification, which is easy but requires energy to accomplish.

Energy from coal and oil made modern civilization possible and these fuels were no threat as long as their use was limited. Now use has grown to a level causing destruction of many environments. Cheap energy based on fission of uranium was promised but in practice is too dangerous and too costly. Energy from hot fusion was promised as the energy of the future but research costing billions of dollars has revealed this expectation is false, yet the studies continue to drain money from development of better energy sources. The electric distribution system has become vulnerable to weather, radiation bursts from the sun, and cyber threats, making centralized electric power unreliable. The problems continue to grow more threatening without useful response from society or governments.

In other words, mankind is at the mercy of energy sources that are poisoning the environment, are too dangerous to use in many countries, are unreliable, and have limited life. Cold fusion is looking good about now, wouldn't you say? Given these problems, why would a sane person not make every possible effort to learn how LENR works, make it reliable, and develop it as a source of unlimited clean energy? What do we have to lose?

The future of LENR depends on the people who will suffer as a result of flawed energy sources because only they have self-interest to develop a new source. How many of these people are willing to encourage interest in LENR as a source of ideal power? Unfortunately, their influence is limited because the media and social leaders are unwilling to challenge the myth about LENR even though this myth has very little relationship to reality. It is not true that LENR was found to be error. It is not true the effect cannot be replicated. Moreover, it is not true that it violates known laws of nature. In contrast, the effect is real, it is difficult to replicate, and reveals a new law of nature. The future depends on these facts being accepted by the scientific and political leadership.

6.0.0 SCIENTIFIC

Many conventional scientists are arrogant — not a large number but enough to be a problem. Such people believe they understand nature and any new idea that conflicts with what they were taught must be wrong. The claims about cold fusion conflicted with what is known and taught about hot fusion. Therefore, the claims about cold fusion must be wrong. Fleischmann and Pons must have made a mistake. The fact that the mistake cannot be found is not important.

Now nearly a thousand papers show that Fleischmann and Pons were correct. These papers are largely unknown and unread because conventional scientific journals refuse their publication. Books have summarized this work, but remain largely ignored. As a result, the scientific profession is uninformed about what is actually known about LENR. Rejection is now based mostly on ignorance.

LENR is a complex phenomenon lacking funding to hire skilled scientists to fully investigate its behavior, resulting in partial understanding. The main problems involve the difficulty in making the effect work at every attempt, achieving total control of generated power, and extending the lifetime for energy production. In addition, the present energy density is not as great as might be possible if better knowledge were available. Solving these problems would not be difficult when skill and money are applied to the problem.

In addition to energy production, the phenomenon has two other important features. First, it gives the promise of reducing radioactive contamination created at today's nuclear fission plants. Second, a new kind of interaction between nuclei has been revealed. Such new discoveries always lead to unanticipated applications. Exploring these features has barely started.

Three problems require solution before commercial application is practical. First, the NAE must be made in large quantity with reliability. Second, power production needs to be controlled into a variable load. Third, the lifetime for power production needs to be increased by making the NAE more robust.

Presently, the NAE is made by using a chemical reaction to produce stress, resulting in a variable number of cracks having the critical gap and with uncertain location. A better but more challenging method would be to use nano-machining to create the NAE with high density and reproducible behavior. This method would allow the NAE to be applied to various surfaces and materials in a form more suited to the application and with a more durable form compared to the present methods.

Control is important because LENR has the potential to experience run-away during which the temperature can rise until the NAE is destroyed. This temperature may also generate enough pressure to cause local physical damage. However, a runaway cannot produce a nuclear event, harmful radiation, or radioactive contamination, as is characteristic of other nuclear energy sources. In this case, the damage results from only excessive physical pressure, which can be kept small. In fact, this is one of the safest energy sources presently known.

The lifetime for energy production will be temperature dependent, with a shorter lifetime resulting at a higher temperature. This relationship needs to be understood and extended to the highest possible temperature, the limit of which is presently unknown, to make conversion of heat to electric power most efficient. Very long lifetimes appear to be possible at lower temperatures suitable for space heating, cooking, and driving some chemical processes.

6.1.0 COMMERCIAL

Because the source of power comes from hydrogen, the main element in water, availability of this energy is unlimited. Because water is everywhere, no country can be denied access. Consequently, the energy cannot be controlled or metered. While this limits how profits can be made from selling the fuel, it also reduces conflicts over access to energy. Is not avoiding war as useful as making money? The commercial profit must come from selling the special material in which the nuclear reactions take place. The ability to make such material has yet to be mastered.

Because rare palladium has been found to support the process, concerns have been expressed about whether enough palladium is available to make power for worldwide use. This concern is not justified because very little palladium is required and because many other common metals are expected to work just as well. Besides, once unlimited energy becomes available, extraction of even rare metals becomes much cheaper. Nickel has now been made nuclear active using a special treatment. While nickel is cheaper and more plentiful than palladium, these factors are not important because so little of the nuclear-active material is required. In fact, once the phenomenon has been understood and mastered, even cheaper materials can be expected to function even more effectively.

Although money made from supplying LENR energy will be limited, the profit resulting from using this cheap energy in a plethora of applications and useful technologies not otherwise practical at the present time will be great.

An additional problem results from the difficulty of obtaining patent protection in the US. Not only are claims for such energy normally rejected but the large number of denied or unexamined patents creates a potential legal nightmare. Other countries have been much more willing to grant patent protection. Presently, methods for achieving success are being kept as trade secrets. This denies other people the ability to build on this success and advance the technology. The US government needs to be held accountable for slowing commercial development of this technology.

6.2.0 POLITICAL

Oil, its extraction, refinement, and distribution, controls politics in many countries. LENR is a serious threat to this wealth and political control. Consequently, political barriers against use of LENR energy are expected in oil-rich countries. In contrast, countries without this resource are expected to welcome this energy and use it to become free of influence by the petro-countries. The resulting conflict is expected to have serious and potentially devastating consequences in some parts of the world.

Even though LENR will provide an ideal energy for the future, reaching this eventual goal may be difficult and painful. The implications are as profound as was the discovery of fire. Transition from conventional energy sources to this new one will test the ability of the human race to survive. Not making the transition will doom mankind. That is our present choice.

References

1. E. K. Storms, *The science of low energy nuclear reaction*. (World Scientific, Singapore, 2007), 312 pages.
2. E. K. Storms, T. W. Grimshaw, Judging the validity of the Fleischmann-Pons effect. *J. Cond. Matter Nucl. Sci.* **3**, 9-30 (2010).
3. M. Fleischmann, S. Pons, M. Hawkins, Electrochemically induced nuclear fusion of deuterium. *J. Electroanal. Chem.* **261**, 301-308 and errata in **263**, 187-188 (1989).
4. B. Simon, *Undead science: Science studies and the afterlife of cold fusion*. (Rutgers University Press, New Brunswick, NJ, 2002), pp. 252.
5. C. G. Beaudette, *Excess heat: Why cold fusion research prevailed*. (Oak Grove Press/Infinite Energy, Concord, NH, 2000), 365 pages.
6. E. Mallove, *Fire from ice*. (John Wiley, NY, 1991).
7. P. A. Mosier-Boss, F. P. G. Forsley, F. Gordon, How the flawed journal review process impedes paradigm shifting discoveries. *J. Cond. Matter Nucl. Sci.* **12**, 1-12 (2013).
8. C. L. Frazier, Proceedings of new energy conference rejected by publisher. *Infinite Energy* **16**, 95, 15-16 (2011).
9. V. A. Chechin, V. A. Tsarev, M. Rabinowitz, Y. E. Kim, Critical review of theoretical models for anomalous effects in deuterated metals. *Int. J. Theo. Phys.* **33**, 617-670 (1994).
10. J. Rothwell, *Cold fusion and the future*. (www.LENR.org, 2007).
11. E. Storms, paper presented at the 6th International Conference on Cold Fusion, Progress in New Hydrogen Energy, Lake Toya, Hokkaido, Japan, October 13-18, 1996.
12. E. K. Storms, The nature of the energy-active state in Pd-D. *Infinite Energy*, **1**, 5/6, 77 (1995).
13. E. Storms, paper presented at the American Physical Society Winter Meeting, Austin Convention Center, Austin, TX, 2003.
14. K. C. Jordan, B. C. Blanke, W. A. Dudley, Half-life of tritium. *J. Inorg. Nucl. Chem.* **29**, 2129 (1967).
15. R. R. Adzic, D. Gervasio, I. Bae, B. Cahan, E. Yeager, paper presented at the 1st Annual Conference on Cold Fusion, University of Utah Research Park, Salt Lake City, UT, March 28-31, 1990.
16. D. A. Corrigan, E. W. Schneider, Tritium separation effects during heavy water electrolysis: implications for reported observations of cold fusion. *J. Electroanal. Chem.* **281**, 305 (1990).
17. P. G. Sona, F. Parmigiani, F. Barberis, A. Battaglia, R. Berti, G. Buzzanca, A. Capelli, D. Capra, M. Ferrari, Preliminary tests on tritium and neutrons in cold nuclear fusion within palladium cathodes. *Fusion Technol.* **17**, 713 (1990).

18. E. Brillas, G. Sardin, J. Casado, X. Doménech, J. Sánchez, paper presented at the 2nd Annual Conference on Cold Fusion, The Science of Cold Fusion, Como, Italy, June 29-July 4, 1991.

19. J. Sevilla, B. Escarpizo, F. Fernandez, F. Cuevas, C. Sanchez, paper presented at the 3rd International Conference on Cold Fusion, Frontiers of Cold Fusion, Nagoya, Japan, October 21-25, 1992.

20. G. R. Boucher, F. E. Collins, R. L. Matlock, Separation factors for hydrogen isotopes on palladium. *Fusion Technol.* **24**, 200 (1993).

21. D. Hodko, J. Bockris, Possible excess tritium production on Pd codeposited with deuterium. *J. Electroanal. Chem.* **353**, 33 (1993).

22. S. Szpak, P. A. Mosier-Boss, R. D. Boss, J. J. Smith, On the behavior of the Pd/D system: evidence for tritium production. *Fusion Technol.* **33**, 38 (1998).

23. Y. Iwamura, T. Itoh, I. Toyoda, Observation of anomalous nuclear effects in D_2-Pd system. *Trans. Fusion Technol.* **26**, 160-164 (1994).

24. T. N. Claytor, M. J. Schwab, D. J. Thoma, D. F. Teter, D. G. Tuggle, paper presented at the 7th International Conference on Cold Fusion, Vancouver, Canada, April 19-24, 1998.

25. D. D. Afonichev, E. G. Galkin, paper presented at the 15th International Conference on Condensed Matter Nuclear Science, Rome, Italy, October 5-9, 2009.

26. P. K. Iyengar, M. Srinivasan, *BARC studies in cold fusion*. (BARC, India, Bombay, 1989).

27. V. A. Romodanov, V. Savin, V. Elksnin, Y. Skuratnik, paper presented at the 4th International Conference on Cold Fusion, Lahaina, HI December 6-9, 1993.

28. V. A. Romodanov, Y. B. Skuratnik, A. K. Pokrovsky, paper presented at the 8th International Conference on Cold Fusion, Lerici (La Spezia), Italy, May 21-26, 2000.

29. V. A. Romodanov, paper presented at the 10th International Conference on Cold Fusion, Cambridge, MA, August 24-29, 2003.

30. T. Itoh, Y. Iwamura, N. Gotoh, I. Toyoda, paper presented at the 5th International Conference on Cold Fusion, Monte-Carlo, Monaco, April 9-13, 1995.

31. B. W. Clarke, B. M. Oliver, M. C. H. McKubre, F. L. Tanzella, P. Tripodi, Search for ^{3}He and ^{4}He in Arata-style palladium cathodes II: Evidence for tritium production. *Fusion Sci. & Technol.* **40**, 152-167 (2001).

32. W. B. Clarke, B. M. Oliver, Response to "Comments on 'Search for ^{3}He and ^{4}He in Arata-Style Palladium Cathodes II: Evidence for tritium production" (Lett. to Ed;). *Fusion Sci. Technol.* **41**, 153 (2002).

33. W. B. Clarke, B. M. Oliver, Response to "Comments on 'Search for ^{3}He and ^{4}He in Arata-style palladium cathodes I: A negative result' and

'search for ^{3}He and ^{4}He in Arata-style palladium cathodes II: Evidence for tritium production'". *Fusion Sci. & Technol.* **43**, 135 (2003).

34. M. C. H. McKubre, F. L. Tanzella, P. Tripodi, P. L. Hagelstein, paper presented at the 8th International Conference on Cold Fusion, Lerici (La Spezia), Italy, May 21-26, 2000.
35. Y. Arata, Y. C. Zhang, Helium (^{4}He, ^{3}He) within deuterated Pd-black. *Proc. Jpn. Acad., Ser. B* **73**, 1 (1997).
36. N. J. C. Packham, K. L. Wolf, J. C. Wass, R. C. Kainthla, J. O. M. Bockris, Production of tritium from D_2O electrolysis at a palladium cathode. *J. Electroanal. Chem.* **270**, 451 (1989).
37. J. O. M. Bockris, Addition to 'A review of the investigations of the Fleischmann-Pons phenomena'. *Fusion Technol.* **18**, 523 (1990).
38. J. O. M. Bockris, G. H. Lin, R. C. Kainthla, N. J. C. Packham, O. Velev, paper presented at the 1st Annual Conference on Cold Fusion, University of Utah Research Park, Salt Lake City, UT, March 28-31, 1990.
39. G. H. Lin, R. C. Kainthla, N. J. C. Packham, J. O. M. Bockris, Electrochemical fusion: a mechanism speculation. *J. Electroanal. Chem.* **280**, 207 (1990).
40. G. H. Lin, R. C. Kainthla, N. J. C. Packham, O. A. Velev, J. Bockris, On electrochemical tritium production. *Int. J. Hydrogen Energy* **15**, 537 (1990).
41. J. Bockris, C. Chien, D. Hodko, Z. Minevski, paper presented at the 3rd International Conference on Cold Fusion, Frontiers of Cold Fusion, Nagoya, Japan, October 21-25, 1992.
42. C.-C. Chien, D. Hodko, Z. Minevski, J. O. M. Bockris, On an electrode producing massive quantities of tritium and helium. *J. Electroanal. Chem.* **338**, 189-212 (1992).
43. E. Storms, C. L. Talcott, Electrolytic tritium production. *Fusion Technol.* **17**, 680 (1990).
44. G. Taubes, *Bad science: The short life and weird times of cold fusion*. (Random House, NY, 1993), pp. 503.
45. O. Reifenschweiler, Cold fusion and decrease of tritium radioactivity. *LENR-CANR.org* (2003).
46. O. Reifenschweiler, Reduced radioactivity of tritium in small titanium particles. *Phys. Lett. A* **184**, 149 (1994).
47. O. Reifenschweiler, paper presented at the 5th International Conference on Cold Fusion, Monte-Carlo, Monaco, April 9-13, 1995.
48. O. Reifenschweiler, Some experiments on the decrease of tritium radioactivity. *Fusion Technol.* **30**, 261 (1996).
49. O. Reifenschweiler, About the possibility of decreased radioactivity of heavy nuclei. *Fusion Technol.* **31**, 291 (1997).
50. E. K. Storms. (1990).

51. N. Hoffman, *A dialogue on chemically induced nuclear effects: A guide for the perplexed about cold fusion*. (American Nuclear Society, La Grange Park, IL, 1995).
52. E. K. Storms, C. Talcott-Storms, The effect of hydriding on the physical structure of palladium and on the release of contained tritium. *Fusion Technol*. **20**, 246 (1991).
53. J. Kopecky, *Atlas of neutron capture cross-sections*. (International Nuclear Data Committee, IAEA, Petten, Netherlands, 1997).
54. P. Jung, Ed., *Diffusion and Clustering of Helium in Noble Metals*, (Plenum Press, NY, 1991), pp. 59.
55. J. Xia, W. Hu, J. Yang, B. Ao, X. Wang, A comparative study of helium atom diffusion via an interstitial mechanism in nickel and palladium. *Phys. Stat. Sol. B* **243B**, 579-583 (2006).
56. G. C. Abell, L. K. Matson, R. H. Steinmeyer, R. C. Bowman Jr., B. M. Oliver, Helium release from aged palladium tritide. *Phys. Rev. B: Mater. Phys*. **41**, 1220 (1990).
57. W. J. Camp, Helium detrapping and release from metal tritides. *J. Vac. Sci. Technol*. **14**, 514-517 (1977).
58. M. Miles, B. F. Bush, J. J. Lagowski, Anomalous effects involving excess power, radiation, and helium production during D_2O electrolysis using palladium cathodes. *Fusion Technol*. **25**, 478 (1994).
59. M. C. McKubre, paper presented at the ARL Workshop, Adelphi, MD, June 29, 2010.
60. E. K. Storms, The status of cold fusion. *Naturwissenschaften* **97**, 861 (2010).
61. T. O. Passell, paper presented at the 6th International Conference on Cold Fusion, Progress in New Hydrogen Energy, Lake Toya, Hokkaido, Japan, October 13-18, 1996.
62. S. Krivit, Nuclear phenomena in low-energy nuclear reaction research. *Naturwissenschaften* **100**, 899-900 (2013).
63. E. K. Storms, Efforts to explain low-energy nuclear reactions. *Naturwissenschaften* **100**, 1103 (2013).
64. A. Widom, L. Larsen, Theoretical standard model rates of proton to neutron conversions near metallic hydride surfaces. *arXiv:nucl-th/0608059v2*, (2007).
65. Y. N. Srivastava, A. Widom, L. Larsen, in *Low-Energy Nuclear Reactions Sourcebook Volume 2,* J. Marwan, S. Krivit, Eds. (Oxford University Press, 2009).
66. Y. Arata, Y. C. Zhang, Anomalous production of gaseous ^{4}He at the inside of 'DS cathode' during D_2O-electrolysis. *Proc. Jpn. Acad., Ser. B* **75**, 281 (1999).
67. M. C. McKubre, F. Tanzella, Cold Fusion, LENR, CMNS, FPE: One perspective on the state of the science based on measurements made at SRI. *J. Cond. Matter Nucl. Sci*. **4**, 32-44 (2011).

68. T. O. Passell, paper presented at the 10th International Conference on Cold Fusion, Cambridge, MA, August 24-29, 2003.
69. Y. Arata, Y. C. Zhang, paper presented at the 8th International Conference on Cold Fusion, Lerici (La Spezia), Italy, May 21-26, 2000.
70. B. F. Bush, J. J. Lagowski, "Trace elements added to palladium by electrolysis in heavy water," *EPRI TP-108743* (EPRI, 1999).
71. M. Miles, paper presented at the 10th International Conference on Cold Fusion, Cambridge, MA, August 24-29, 2003.
72. G. L. Wendt, C. E. Irion, Experimental attempts to decompose tungsten at high temperatures. *Science* **55**, 422 (1922).
73. F. Paneth, K. Peters, On the transmutation of hydrogen to helium. *Naturwissenschaften* **14**, 956-962 (in German) (LA-TR-989-914) (1926).
74. F. Paneth, Neuere Versuche über die Verwandlung von Wasserstoff in Helium. *Naturwissenschaften* **15**, 379-379 (1927).
75. D. Albagli, R. Ballinger, V. Cammarata, X. Chen, R. M. Crooks, C. Fiore, M. P. J. Gaudreau, I. Hwang, C. K. Li, P. Linsay, S. C. Luckhardt, R. R. Parker, R. D. Petrasso, M. O. Schloh, K. W. Wenzel, M. S. Wrighton, Measurement and analysis of neutron and gamma-ray emission rates, other fusion products, and power in electrochemical cells having Pd cathodes. *J. Fusion Energy* **9**, 133 (1990).
76. A. Alessandrello, E. Bellotti, C. Cattadori, C. Antonione, G. Bianchi, S. Rondinini, S. Torchio, E. Fiorini, A. Guiliani, S. Ragazzi, L. Zanotti, C. Gatti, Search for cold fusion induced by electrolysis in palladium. *Il Nuovo Cimento* **A103**, 1617 (1990).
77. V. B. Brudanin, V. M. Bystritsky, V. G. Egorov, S. G. Stetsenko, I. A. Yutlandov, Search for the cold fusion d(d,^{4}He) in electrolysis of D_2O. *Phys. Lett. A* **151**, 543 (1990).
78. J. I. Matsuda, T. Matsumoto, K. Nagao, An attempt to detect (3)He from the cold nuclear fusion. *J. Geochem.* **24**, 379 (1990).
79. J. R. Morrey, M. W. Caffee, H. Farrar IV, N. J. Hoffman, G. B. Hudson, R. H. Jones, M. D. Kurz, J. Lupton, B. M. Oliver, B. V. Ruiz, J. F. Wacker, A. van Veen, Measurements of helium in electrolyzed palladium. *Fusion Technol.* **18**, 659 (1990).
80. B. F. Bush, J. J. Lagowski, M. H. Miles, G. S. Ostrom, Helium production during the electrolysis of D_2O in cold fusion experiments. *J. Electroanal. Chem.* **304**, 271-278 (1991).
81. M. Miles, B. F. Bush, G. S. Ostrom, J. J. Lagowski, paper presented at the 2nd Annual Conference on Cold Fusion, The Science of Cold Fusion, Como, Italy, June 29-July 4, 1991.
82. M. Miles, B. F. Bush, paper presented at the 3rd International Conference on Cold Fusion, Frontiers of Cold Fusion, Nagoya, Japan, October 21-25, 1992.

83. M. H. Miles, R. A. Hollins, B. F. Bush, J. J. Lagowski, R. E. Miles, Correlation of excess power and helium production during D_2O and H_2O electrolysis using palladium cathodes. *J. Electroanal. Chem.* **346**, 99-117 (1993).
84. M. H. Miles, B. F. Bush, paper presented at the 4th International Conference on Cold Fusion, Lahaina, HI, December 6-9, 1993.
85. S. E. Jones, Current issues in cold fusion research: heat, helium, tritium, and energetic particles. *Surf. Coatings Technol.* **51**, 283 (1992).
86. M. Miles, C. P. Jones, Cold fusion experimenter Miles responds to critic. *21st Century Sci. & Technol.* **Spring**, 75 (1992).
87. M. H. Miles, B. F. Bush, D. E. Stilwell, Calorimetric principles and problems in measurements of excess power during Pd-D_2O electrolysis. *J. Phys. Chem.* **98**, 1948 (1994).
88. S. E. Jones, L. D. Hansen, Examination of claims of Miles et al. in Pons-Fleischmann-type cold fusion experiments. *J. Phys. Chem.* **99**, 6966 (1995).
89. M. H. Miles, Reply to 'An assessment of claims of excess heat in cold fusion calorimetry'. *J. Phys. Chem. B* **102**, 3648 (1998).
90. M. H. Miles, Reply to 'Examination of claims of Miles et al. in Pons-Fleischmann-type cold fusion experiments'. *J. Phys. Chem. B* **102**, 3642 (1998).
91. B. Y. Liaw, B. E. Liebert, paper presented at the 3rd International Conference on Cold Fusion, Frontiers of Cold Fusion, Nagoya, Japan, October 21-25, 1992.
92. B. Y. Liaw, P.-L. Tao, P. Turner, B. E. Liebert, paper presented at the 8th World Hydrogen Energy Conf., Honolulu, HI, July 22-27, 1990.
93. B. Y. Liaw, P.-L. Tao, B. E. Liebert, paper presented at the 2nd Annual Conference on Cold Fusion, The Science of Cold Fusion, Como, Italy, June 29-July 4, 1991.
94. B. Y. Liaw, P.-L. Tao, P. Turner, B. E. Liebert, Elevated-temperature excess heat production in a Pd + D system. *J. Electroanal. Chem.* **319**, 161 (1991).
95. B. Y. Liaw, P.-L. Tao, B. E. Liebert, Helium analysis of palladium electrodes after molten salt electrolysis. *Fusion Technol.* **23**, 92-97 (1993).
96. Y. Ding, B. Y. Liaw, paper presented at the 9th International Conference on Solid State Ionics, The Hague, The Netherlands, Sept. 12-17, 1993.
97. A. Zywocinski, H.-L. Li, A. A. Tuinman, P. Campbell, J. Q. Chambers, W. A. Van Hook, Analysis for light atoms produced in the bulk phase of a tubular palladium/ silver alloy cathode working electrode. *J. Electroanal. Chem.* **319**, 195 (1991).

98. H. Sakaguchi, G. Adachi, K. Nagao, paper presented at the 3rd International Conference on Cold Fusion, Frontiers of Cold Fusion, Nagoya, Japan, October 21-25, 1992.
99. E. Yamaguchi, T. Nishioka, paper presented at the 3rd International Conference on Cold Fusion, Frontiers of Cold Fusion, Nagoya, Japan, October 21-25, 1992.
100. E. Yamaguchi, T. Nishioka, Helium-4 production and its correlation with heat evolution. *Oyo Butsuri* **62**, 712 (in Japanese) (1993).
101. E. Yamaguchi, T. Nishioka, Helium-4 production from deuterated palladium. *Kakuyuogo Kenkyo* **69**, 743 (in Japanese) (1993).
102. Q. F. Zhang, Q. Q. Gou, Z. H. Zhu, B. L. Xio, J. M. Lou, F. S. Liu, J. X. S., Y. G. Ning, H. Xie, Z. G. Wang, paper presented at the 3rd International Conference on Cold Fusion, Frontiers of Cold Fusion, Nagoya, Japan, October 21-25, 1992.
103. D. Gozzi, R. Caputo, P. L. Cignini, M. Tomellini, G. Gigli, G. Balducci, E. Cisbani, S. Frullani, F. Garibaldi, M. Jodice, G. M. Urciuoli, paper presented at the 4th International Conference on Cold Fusion, Lahaina, HI, December 6-9, 1993.
104. D. Gozzi, R. Caputo, P. L. Cignini, M. Tomellini, G. Gigli, G. Balducci, E. Cisbani, S. Frullani, F. Garibaldi, M. Jodice, G. M. Urciuoli, paper presented at the 4th International Conference on Cold Fusion, Lahaina, HI, December 6-9, 1993.
105. D. Gozzi, R. Caputo, P. L. Cignini, M. Tomellini, G. Gigli, G. Balducci, E. Cisbani, S. Frullani, F. Garibaldi, M. Jodice, G. M. Urciuoli, Calorimetric and nuclear byproduct measurements in electrochemical confinement of deuterium in palladium. *J. Electroanal. Chem.* **380**, 91-107 (1995).
106. D. Gozzi, R. Caputo, P. L. Cignini, M. Tomellini, G. Gigli, G. Balducci, E. Cisbani, S. Frullani, F. Garibaldi, M. Jodice, G. M. Urciuoli, Quantitative measurements of helium-4 in the gas phase of Pd + D2O electrolysis. *J. Electroanal. Chem.* **380**, 109-116 (1995).
107. D. Gozzi, P. L. Cignini, R. Caputo, M. Tomellini, G. Balducci, G. Gigli, E. Cisbani, S. Frullani, F. Garibaldi, M. Jodice, G. M. Urciuoli, Excess heat and nuclear byproduct measurements in electrochemical confinement of deuterium in palladium. *J. Electroanal. Chem.* **380**, 91 (1995).
108. T. Aoki, Y. Kurata, H. Ebihara, N. Yoshikawa, Helium and tritium concentrations in electrolytic cells. *Trans. Fusion Technol.* **26**, 214-220 (1994).
109. M. H. Miles, B. F. Bush, Heat and helium measurements in deuterated palladium. *Trans. Fusion Technol.* **26**, 156-159 (1994).
110. M. Miles, K. B. Johnson, M. A. Imam, paper presented at the 6th International Conference on Cold Fusion, Progress in New Hydrogen Energy, Lake Toya, Hokkaido, Japan, October 13-18, 1996.

111. E. Botta, T. Bressani, D. Calvo, C. Fanara, F. Lazzi, paper presented at the 6th International Conference on Cold Fusion, Progress in New Hydrogen Energy, Lake Toya, Hokkaido, Japan, October 13-18, 1996.
112. E. Botta, R. Bracco, T. Bressani, D. Calvo, V. Cela, C. Fanara, U. Ferracin, F. Iazzi, paper presented at the 5th International Conference on Cold Fusion, Monte-Carlo, Monaco, April 9-13, 1995.
113. A. Coehn, Nachweis Von Protonen in Metallen. *Z. Elektrochem.* **35**, 676 (1929).
114. M. Cola, E. Del Gindice, A. De Ninno, G. Preparata, paper presented at the 8th International Conference on Cold Fusion, Lerici (La Spezia), Italy, May 21-26, 2000.
115. D. Gozzi, F. Cellucci, P. L. Cignini, G. Gigli, M. Tomellini, E. Cisbani, S. Frullani, G. M. Urciuoli, X-ray, heat excess and ^4He in the D/Pd system. *J. Electroanal. Chem.* **452**, 251-271 (1998).
116. Y. Arata, Y.-C. Zhang, Deuterium nuclear reaction process within solid. *J. New Energy* **2**, 27 (1997).
117. Y. Arata, C. Zhang, Presence of helium ($^2He_4$, $^2He_3$) confirmed in deuterated Pd-black by the "vi-effect" in a "closed QMS" environment. *Proc. Japan. Acad. B* **73**, 62 (1997).
118. Y. Arata, Y.-C. Zhang, Solid-state plasma fusion ('cold fusion'). *J. High Temp. Soc.* **23** (special volume), 1-56 (1997).
119. Y. Arata, C. Zhang, Presence of helium ($^2He_4$, $^2He_3$) confirmed in highly deuterated Pd-black by the new detecting methodology. *J. High Temp. Soc.* **23**, 110 (in Japanese) (1997).
120. Y. Arata, Z.-Y. Chang, Presence of helium($^2He_4$, $^2He_3$) confirmed in deuterated Pd-black by the "Vi-effect" in a closed QMS" environment. *Proc. Japan Acad.* **73**, 62 (1997).
121. Y. Arata, Y. C. Zhang, Critical condition to induce 'excess energy' within [DS-H_2O] cell. *Proc. Jpn. Acad., Ser. B* **75 Ser. B**, 76 (1999).
122. Y. Arata, Y. C. Zhang, Definitive difference between [DS-D_2O] and [Bulk-D_2O] cells in 'deuterium-reaction'. *Proc. Jpn. Acad., Ser. B* **75 Ser. B**, 71 (1999).
123. Y. Arata, Y. C. Zhang, Observation of anomalous heat release and helium-4 production from highly deuterated fine particles. *Jpn. J. Appl. Phys. Part 2* **38**, L774 (1999).
124. B. F. Bush, J. J. Lagowski, paper presented at the 7th International Conference on Cold Fusion, Vancouver, Canada, April 19-24, 1998.
125. G. S. Qiao, X. L. Han, L. C. Kong, S. X. Zheng, H. F. Huang, Y. J. Yan, Q. L. Wu, Y. Deng, S. L. Lei, X. Z. Li, paper presented at the 7th International Conference on Cold Fusion, Vancouver, Canada, April 19-24, 1998.
126. X. Z. Li, W. Z. Yue, G. S. Huang, H. Shi, L. Gao, M. L. Liu, F. S. Bu, paper presented at the 6th International Conference on Cold Fusion,

Progress in New Hydrogen Energy, Lake Toya, Hokkaido, Japan, October 13-18, 1996.

127. X. Z. Li, W. Z. Yue, G. S. Huang, H. Shi, L. Gao, M. L. Liu, F. S. Bu, "Excess heat" measurement in gas-loading D/Pd system. *J. New Energy* **1**, 34 (1996).

128. A. Takahashi, paper presented at the 7th International Conference on Cold Fusion, Vancouver, Canada, April 19-24, 1998.

129. M. C. H. McKubre, F. L. Tanzella. (1999).

130. M. C. McKubre, F. Tanzella, P. Tripodi, Evidence of d-d fusion products in experiments conducted with palladium at near ambient temperatures. *Trans. Am. Nucl. Soc.* **83**, 367 (2000).

131. Y. Isobe, S. Uneme, K. Yabuta, H. Mori, T. Omote, S. Ucda, K. Ochiai, H. Miyadera, A. Takahashi, paper presented at the 8th International Conference on Cold Fusion, Lerici (La Spezia), Italy, May 21-26, 2000.

132. M. Matsunaka, Y. Isobe, S. Ueda, K. Yabuta, T. Ohishi, H. Mori, A. Takahashi, paper presented at the 9th International Conference on Cold Fusion, Condensed Matter Nuclear Science, Tsinghua Univ., Beijing, China, May 19-24, 2002.

133. M. C. H. McKubre, paper presented at the 10th International Conference on Cold Fusion, Cambridge, MA, August 24-29, 2003.

134. M. Miles, M. A. Imam, M. Fleischmann, Excess heat and helium production in the palladium-boron system. *Trans. Am. Nucl. Soc.* **83**, 371-372 (2000).

135. M. Miles, M. A. Imam. (United States, 6764561 (2004)).

136. A. B. Karabut, E. A. Karabut, Experimental results on Excess Heat Power, Impurity Nuclides and X-ray Production in Experiments with a High-Voltage Electric Discharge System. *J. Cond. Matter Nucl. Sci.* **6**, 199-216 (2012).

137. A. DeNinno, A. Franttolillo, A. Rizzo, F. Scaramuzzi, C. Alessandrini, paper presented at the 8th International Conference on Cold Fusion, Lerici (La Spezia), Italy, May 21-26, 2000.

138. A. DeNinno, A. Frattolillo, A. Rizzo, E. Del Giudice, G. Preparata, "Experimental evidence of ^{4}He production in a cold fusion experiment," (ENEA - Unita Tecnico Scientfica Fusione Centro Ricerche Frascati, Roma, 2002).

139. A. DeNinno, A. Frattolillo, A. Rizzo, E. Del Gindice, paper presented at the 10th International Conference on Cold Fusion, Cambridge, MA, August 24-29, 2003.

140. A. DeNinno, A. Frattolillo, A. Rizzo, E. Del Giudice, paper presented at the 5th International Workshop on Anomalies in Hydrogen/Deuterium Loaded Metals, Asti, Italy, March 19-21, 2004.

141. R. Stringham, paper presented at the Am. Phys. Soc., Seattle, WA, March 11-15, 2001.

142. R. Stringham, paper presented at the 10th International Conference on Cold Fusion, Cambridge, MA, August 24-29, 2003.

143. R. George, paper presented at the 6th International Workshop on Anomalies in Hydrogen/Deuterium Loaded Metals, Siena, Italy, May 13-15, 2005.

144. M. Apicella, E. Castagna, L. Capobianco, L. D'Aulerio, G. Mazzitelli, F. Sarto, A. Rosada, E. Santoro, V. Violante, M. C. McKubre, F. Tanzella, C. Sibilia, paper presented at the 12th International Conference on Condensed Matter Nuclear Science, Yokohama, Japan, November 27-December 2, 2005.

145. Y. Arata, Y. C. Zhang, X. F. Wang, paper presented at the 15th International Conference on Condensed Matter Nuclear Science, Rome, Italy, October 5-9, 2009.

146. E. K. Storms, paper presented at the American Physical Society, Denver, CO, March 5-9, 2007.

147. W. J. M. F. Collis, paper presented at the 6th International Workshop on Anomalies in Hydrogen/Deuterium loaded Metals, Siena, Italy, May 13-16, 2005.

148. G. Miley, P. J. Shrestha, paper presented at the 12th International Conference on Condensed Matter Nuclear Science, Yokohama, Japan, November 27-December 2, 2005.

149. G. Miley, paper presented at the 14th International Conference on Condensed Matter Nuclear Science, Washington DC, August 10-15, 2008.

150. A. Kitamura, Y. T., T. Nohmi, S. Y., Y. Miyoshi, A. Taniike, Y. Furuyama, A. Takahashi, "MIDE (Metal Deuteride Energy) Project 2009," (2009).

151. G. Miley, X. Yang, H. Hora, Ultra-High Density Deuteron-cluster Electrode for Low-energy Nuclear Reactions. *J. Cond. Matter Nucl. Sci.* **4**, 256-268 (2011).

152. M. Srinivasan, G. Miley, E. K. Storms, in *Nuclear Energy Encyclopedia: Science, Technology, and Applications,* S. Krivit, J. H. Lehr, T. B. Kingery, Eds. (John Wiley & Sons, Hoboken, NJ, 2011), pp. 503-539.

153. J.-P. Biberian, Biological transmutations: historical perspective. *J. Cond. Matter Nucl. Sci.* **7**, 11-15 (2012).

154. P. A. Mosier-Boss, A Review on nuclear products generated during low-energy nuclear reactions (LENR). *J. Cond. Matter Nucl. Sci.* **6**, 135-148 (2012).

155. D. Nagel, Characteristics and energetics of craters in LENR experimental materials. *J. Cond. Matter Nucl. Sci.* **10**, 1-14 (2013).

156. T. Hioki, N. Sugimoto, T. Nishi, A. Itoh, T. Motohiro, paper presented at the 17th International Conference on Condensed Matter Nuclear Science, Daejeon, Korea, August 12-17, 2012.

157. D. Afonichev, paper presented at the 10th International Conference on Cold Fusion, Cambridge, MA, August 24-29, 2003.
158. L. Daddi, paper presented at the Asti Workshop on Anomalies in Hydrogen/Deuterium Loaded Metals, Rocca d'Arazzo, Italy, November 27-30, 1997.
159. G. Miley, G. Narne, T. Woo, Use of combined NAA and SIMS analyses for impurity level isotope detection. *J. Radioanalytical and Nuclear Chemistry* **263**, 691-696 (2005).
160. T. O. Passell, paper presented at the 7th International Conference on Cold Fusion, Vancouver, Canada, April 19-24, 1998.
161. T. O. Passell, R. George, paper presented at the 8th International Conference on Cold Fusion, Lerici (La Spezia), Italy, May 21-26, 2000.
162. A. Rosada, E. Santoro, F. Sarto, V. Violante, P. Avino, paper presented at the 15th International Conference on Condensed Matter Nuclear Science, Rome, Italy, October 5-9, 2009.
163. H. Kozima, paper presented at the 10th International Conference on Cold Fusion, Cambridge, MA, August 24-29, 2003.
164. G. Miley, H. Hora, A. G. Lipson, H. Leon, P. J. Shrestha, paper presented at the 8th International Workshop on Anomalies in Hydrogen/Deuterium Loaded Metals, Catania, Italy, October 13-18, 2007.
165. A. Karabut, paper presented at the 11th International Conference on Cold Fusion, Marseilles, France, October 31-November 5, 2004.
166. M. Ohta, A. Takahashi, paper presented at the 9th International Conference on Cold Fusion, Condensed Matter Nuclear Science, Tsinghua Univ., Beijing, China, May 19-24, 2002.
167. A. Takahashi, M. Ohta, T. Mizuno, Radiation-less fission products by selective channel low-energy photofission for A>100 elements. *Trans. Am. Nucl. Soc.* **83**, 369 (2000).
168. A. Takahashi, M. Ohta, T. Mizuno, paper presented at the 8th International Conference on Cold Fusion, Lerici (La Spezia), Italy, May 21-26, 2000.
169. G. Miley, paper presented at the Asti Workshop on Anomalies in Hydrogen/Deuterium Loaded Metals, Rocca d'Arazzo, Italy, November 27-30, 1997.
170. B. F. Bush, Cold fusion/cold fission to account for radiation remediation. *J. New Energy* **2**, 32 (1997).
171. I. B. Savvatimova, A. B. Karabut, Nuclear reaction products detected at the cathode after a glow discharge in deuterium. *Poverkhnost (Surface)*, 63 (in Russian) (1996).
172. T. Mizuno, T. Ohmori, M. Enyo, Isotopic changes of the reaction products induced by cathodic electrolysis in Pd. *J. New Energy* **1**, 31 (1996).

173. H. Hora, J. A. Patterson, The d and p reactions in low-energy nuclear fusion, transmutation, and fission. *Trans. Amer. Nucl. Soc.* **76**, 144 (1996).

174. I. Savvatimova, Y. Kucherov, A. Karabut, Cathode material change after deuterium glow discharge experiments. *Trans. Fusion Technol.* **26**, 389-394 (1994).

175. I. Savvatimova, Y. Kucherov, A. Karabut, paper presented at the 4th International Conference on Cold Fusion, Lahaina, HI, December 6-9, 1993.

176. S. Szpak, P. A. Mosier-Boss, F. Gordon, paper presented at the 11th International Conference on Cold Fusion, Marseilles, France, October 31-November 5, 2004.

177. G. H. Miley, P. J. Shrestha, paper presented at the 10th International Conference on Cold Fusion, Cambridge, MA, August 24-29, 2003.

178. G. H. Miley, paper presented at the 7th International Conference on Cold Fusion, Vancouver, Canada, April 19-24, 1998.

179. H. Hora, G. H. Miley, J. C. Kelly, G. Salvaggi, A. Tate, F. Osman, R. Castillo, Proton-metal reactions in thin films with Boltzmann distribution similar to nuclear astrophysics. *Fusion Technol.* **36**, 331 (1999).

180. M. R. Swartz, Three physical regions of anomalous activity in deuterated palladium. *Infinite Energy* **14**, 81, 19-31 (2008).

181. D. Baranov, Y. Bazhutov, N. Khokhov, V. P. Koretsky, A. B. Kuznetsov, Y. Skuratnik, N. Sukovatkin, paper presented at the 4th International Conference on Cold Fusion, Lahaina, HI, December 6-9, 1993.

182. I. Savvatimova, paper presented at the 13th International Conference on Condensed Matter Nuclear Science, Sochi, Russia, June 25-July 1, 2007.

183. I. Savvatimova, G. Savvatimov, A. A. Kornilova, paper presented at the 13th International Conference on Condensed Matter Nuclear Science, Sochi, Russia, June 25-July 1, 2007.

184. A. Takahashi, F. Celani, Y. Iwamura, paper presented at the 12th Condensed Matter Nuclear Science, Yokohama, Japan, November 27-December 2, 2005.

185. V. Vysotskii, A. Odintsov, V. N. Pavlovich, A. Tashirev, A. A. Kornilova, paper presented at the 11th International Conference on Cold Fusion, Marseilles, France, October 31-November 5, 2004.

186. D. V. Filippov, L. I. Urutskoev, A. A. Rukhadze, paper presented at the 11th International Conference on Cold Fusion, Marseilles, France, October 31-November 5, 2004.

187. J. Dash, D. Chicea, paper presented at the 10th International Conference on Cold Fusion, Cambridge, MA, August 24-29, 2003.

188. J. Dash, I. Savvatimova, S. Frantz, E. Weis, H. Kozima, paper presented at the 9th International Conference on Cold Fusion, Beijing, China, May 19-24, 2002.
189. H. Yamada, K. Uchiyama, N. Kawata, Y. Kurisawa, M. Nakamura, Producing a radioactive source in a deuterated palladium electrode under direct-current glow discharge. *Fusion Technol.* **39**, 253 (2001).
190. M. Bernardini, C. Manduchi, G. Mengoli, G. Zannoni, paper presented at the 8th International Conference on Cold Fusion, Lerici (La Spezia), Italy, May 21-26, 2000.
191. K. Nakamura, Y. Kishimoto, I. Ogura, Element conversion by arcing in aqueous solution. *J. New Energy* **2**, 53-55 (1997).
192. R. Notoya, T. Ohnishi, Y. Noya, Nuclear reactions caused by electrolysis in light and heavy water solutions. *J. New Energy* **1**, 40-45 (1996).
193. A. Michrowski, Advanced transmutation processes and their application for the decontamination of radioactive nuclear waste. *J. New Energy* **1**, 122 (1996).
194. R. T. Bush, Electrolytic stimulated cold nuclear synthesis of strontium from rubidium. *J. New Energy* **1**, 28 (1996).
195. I. Savvatimova, A. Karabut, paper presented at the 5th International Conference on Cold Fusion, Monte-Carlo, Monaco, April 9-13, 1995.
196. R. Notoya, paper presented at the 5th International Conference on Cold Fusion, Monte-Carlo, Monaco, April 9-13, 1995.
197. T. Mizuno, K. Inoda, T. Akimoto, K. Azumi, M. Kitaichi, K. Kurokawa, T. Ohmori, M. Enyo, Formation of ^{197}Pt radioisotopes in solid state electrolyte treated by high temperature electrolysis in D_2 gas. *Infinite Energy* **1**, 4, 9 (1995).
198. R. T. Bush, R. D. Eagleton, paper presented at the 4th International Conference on Cold Fusion, Lahaina, HI, December 6-9, 1993.
199. E. G. Campari, S. Focardi, V. Gabbani, V. Montalbano, F. Piantelli, E. Porcu, E. Tosti, S. Veronesi, paper presented at the 8th International Conference on Cold Fusion, Lerici (La Spezia), Italy, May 21-26, 2000.
200. S. Focardi, V. Gabbani, V. Montalbano, F. Piantelli, S. Veronesi, paper presented at the Asti Workshop on Anomalies in Hydrogen/Deuterium Loaded Metals, Rocca d'Arazzo, Italy, November 27-30, 1997.
201. A. Rossi. (WO2011/0005506 A1).
202. R. E. Godes. (Profusion Energy, Inc., USA, US 2007/0206715 A1).
203. A. Rossi. (WO 2009/125444 A1).
204. R. Kurup, P. A. Kurup, Actinidic archaea mediates biological transmutation in human systems: experimental evidence. *Adv. in Nat. Sci.* **5**, 47-49 (2012).

205. V. Vysotskii, A. A. Kornilova, Low-energy nuclear reactions and transmutation of stable and radioactive isotopes in growing biological systems. *J. Cond. Matter Nucl. Sci.* **4**, 146-160 (2011).
206. V. Vysotskii, A. B. Tashyrev, A. A. Kornilova, in *ACS Symposium Series 998, Low-Energy Nuclear Reactions Sourcebook,* J. Marwan, S. B. Krivit, Eds. (American Chemical Society, Washington, DC, 2008), pp. 295-309.
207. V. Vysotskii, A. A. Kornilova, A. Tashirev, J. Kornilova, paper presented at the 12th Condensed Matter Nuclear Science, Yokohama, Japan, November 27-December 2, 2005.
208. A. Triassi, paper presented at the 11th International Conference on Cold Fusion, Marseilles, France, October 31-November 5, 2004.
209. V. Vysotskii, A. A. Kornilova, *Nuclear fusion and transmutation of isotopes in biological systems.* (MIR Publishing House, Russia, 2003).
210. V. Vysotskii, V. Shevel, A. Tashirev, A. A. Kornilova, paper presented at the 10th International Conference on Cold Fusion, Cambridge, MA, August 24-29, 2003.
211. V. I. Vysotskii, A. A. Kornilova, I. I. Samoylenko, G. A. Zykov, paper presented at the 9th International Conference on Cold Fusion, Condensed Matter Nuclear Science, Tsinghua Univ., Beijing, China, May 19-24, 2002.
212. A. A. Kornilova, V. I. Vysotskii, G. A. Zykov, paper presented at the 9th International Conference on Cold Fusion, Condensed Matter Nuclear Science, Tsinghua Univ., Beijing, China, May 19-24, 2002.
213. V. Vysotskii, A. A. Kornilova, I. I. Samoylenko, Z. G. A., Observation and mass-spectrometry. Study of controlled transmutation of intermediate mass isotopes in growing biological cultures. *Infinite Energy* **6**, 36, 64-68 (2001).
214. V. Vysotskii, A. A. Kornilova, I. I. Samoylenko, G. A. Zykov, paper presented at the 8th International Conference on Cold Fusion, Lerici (La Spezia), Italy, May 21-26, 2000.
215. M. S. Benford, Biological nuclear reactions: empirical data describes unexplained SHC phenomenon. *J. New Energy* **3**, 19 (1999).
216. V. Vysotskii, A. A. Kornilova, I. I. Samoylenko, Experimental discovery and investigation of the phenomenon of nuclear transmutation of isotopes in growing biological cultures. *Infinite Energy* **2**, 10, 63-66 (1996).
217. V. I. Vysotskii, A. A. Kornilova, I. I. Samoylenko, paper presented at the 6th International Conference on Cold Fusion, Progress in New Hydrogen Energy, Lake Toya, Hokkaido, Japan, October 13-18, 1996.
218. H. Kozima, K. Hiroe, M. Nomura, M. Ohta, On the elemental transmutation in biological and chemical systems. *Cold Fusion* **17**, (1996).

219. P. Thompkins, C. Byrd, *The secret life of plants*. (Penguin Books, New York, 1993).
220. H. Komaki, paper presented at the 4th International Conference on Cold Fusion, Lahaina, December 6-9, 1993.
221. H. Komaki, paper presented at the 3rd International Conference on Cold Fusion, Frontiers of Cold Fusion, Nagoya, Japan, October 21-25, 1992.
222. C. L. Kervran, *Biological transmutation*. (Beekman Publishers, Inc, 1980).
223. C. L. Kervran, *Biological transmutations*. (Swan House Publishing Co., 1972).
224. H. Komaki, Formation de protines et variations minerales par des microorganismes en milieu de culture, sort avec or sans potassium, sort avec ou sans phosphore. *Revue de Pathologie Comparee* **69**, 83 (1969).
225. H. Komaki, Production de protein par 29 souches de microorganismes et augmentation du potassium en milieu de culture sodique sans potassium. *Revue de Pathologie Comparee* **67**, 213 (1967).
226. C. L. Kervran, Transmutations biologiques, metabolismes aberrants de l'asote, le potassium et le magnesium. *Librairie Maloine S. A, Paris*, (1963).
227. Y. Iwamura, H. Itoh, N. Gotoh, M. Sakano, I. Toyoda, H. Sakata, Detection of anomalous elements, X-ray and excess heat induced by continuous diffusion of deuterium through multi-layer cathode (Pd/CaO/Pd). *Infinite Energy* **4**, 20, 56 (1998).
228. Y. Iwamura, T. Itoh, N. Gotoh, M. Sakano, I. Toyoda, H. Sakata, paper presented at the 7th International Conference on Cold Fusion, Vancouver, Canada, April 19-24, 1998.
229. Y. Iwamura, T. Itoh, N. Gotoh, I. Toyoda, Detection of anomalous elements, X-ray, and excess heat in a D_2-Pd system and its interpretation by the electron-induced nuclear reaction model. *Fusion Technol.* **33**, 476 (1998).
230. Y. Iwamura, T. Itoh, N. Yamazaki, H. Yonemura, K. Fukutani, D. Sekiba, Recent advances in deuterium permeation transmutation experiments. *J. Cond. Matter Nucl. Sci.* **10**, 63-71 (2013).
231. Y. Iwamura, T. Itoh, M. Sakano, paper presented at the 8th International Conference on Cold Fusion, Lerici (La Spezia), Italy, May 21-26, 2000.
232. Y. Iwamura, M. Sakano, T. Itoh, Elemental analysis of Pd complexes: effects of D_2 gas permeation. *Jpn. J. Appl. Phys. A* **41**, 4642-4650 (2002).
233. Y. Iwamura, T. Itoh, M. Sakano, S. Sakai, paper presented at the 9th International Conference on Cold Fusion, Beijing, China, May 19-25, 2002.
234. Y. Iwamura, T. Itoh, M. Sakano. (Mitsubishi Heavy Industries, Ltd., U.S.A., US 2002/0080903 A1).

235. Y. Iwamura, T. Itoh, M. Sakano, S. Sakai, S. Kuribayashi, paper presented at the 10th International Conference on Cold Fusion, Cambridge, MA, August 24-29, 2003.

236. Y. Iwamura, T. Itoh, M. Sakano, N. Yamazaki, S. Kuribayashi, Y. Terada, T. Ishikawa, J. Kasagi, paper presented at the 11th International Conference on Condensed Matter Nuclear Science, Marseilles, France, October 31-November 5, 2004.

237. Y. Iwamura, T. Itoh, N. Yamazaki, J. Kasagi, Y. Terada, T. Ishikawa, D. Sekiba, H. Yonemura, K. Fukutani, Observation of low energy nuclear transmutation reactions induced by deuterium permeation through multilayer Pd and CaO thin film. *J. Cond. Matter Nucl. Sci.* **4**, 132-144 (2011).

238. A. Kitamura, R. Nishio, H. Iwai, R. Satoh, A. Taniike, Y. Furuyama, paper presented at the 12th International Conference on Condensed Matter Nuclear Science, Yokohama, Japan, November 27-December 2, 2005.

239. A. B. Karabut, Y. R. Kucherov, I. B. Savvatimova, paper presented at the 3rd International Conference on Cold Fusion, Frontiers of Cold Fusion, Nagoya, Japan, October 21-25, 1992.

240. A. B. Karabut, Y. R. Kucherov, I. B. Savvatimova, Nuclear product ratio for glow discharge in deuterium. *Phys. Lett. A* **170**, 265-272 (1992).

241. I. B. Savvatimova, A. B. Karabut, Radioactivity of palladium cathodes after irradiation in a glow discharge. *Poverkhnost (Surface)*, 76 (in Russian) (1996).

242. A. Karabut, Y. Kucherov, I. Savvatimova, Possible nuclear reactions mechanisms at glow discharge in deuterium. *J. New Energy* **1**, 20 (1996).

243. A. B. Karabut, paper presented at the 8th International Conference on Cold Fusion, Lerici (La Spezia), Italy, May 21-26, 2000.

244. A. B. Karabut, A. G. Lipson, A. S. Roussetsky, paper presented at the 8th International Conference on Cold Fusion, Lerici (La Spezia), Italy, May 21-26, 2000.

245. A. B. Karabut, paper presented at the 9th International Conference on Cold Fusion, Condensed Matter Nuclear Science, Tsinghua Univ., Beijing, China, May 19-24, 2002.

246. A. G. Lipson, A. B. Karabut, A. S. Roussetsky, paper presented at the 9th International Conference on Cold Fusion, Condensed Matter Nuclear Science, Tsinghua Univ., Beijing, China, May 19-24, 2002.

247. A. Karabut, paper presented at the 11th International Conference on Emerging Nuclear Energy Systems, Albuquerque, NM, September 29-October 4, 2004.

248. A. B. Karabut, paper presented at the 10th International Conference on Cold Fusion, Cambridge, MA, August 24-29, 2003.

249. A. Karabut, paper presented at the 12th International Conference on Condensed Matter Nuclear Science, Yokohama, Japan, November 27-December 2, 2005.
250. A. B. Karabut, paper presented at the 13th International Conference on Condensed Matter Nuclear Science, Sochi, Russia, June 25-July 1, 2007.
251. I. Savvatimova, paper presented at the 7th International Conference on Cold Fusion, Vancouver, Canada, April 19-24, 1998.
252. I. Savvatimova, D. V. Gavritenkov, paper presented at the 11th International Conference on Cold Fusion, Marseilles, France, October 31-November 5, 2004.
253. I. B. Savvatimova, A. D. Senchukova, I. P. Chernov, paper presented at the 6th International Conference on Cold Fusion, Lake Toya, Japan, October 13-18, 1996.
254. I. Savvatimova, D. V. Gavritenkov, paper presented at the 12th International Conference on Condensed Matter Nuclear Science, Yokohama, Japan, November 27-December 2, 2005.
255. V. Muromtsev, V. Platonov, I. B. Savvatimova, paper presented at the 12th International Conference on Condensed Matter Nuclear Science, Yokohama, Japan, November 27-December 2, 2005.
256. I. Savvatimova, paper presented at the 8th International Workshop on Anomalies in Hydrogen/Deuterium Loaded Metals, Catania, Italy, October 13-18, 2007.
257. I. B. Savvatimova, Transmutation of Elements in Low-energy Glow Discharge and the Associated Processes. *J. Cond. Matter Nucl. Sci.* **6**, 181-198 (2012).
258. E. K. Storms, B. Scanlan, paper presented at the 8th International Workshop on Anomalies in Hydrogen/Deuterium Loaded Metals, Catania, Sicily, October 13-18, 2007.
259. R. T. Bush, R. D. Eagleton, paper presented at the 3rd International Conference on Cold Fusion, Frontiers of Cold Fusion, Nagoya, Japan, October 21-25, 1992.
260. R. Notoya, M. Enyo, paper presented at the 3rd International Conference on Cold Fusion, Frontiers of Cold Fusion, Nagoya, Japan, October 21-25, 1992.
261. R. Sundaresan, J. O. M. Bockris, Anomalous reactions during arcing between carbon rods in water. *Fusion Technol.* **26**, 261 (1994).
262. M. Singh, M. D. Saksena, V. S. Dixit, V. B. Kartha, Verification of the George Oshawa experiment for anomalous production of iron from carbon arc in water. *Fusion Technol.* **26**, 266 (1994).
263. T. Grotz, Investigation of reports of the synthesis of iron via arc discharge through carbon compounds. *J. New Energy* **1**, 106 (1996).
264. X. L. Jiang, L. J. Han, W. Kang, paper presented at the 7th International Conference on Cold Fusion, Vancouver, Canada, April 19-24, 1998.

265. H. E. Ransford, Non-Stellar nucleosynthesis: transition metal production by DC plasma-discharge electrolysis using carbon electrodes in a non-metallic cell. *Infinite Energy* **4**, 23, 16 (1999).
266. T. Hanawa, paper presented at the 8th International Conference on Cold Fusion, Lerici (La Spezia), Italy, May 21-26, 2000.
267. T. Matsumoto, paper presented at the 5th International Conference on Cold Fusion, Monte-Carlo, Monaco, April 9-13, 1995.
268. T. Matsumoto, Experiments of underwater spark discharge with pinched electrodes. *J. New Energy* **1**, 79 (1996).
269. T. Matsumoto, paper presented at 7th International Conference on Cold Fusion, Vancouver, Canada, April 19-24, 1998.
270. T. Mizuno, T. Ohmori, T. Akimoto, paper presented at the 7th International Conference on Cold Fusion, Vancouver, Canada, April 19-24, 1998.
271. T. Mizuno, T. Ohmori, K. Azumi, T. Akimoto, A. Takahashi, paper presented at the 8th International Conference on Cold Fusion, Lerici (La Spezia), Italy, May 21-26, 2000.
272. T. Mizuno, T. Ohmori, T. Akimoto, A. Takahashi, Production of heat during plasma electrolysis. *Jpn. J. Appl. Phys. A* **39**, 6055 (2000).
273. D. Cirillo, R. Germano, V. Tontodonato, A. Widom, Y. N. Srivastava, E. Del Giudice, G. Vitiello, Experimental evidence of a neutron flux generation in a plasma discharge electrolytic cell. *Key Engineering Materials* **495**, 104-107 (2012).
274. V. Nassisi, Incandescent Pd and anomalous distribution of elements in deuterated samples processed by an excimer laser. *J. New Energy* **2**, 14-19 (1997).
275. V. Nassisi, Transmutation of elements in saturated palladium hydrides by an XeCl excimer laser. *Fusion Technol.* **33**, 468 (1998).
276. Castellano, M. Di Giulio, M. Dinescu, V. Nassisi, A. Conte, P. P. Pompa, paper presented at the 8th International Conference on Cold Fusion, Lerici (La Spezia), Italy, May 21-26, 2000.
277. M. Di Giulio, E. Filippo, D. Manno, V. Nassisi, Analysis of nuclear transmutations observed in D- and H-loaded films. *J. Hydrogen Eng.* **27**, 527 (2002).
278. V. Nassisi, G. Caretto, A. Lorusso, D. Manno, L. Fama, G. Buccolieri, A. Buccolieri, U. Mastromatteo, Modification of Pd-H_2 and Pd-D_2 thin films processed by He-Ne laser. *J. Cond. Matter. Nucl. Sci.* **5**, 1-6 (2011).
279. V. Nassisi, M. L. Longo, Experimental results of transmutation of elements observed in etched palladium samples by an excimer laser. *Fusion Technol.* **37**, 247 (2000).
280. V. Violante, E. Castagna, C. Sibilia, S. Paoloni, F. Sarto, paper presented at the 10th International Conference on Cold Fusion, Cambridge, MA, August 24-29, 2003.

281. J. Tian, L. H. Jin, B. J. Shen, Q. S. Wang, J. Dash, paper presented at the 13th International Conference on Condensed Matter Nuclear Science, Sochi, Russia, June 25-July 1, 2007.
282. Y. Iwamura, S. Tsuruga, T. Itoh, paper presented at the 13th Meeting of the Japan Cold Fusion Research Society, Nagoya, Japan, December 8-9, 2012.
283. T. O. Passell, paper presented at the 9th International Conference on Cold Fusion, Condensed Matter Nuclear Science, Tsinghua Univ., Beijing, China, May 19-24, 2002.
284. P. L. Hagelstein, Neutron yield for energetic deuterons in PdD and in D_2O. *J. Cond. Matter Nucl. Sci.* **3**, 35-40 (2010).
285. P. L. Hagelstein, On the connection between Ka X-rays and energetic alpha particles in Fleischmann-Pons experiments. *J. Cond. Matter Nucl. Sci.* **3**, 50-58 (2010).
286. P. L. Hagelstein, Secondary neutron yield in the presence of energetic alpha particles in PdD. *J. Cond. Matter Nucl. Sci.* **3**, 41-49 (2010).
287. L. Kowalski, Comment on "The use of CR-39 in Pd/D co-deposition experiments" by P.A. Mosier-Boss, S. Szpak, F.E. Gordon and L.P.G. Forsley. *Cur. Phys. J. Appl. Phys.* **44**, 287 (2008).
288. L. Kowalski, Chemically-induced nuclear activity or an illusion? (2009).
289. L. Kowalski, paper presented at the 14th International Conference on Condensed Matter Nuclear Science, Washington DC, August 10-15, 2008.
290. L. Kowalski, Comments on codeposition electrolysis results. *J. Cond. Matter Nucl. Sci.* **3**, 1-3 (2010).
291. S. Szpak, J. Dea, Evidence for the induction of nuclear activity in polarized Pd/H-H_2O system. *J. Cond. Matter Nucl. Sci.* **9**, 21-29 (2012).
292. P. A. Mosier-Boss, paper presented at the 17th International Conference on Condensed Matter Nuclear Science, Daejeon, Korea, August 12-17, 2012.
293. P. A. Mosier-Boss, F. E. Gordon, F. P. G. Forsley, Characterization of neutrons emitted during Pd/D co-deposition. *J. Cond. Matter Nucl. Sci.* **6**, 13-23 (2012).
294. P. A. Mosier-Boss, J. Dea, F. Gordon, L. Forsley, M. Miles, Review of twenty years of LENR research using Pd/D co-deposition. *J. Cond. Matter Nucl. Sci.* **4**, 173-187 (2011).
295. P. A. Mosier-Boss, L. Forsley, F. Gordon, Comments on co-deposition electrolysis results: a response to Kowalski. *J. Cond. Matter Nucl. Sci.* **3**, 4-8 (2010).
296. P. A. Mosier-Boss, J. Y. Dea, L. P. G. Forsley, M. S. Morey, J. R. Tinsley, J. P. Hurley, F. E. Gordon, Comparison of Pd/D co-deposition

and DT neutron generated triple tracks observed in CR-39 detectors. *Eur. Phys. J. Appl. Phys.* **51**, 20901-20911 (2010).

297. S. Szpak, P. A. Mosier-Boss, F. Gordon, Further evidence of nuclear reactions in the Pd/D lattice: emission of charged particles. *Naturwissenschaften* **94**, 515 (2009).

298. P. A. Mosier-Boss, L. Forsley, F. Gordon. (http://chiefio.wordpress.com/2012/05/26/spawar-space-and-naval-warfare-lenr-proof/, Univ. of Missouri Talk, 2009).

299. P. A. Mosier-Boss, S. Szpak, F. E. Gordon, L. P. G. Forsley, Triple tracks in CR-39 as the result of Pd/D co-deposition: evidence of energetic neutrons. *Naturwissenschaften* **96**, 135-142 (2009).

300. P. A. Mosier-Boss, F. Gordon, L. Forsley, in *Low-Energy Nuclear Reactions Sourcebook Volume 2,* J. Marwan, S. Krivit, Eds. (Oxford University Press, 2009).

301. P. A. Mosier-Boss, S. Szpak, F. Gordon, L. Forsley, Characterization of tracks in CR-39 detectors obtained as a result of Pd/D Co-deposition. *Eur. Phys. J. Appl. Phys.* **46**, 30901 (2009).

302. P. A. Mosier-Boss, S. Szpak, F. Gordon, L. Forsley, Reply to comment on “The use of CR-39 in Pd/D co-deposition experiments”: a response to Kowalski. *Eur. Phys. J. Appl. Phys.* **44**, 291 (2008).

303. P. A. Mosier-Boss, S. Szpak, F. Gordon, L. Forsley, in *ACS Symposium Series 998, Low-Energy Nuclear Reactions Sourcebook,* J. Marwan, S. B. Krivit, Eds. (American Chemical Society, Washington, DC, 2008), pp. 311-334.

304. L. Forsley, P. A. Mosier-Boss, paper presented at the American Physical Society, New Orleans, LA, March 10, 2008.

305. P. A. Mosier-Boss, S. Szpak, F. E. Gordon, L. P. G. Forsley, Use of CR-39 in Pd/D co-deposition experiments. *Eur. Phys. J. Appl. Phys.* **40**, 293-303 (2007).

306. A. Lipson, I. Chernov, V. Sokhoreva, V. Mironchik, A. Roussetski, A. Tsivadze, Y. Cherdantsev, B. Lyakhov, E. Saunin, M. Melich, paper presented at the 15th International Conference on Condensed Matter Nuclear Science, Rome, Italy, October 5-9, 2009.

307. A. G. Lipson, G. Miley, A. S. Roussetski, B. F. Lyakhov, E. I. Saunin, paper presented at the 12th International Conference on Condensed Matter Nuclear Science, Yokohama, Japan, November 27-December 2, 2005.

308. A. G. Lipson, B. F. Lyakhov, V. A. Kuznetsov, T. S. Ivanova, B. V. Deryagin, The nature of excess energy liberated in a Pd/PdO heterostructure electrochemically saturated with hydrogen (deuterium). *Russ. J. Phys. Chem.* **69**, 1810 (1995).

309. A. G. Lipson, B. F. Lyakhov, A. S. Rousstesky, N. Asami, paper presented at the 8th International Conference on Cold Fusion, Lerici (La Spezia), Italy, May 21-26, 2000.

310. A. G. Lipson, B. F. Lyakhov, A. S. Roussetski, T. Akimoto, T. Mizuno, N. Asami, R. Shimada, S. Miyashita, A. Takahashi, Evidence for low-intensity D-D reaction as a result of exothermic deuterium desorption from Au/Pd/PdO:D heterostructure. *Fusion Technol.* **38**, 238 (2000).
311. A. G. Lipson, A. S. Roussetsky, G. H. Miley, C. H. Castano, paper presented at the 9th International Conference on Cold Fusion, Condensed Matter Nuclear Science, Tsinghua Univ., Beijing, China, May 19-24, 2002.
312. A. G. Lipson, A. S. Roussetski, G. Miley, Energetic alpha and proton emissions on the electrolysis of thin-Pd films. *Trans. Am. Nucl. Soc.* **88**, 638-639 (2003).
313. A. G. Lipson, A. S. Roussetski, A. B. Karabut, G. H. Miley, paper presented at the 10th International Conference on Cold Fusion, Cambridge, MA, August 24-29, 2003.
314. E. K. Storms, B. Scanlan, paper presented at the 14th International Conference on Condensed Matter Nuclear Science, Washington DC, August 10-15, 2008.
315. A. G. Lipson, A. S. Roussetski, G. H. Miley, E. I. Saunin, paper presented at the 10th International Conference on Cold Fusion, Cambridge, MA, August 24-29, 2003.
316. A. S. Roussetski, A. G. Lipson, V. P. Andreanov, paper presented at the 10th International Conference on Cold Fusion, Cambridge, MA, August 24-29, 2003.
317. A. G. Lipson, G. Miley, B. F. Lyakhov, A. S. Roussetski, paper presented at the 11th International Conference on Cold Fusion, Marseilles, France, October 31-November 5, 2004.
318. A. S. Roussetski, A. G. Lipson, B. F. Lyakhov, E. I. Saunin, paper presented at the 12th International Conference on Condensed Matter Nuclear Science, Yokohama, Japan, November 27-December 2, 2005.
319. A. G. Lipson, A. S. Roussetski, G. Miley, paper presented at the 13th International Conference on Condensed Matter Nuclear Science, Sochi, Russia, June 25-July 1, 2007.
320. A. G. Lipson, I. P. Chernov, A. S. Roussetski, Y. Chardantsc, B. F. Lyakhov, E. I. Saunin, M. E. Melich, paper presented at the 14th International Conference on Condensed Matter Nuclear Science, Washington DC, August 10-15, 2008.
321. A. G. Lipson, I. P. Chernov, A. S. Roussetski, Y. P. Cherdantsev, A. Tsivadze, B. Lyakohov, E. I. Saunin, M. E. Melich, in *Low-Energy Nuclear Reactions Sourcebook Volume 2,* J. Marwan, S. Krivit, Eds. (Oxford University Press, 2009).
322. A. S. Roussetski, A. G. Lipson, F. Tanzella, E. I. Saunin, M. C. McKubre, paper presented at the 15th International Conference on Condensed Matter Nuclear Science, Rome, Italy, October 5-9, 2009.

323. A. G. Lipson, A. S. Roussetski, E. I. Saunin, F. Tanzella, B. Earle, M. C. McKubre, paper presented at the 8th International Workshop on Anomalies in Hydrogen/Deuterium Loaded Metals, Catania, Italy, October 13-18, 2007.
324. A. G. Lipson, A. S. Roussetski, E. I. Saunin, paper presented at the 8th International Workshop on Anomalies in Hydrogen/Deuterium Loaded Metals, Catania, Italy, October 13-18, 2007.
325. R. A. Oriani, J. C. Fisher, paper presented at the 10th International Conference on Cold Fusion, Cambridge, MA, August 24-29, 2003.
326. R. A. Oriani, J. C. Fisher, Generation of nuclear tracks during electrolysis. *Jpn. J. Appl. Phys. A* **41**, 6180-6183 (2002).
327. R. A. Oriani, J. C. Fisher, paper presented at the 10th International Conference on Cold Fusion, Cambridge, MA, August 24-29, 2003.
328. J. C. Fisher, paper presented at the 8th International Workshop on Anomalies in Hydrogen/Deuterium Loaded Metals, Catania, Italy, October 13-18, 2007.
329. Y. Iwamura, N. Gotoh, T. Itoh, I. Toyoda, paper presented at the 5th International Conference on Cold Fusion, Monte-Carlo, Monaco, April 9-13, 1995.
330. S. Focardi, F. Piantelli. (WO 95/20816 (1995)).
331. S. Focardi, R. Habel, F. Piantelli, Anomalous heat production in Ni-H systems. *Nuovo Cimento* **107A**, 163 (1994).
332. S. Focardi, V. Gabbani, V. Montalbano, F. Piantelli, S. Veronesi, Large excess heat production in Ni-H systems. *Nuovo Cimento* **111A**, 1233 (1998).
333. A. Battaglia, L. Daddi, S. Focardi, V. Gabbani, V. Montalbano, F. Piantelli, P. G. Sona, S. Veronesi, Neutron emission in Ni-H systems. *Nuovo Cimento* **112 A**, 921 (1999).
334. E. G. Campari, S. Focardi, V. Gabbani, V. Montalbano, F. Piantelli, S. Veronesi, paper presented at the 5th Asti Workshop on Anomalies in Hydrogen/Deuterium Loaded Metals, Asti, Italy, March 19-21, 2004.
335. S. Focardi, V. Gabbani, V. Montalbano, F. Piantelli, F. Veronesi, paper presented at the 11th International Conference on Cold Fusion, Marseilles, France, October 31-November 5, 2004.
336. E. G. Campari, G. Fasano, S. Focardi, G. Lorusso, V. Gabbani, V. Montalbano, F. Piantelli, C. Stanghini, S. Veronesi, paper presented at the 11th International Conference on Cold Fusion, Marseilles, France, October 31-November 5, 2004.
337. E. G. Campari, S. Focardi, V. Gabbani, V. Montalbano, F. Piantelli, S. Veronesi, paper presented at the 11th International Conference on Cold Fusion, Marseilles, France, October 31-November 5, 2004.
338. E. G. Campari, S. Focardi, V. Gabbani, V. Montalbano, F. Piantelli, C. Stanghini, paper presented at the 6th International Workshop on

Anomalies in Hydrogen/Deuterium Loaded Metals, Siena, Italy, May 13-15, 2005.
339. F. Piantelli, paper presented at the 9th International Workshop on Anomalies in Hydrogen/Deuterium Loaded Metals, Pontignano, Italy, September 17-19, 2010.
340. F. Piantelli. (World Property Organization, PCT/IB2009/007549 (2010)).
341. F. Piantelli. (USA, US 2011/0249763 A1).
342. V. Violante, P. Tripodi, D. Di Gioacchino, R. Borelli, L. Bettinali, E. Santoro, A. Rosada, F. Sarto, A. Pizzuto, M. C. H. McKubre, F. Tanzella, paper presented at the 9th International Conference on Cold Fusion, Condensed Matter Nuclear Science, Tsinghua Univ., Beijing, China, May 19-24, 2002.
343. R. T. Bush, R. D. Eagleton, paper presented at the 3rd International Conference on Cold Fusion, Frontiers of Cold Fusion, Nagoya, Japan, October 21-25, 1992.
344. E. K. Storms, B. Scanlan, Nature of energetic radiation emitted from a metal exposed to H_2. *J. Cond. Matter Nucl. Sci.* **11**, 142-156 (2013).
345. R. T. Bush, R. D. Eagleton, Evidence for electrolytically induced transmutation and radioactivity correlated with excess heat in electrolytic cells with light water rubidium salt electrolytes. *Trans. Fusion Technol.* **26**, 344-354 (1994).
346. P. K. Iyengar, M. Srinivasan, S. K. Sikka, A. Shyam, V. Chitra, L. V. Kulkarni, R. K. Rout, M. S. Krishnan, S. K. Malhotra, D. G. Gaonkar, H. K. Sadhukhan, V. B. Nagvenkar, M. G. Nayar, S. K. Mitra, P. Raghunathan, S. B. Degwekar, T. P. Radhakrishnan, R. Sundaresan, J. Arunachalam, V. S. Raju, R. Kalyanaraman, S. Gangadharan, G. Venkateswaran, P. N. Moorthy, K. S. Venkateswarlu, B. Yuvaraju, K. Kishore, S. N. Guha, M. S. Panajkar, K. A. Rao, P. Raj, P. Suryanarayana, A. Sathyamoorthy, T. Datta, H. Bose, L. H. Prabhu, S. Sankaranarayanan, R. S. Shetiya, N. Veeraraghavan, T. S. Murthy, B. K. Sen, P. V. Joshi, K. G. B. Sharma, T. B. Joseph, T. S. Iyengar, V. K. Shrikhande, K. C. Mittal, S. C. Misra, M. Lal, P. S. Rao, Bhabha Atomic Research Centre studies on cold fusion. *Fusion Technol.* **18**, 32-94 (1990).
347. D. D. Afonichev, paper presented at the 10th International Conference on Cold Fusion, Cambridge, MA, August 24-29, 2003.
348. A. B. Karabut, paper presented at the 9th International Conference on Cold Fusion, Condensed Matter Nuclear Science, Tsinghua Univ., Beijing, China, May 19-24, 2002.
349. A. B. Karabut, E. A. Karabut, P. L. Hagelstein, Spectral and temporal characteristics of X-ray emission from metal electrodes in a high-current glow discharge. *J. Cond. Matter Nucl. Sci.* **6**, 217-240 (2012).

350. F. Keeney, S. E. Jones, A. Johnson, D. B. Buehler, F. E. Cecil, G. K. Hubler, P. L. Hagelstein, M. Scott, J. Ellsworth, paper presented at the 10th International Conference on Cold Fusion, Cambridge, MA, August 24-29, 2003.
351. S. E. Jones, paper presented at the Riken Conference on Muon-Catalyzed and Cold Fusion, Tokyo, Japan, November 1989.
352. K. Yi, D. Jiang, X. Qian, J. Lin, Y. Ye, A study of D-D fusion in TiD target induced by ^{197}Au bombardment. *Nucl. Techniques (China)* **17**, 722 (in Chinese) (1994).
353. D. H. Beddingfield, F. E. Cecil, C. S. Galovich, H. Liu, S. Asher, paper presented at the 2nd Annual Conference on Cold Fusion, The Science of Cold Fusion, Como, Italy, June 29-July 4, 1991.
354. T. Wang, Y. Piao, J. Hao, X. Wang, G. Jin, Z. Niu, paper presented at the 6th International Conference on Cold Fusion, Progress in New Hydrogen Energy, Lake Toya, Hokkaido, Japan, October 13-18, 1996.
355. T. Wang, Z. Wang, J. Chen, G. Jin, Y. Piao, Investigating the unknown nuclear reaction in a low-energy (E<300 keV) p + T_2H_x experiment. *Fusion Technol.* **37**, 146 (2000).
356. T. Wang, K. Ochiai, Z. Wang, G. Jing, T. Iida, A. Takahashi, paper presented at the 7th International Conference on Cold Fusion, Vancouver, Canada, April 19-24, 1998.
357. K. Ochiai, K. Maruta, H. Miyamaru, A. Takahashi, paper presented at the 7th International Conference on Cold Fusion, Vancouver, Canada, April 19-24, 1998.
358. H. O. Menlove, M. M. Fowler, E. Garcia, A. Mayer, M. C. Miller, R. R. Ryan, paper presented at the Workshop on Cold Fusion Phenomena, Santa Fe, NM, May 23, 1989.
359. A. De Ninno, A. Frattolillo, G. Lollobattista, L. Martinis, M. Martone, L. Mori, S. Podda, F. Scaramuzzi, Emission of neutrons as a consequence of titanium-deuterium interaction. *Nuovo Cimento Soc. Ital. Fis. A* **101**, 841 (1989).
360. M. R. Swartz, G. Verner, Bremsstrahlung in hot and cold fusion. *J. New Energy* **3**, 90-101 (1999).
361. J. L. McKibben, Can cold fusion be catalyzed by fractionally-charged ions that have evaded FC particle searches. *Infinite Energy* **1**, 4, 14-23 (1995).
362. Y. N. Bazhutov, G. M. Vereshkov, R. N. Kuz'min, A. M. Frolov, paper presented at the Fiz. Plazmy Nekotor. Vopr. Obshch. Fiz. M., 1990.
363. Y. N. Bazhutov, paper presented at the Sixth International Conference on Cold Fusion, Progress in New Hydrogen Energy, Lake Toya, Hokkaido, Japan, October 13-18, 1996.
364. Y. N. Bazhutov, V. P. Koretsky, paper presented at the 6th International Conference on Cold Fusion, Progress in New Hydrogen Energy, Lake Toya, Hokkaido, Japan, October 13-18, 1996.

365. Y. N. Bazhutov, paper presented at the 8th International Conference on Cold Fusion, Lerici (La Spezia), Italy, May 21-26, 2000.
366. J. Rafelski, M. Sawicki, M. Gajda, D. Harley, Nuclear reactions catalyzed by a massive negatively charged particle: how cold fusion can be catalyzed. *Fusion Technol.* **18**, 136 (1990).
367. T. Matsumoto, Prediction of new particle emission on cold fusion. *Fusion Technol.* **18**, 647-651 (1990).
368. T. Matsumoto, paper presented at the Anomalous Nuclear Effects in Deuterium/Solid Systems, AIP Conference Proceedings 228, Brigham Young Univ., Provo, UT, October 22-23, 1990.
369. T. Matsumoto, Observation of gravity decays of multiple-neutron nuclei during cold fusion. *Fusion Technol.* **22**, 164 (1992).
370. T. Matsumoto, Mechanisms of cold fusion: comprehensive explanations by the Nattoh model. *Mem. Fac. Eng. Hokkaido Univ.* **19**, 201 (1995).
371. T. Matsumoto, Cold fusion experiments with ordinary water and thin nickel foil. *Fusion Technol.* **24**, 296-306 (1993).
372. T. Matsumoto, Observation of meshlike traces on nuclear emulsions during cold fusion. *Fusion Technol.* **23**, 103 (1993).
373. I. Savvatimova, paper presented at the 8th International Conference on Cold Fusion, Lerici (La Spezia), Italy, May 21-26, 2000.
374. I. Savvatimova, J. Dash, paper presented at the 9th International Conference on Cold Fusion, Condensed Matter Nuclear Science, Tsinghua Univ., Beijing, China, May 19-24, 2002.
375. G. Lochak, L. I. Urutskoev, paper presented at the 11th International Conference on Cold Fusion, Marseilles, France, October 31-November 5, 2004.
376. F. Tanzella, J. Bao, M. C. McKubre, P. L. Hagelstein, paper presented at the 16th International Conference on Cold Fusion, Chennai, India, February 6-11, 2011.
377. F. Tanzella, M. C. McKubre, paper presented at the 15th International Conference on Condensed Matter Nuclear Science, Rome, Italy, October 5-9, 2009.
378. F. L. Tanzella, J. Bao, M. C. H. McKubre, Cryogenic calorimetry of "exploding" PdD_x wires. *J. Cond. Matter Nucl. Sci.* **6**, 90-100 (2012).
379. E. A. Pryakhin, G. A. Tryapitsina, L. L. Urutskoyev, A. V. Akleyev, paper presented at the 11th International Conference on Cold Fusion, Marseilles, France, October 31-November 5, 2004.
380. K. Shoulders, S. Shoulders, Observations on the role of charge clusters in nuclear cluster reactions. *J. New Energy* **1**, 111-121 (1996).
381. K. Shoulders, Projectiles from the dark side. *Infinite Energy* **12**, 70, 39-40 (2006).
382. K. Shoulders, S. Shoulders, paper presented at the Conference on Future Energy, Bethesda, MD, April 29-May 1, 1999.

383. E. H. Lewis, Tracks of Ball Lightning in Apparatus? *J. Cond. Matter Nucl. Sci.* **2**, 13 (2009).
384. R. A. Oriani, J. C. Fisher, paper presented at the 11th International Conference on Cold Fusion, Marseilles, France, October 31-November 5, 2004.
385. V. I. Vysotskii, S. V. Adamenko, Correlated states of interacting particles and problems of the Coulomb barrier transparency at low energies in nonstationary systems. *Technical Phys.* **55**, 613 (2010).
386. K. Kamada, paper presented at the 3rd International Conference on Cold Fusion, Frontiers of Cold Fusion, Nagoya, Japan, October 21-25, 1992.
387. K. Kamada, Electron impact H-H and D-D fusions in molecules embedded in Al. 1. Experimental results. *Jpn. J. Appl. Phys. A* **31**, L1287 (1992).
388. K. Kamada, Y. Katano, N. Ookubo, I. Yoshizawa, paper presented at the 8th International Conference on Cold Fusion, Lerici (La Spezia), Italy, May 21-26, 2000.
389. K. Kamada, H. Kinoshita, H. Takahashi, paper presented at the 5th International Conference on Cold Fusion, Monte-Carlo, Monaco, April 9-13, 1995.
390. I. P. Chernov, Y. M. Koroteev, V. M. Silkin, Y. I. Tyurin, paper presented at the 8th International Workshop on Anomalies in Hydrogen/Deuterium Loaded Metals, Catania, Italy, October 13-18, 2007.
391. I. P. Chernov, Y. M. Koroteev, V. M. Silkin, Y. I. Tyurin, paper presented at the 13th International Conference on Condensed Matter Nuclear Science, Sochi, Russia, June 25-July 1, 2007.
392. S. Adamenko, V. Vysotskii, paper presented at the 10th International Conference on Cold Fusion, Cambridge, MA, August 24-29, 2003.
393. S. Adamenko, V. Vysotskii, paper presented at the 11th International Conference on Cold Fusion, Marseilles, France, October 31-November 5, 2004.
394. S. V. Adamenko, V. I. Vysotskii, Evolution of annular self-controlled electron-nucleus collapse in condensed targets. *Foundations of Phys.* **34**, 1801-1831 (2004).
395. S. V. Adamenko, V. I. Vysotskii, Mechanism of synthesis of superheavy nuclei via the process of controlled electron-nuclear collapse. *Foundations of Phys.* **17**, 203-233 (2004).
396. S. V. Adamenko, A. S. Adamenko, V. I. Vysotskii, Full range nucleosynthesis in the laboratory. *Infinite Energy* **9**, 54, (2004).
397. S. Adamenko, V. Vysotskii, paper presented at the 12th International Conference on Condensed Matter Nuclear Science, Yokohama, Japan, November 27-December 2, 2005.

398. S. V. Adamenko, F. Selleri, A. van der Merwe, Eds., *Controlled nucleosynthesis: Breakthroughs in experiment and theory*, (Springer, Dordrecht, The Netherlands, 2007), pp. 773.

399. V. Adamenko, V. Vysotskii, paper presented at the 14th International Conference on Condensed Matter Nuclear Science, Washington DC, August 10-15, 2008.

400. C. Steinert, Laser-induced 'semicold' fusion. *Fusion Technol.* **17**, 206 (1990).

401. I. L. Beltyukov, N. B. Bondarenko, A. A. Janelidze, M. Y. Gapanov, K. G. Gribanov, S. V. Kondratov, A. G. Maltsev, P. I. Novikov, S. A. Tsvetkov, V. I. Zakharov, Laser-induced cold nuclear fusion in Ti-H_2-D_2-T_2 compositions. *Fusion Technol.* **20**, 234-238 (1991).

402. O. M. Vokhnik, B. I. Goryachev, A. A. Zubrilo, G. P. Kutznetsova, Y. V. Popov, S. I. Svertilov, Search for effects related to nuclear fusion in the optical breakdown of heavy water. *Sov. J. Nucl. Phys.* **55**, 1772 (1992).

403. V. Violante, M. Bertolotti, E. Castagna, I. Dardik, M. C. McKubre, S. Moretti, S. Lesin, F. Sarto, F. Tanzella, T. Zilov, paper presented at the 12th International Conference on Condensed Matter Nuclear Science, Yokohama, Japan, November 27-December 2, 2005.

404. M. Apicella, E. Castagna, L. Capobianco, L. Daulerio, M. C. McKubre, A. Rosada, E. Santoro, F. Sarto, C. Sibilia, F. Tanzella, V. Violante, paper presented at the 5th International Workshop on Anomalies in Hydrogen/Deuterium Loaded Metals, Asti, Italy, March 19-21, 2004.

405. J. Tian, L. H. Jin, B. J. Shen, Z. K. Weng, X. Lu, paper presented at the 14th International Conference on Condensed Matter Nuclear Science, Washington, DC, August 10-15, 2008.

406. E. V. Barmina, P. G. Kuzmin, S. F. Timashev, G. A. Shafeev, L. Y. Karpov, Laser-induced synthesis and decay of Tritium under exposure of solid targets in heavy water. *arXiv:1306.080v1*, (2013).

407. S. Badiei, P. U. Andersson, L. Holmlid, High-energy Coulomb explosions in ultra-dense deuterium: time-of-flight-mass spectrometry with variable energy and flight length. *Int. J. Mass Spectro.* **282**, 70-76 (2009).

408. S. Badiei, L. Holmlid, Experimental studies of fast fragments of H Rydberg matter. *J. Phys. B: At. Mol. Opt. Phys.* **39**, 4191-4212 (2006).

409. K. Tsuchiya, A. Watanabe, M. Ozaki, S. Sasabe, paper presented at the 14th International Conference on Condensed Matter Nuclear Science, Washington, DC, August 10-15, 2008.

410. L. Caneve, paper presented at the 15th International Conference on Condensed Matter Nuclear Science, Rome, Italy, October 5-9, 2009.

411. R. M. Montereali, S. Almaviva, T. Marolo, M. A. Vincenti, F. Sarto, C. Sibilia, E. Castagna, V. Violante, paper presented at the 12th

International Conference on Condensed Matter Nuclear Science, Yokohama, Japan, November 27-December 2, 2005.

412. Y. Arata, Y. Zhang, paper presented at the 10th International Conference on Cold Fusion, Cambridge, MA, August 24-29, 2003.

413. D. Letts, D. Cravens, Laser stimulation of deuterated palladium. *Infinite Energy* **9**, 84, 10 (2003).

414. D. Letts, D. Cravens, paper presented at the 10th International Conference on Cold Fusion, Cambridge, MA, August 24-29, 2003.

415. E. K. Storms, paper presented at the 10th International Conference on Cold Fusion, Cambridge, MA, August 24-29, 2003.

416. M. C. McKubre, paper presented at the 5th International Workshop on Anomalies in Hydrogen/Deuterium Loaded Metals, Asti, Italy, March 19-21, 2004.

417. M. R. Swartz, paper presented at the Tenth International Conference on Cold Fusion, Cambridge, MA, Aug. 24-29, 2003.

418. D. Letts, D. Cravens, P. L. Hagelstein, in *Low-Energy Nuclear Reactions Sourcebook Volume 2,* J. Marwan, S. Krivit, Eds. (Oxford University Press, 2009).

419. P. L. Hagelstein, D. Letts, Analysis of some experimental data from the two-laser experiment. *J. Cond. Matter Nucl. Sci.* **3**, 77-92 (2010).

420. P. L. Hagelstein, D. Letts, D. Cravens, Terahertz difference frequency response of PdD in two-laser experiments. *J. Cond. Matter Nucl. Sci.* **3**, 59-76 (2010).

421. R. W. Bass, paper presented at the 10th International Conference on Cold Fusion, Cambridge, MA, August 24-29, 2003.

422. S. Szpak, F. Gordon, paper presented at the 17th International Conference on Condensed Matter Nuclear Science, Daejeon, Korea, August 12-17, 2012.

423. S. Szpak, P. A. Mosier-Boss, C. Young, F. Gordon, The effect of an external electric field on surface morphology of co-deposited Pd/D films. *J. Electroanal. Chem.* **580**, 284-290 (2005).

424. S. Szpak, P. A. Mosier-Boss, C. Young, F. Gordon, Evidence of nuclear reactions in the Pd lattice. *Naturwissenschaften* **92**, 394 (2005).

425. S. Szpak, P. A. Mosier-Boss, M. Miles, M. Fleischmann, Thermal behavior of polarized Pd/D electrodes prepared by co-deposition. *Thermochim. Acta* **410**, 101 (2004).

426. S. Szpak, P. A. Mosier-Boss, M. H. Miles, Calorimetry of the Pd+D co-deposition. *Fusion Technol.* **36**, 234 (1999).

427. S. Szpak, P. A. Mosier-Boss, J. Dea, F. Gordon, paper presented at the 10th International Conference on Cold Fusion, Cambridge, MA, Aug. 24-29, 2003.

428. S. Szpak, P. A. Mosier-Boss, Nuclear and thermal events associated with Pd + D co-deposition. *J. New Energy* **1**, 54 (1996).

429. S. Szpak, P. A. Mosier-Boss, J. J. Smith, On the behavior of the cathodically polarized Pd/D system: search for emanating radiation. *Physics Lett. A* **210**, 382 (1996).
430. S. Szpak, P. A. Mosier-Boss, J. J. Smith, paper presented at the 2nd Annual Conference on Cold Fusion, The Science of Cold Fusion, Como, Italy, June 29-July 4, 1991.
431. S. Szpak, P. A. Mosier-Boss, J. J. Smith, On the behavior of Pd deposited in the presence of evolving deuterium. *J. Electroanal. Chem.* **302**, 255 (1991).
432. M. R. Swartz, G. Verner, A. Weinberg, paper presented at the 14th International Conference on Condensed Matter Nuclear Science, Washington DC, August 10-15, 2008.
433. O. Shirai, S. Kihara, Y. Sohrin, M. Matsui, Some experimental results relating to cold nuclear fusion. *Bull. Inst. Chem. Res., Kyoto Univ.* **69**, 550 (1991).
434. M. H. Miles, S. Szpak, P. A. Mosier-Boss, M. Fleischmann, paper presented at the 9th International Conference on Cold Fusion, Tsinghua University, Beijing, China, May 19-24, 2002.
435. J. P. Biberian, paper presented at the 5th International Workshop on Anomalies in Hydrogen/Deuterium Loaded Metals, Asti, Italy, March 19-21, 2004.
436. D. Letts, P. L. Hagelstein, Modified Szpak protocol for excess heat. *J. Cond. Matter Nucl. Sci.* **6**, 44-54 (2012).
437. M. C. H. McKubre, S. Crouch-Baker, A. M. Riley, S. I. Smedley, F. L. Tanzella, paper presented at the 3rd International Conference on Cold Fusion, Frontiers of Cold Fusion, Nagoya, Japan, October 21-25, 1992.
438. F. L. Tanzella, S. Crouch-Baker, A. McKeown, M. C. H. McKubre, M. Williams, S. Wing, paper presented at the 6th International Conference on Cold Fusion, Progress in New Hydrogen Energy, Lake Toya, Hokkaido, Japan, October 13-18, 1996.
439. E. K. Storms, paper presented at the 7th International Conference on Cold Fusion, Vancouver, Canada, April 19-24, 1998.
440. Y. Oya, M. Aida, K. Iinuma, M. Okamoto, paper presented at the 7th International Conference on Cold Fusion, Vancouver, Canada, April 19-24, 1998.
441. Y. Oya, H. Ogawa, M. Aida, K. Iinuma, M. Okamoto, paper presented at the 7th International Conference on Cold Fusion, Vancouver, Canada, April 19-24, 1998.
442. R. Dus, E. Nowicka, Segregation of deuterium and hydrogen on surfaces of palladium deuteride and hydride at low temperatures. *Langmuir* **16**, 584 (2000).
443. M. C. McKubre, paper presented at the 15th International Conference on Condensed Matter Nuclear Science, Rome, Italy, October 5-9, 2009.

444. R. A. Oriani, J. C. Nelson, S.-K. Lee, J. H. Broadhurst, Calorimetric measurements of excess power output during the cathodic charging of deuterium into palladium. *Fusion Technol.* **18**, 652 (1990).
445. M. C. H. McKubre, S. Crouch-Baker, F. L. Tanzella, M. Williams, S. Wing, paper presented at the 6th International Conference on Cold Fusion, Progress in New Hydrogen Energy, Lake Toya, Hokkaido, Japan, October 13-18, 1996.
446. A. Czerwinski, Influence of lithium cations on hydrogen and deuterium electrosorption in palladium. *Electrochim. Acta* **39**, 431 (1994).
447. N. Asami, T. Senjuh, H. Kamimura, M. Sumi, E. Kennel, T. Sakai, K. Mori, H. Watanabe, K. Matsui, paper presented at the 6th International Conference on Cold Fusion, Progress in New Hydrogen Energy, Lake Toya, Hokkaido, Japan, October 13-18, 1996.
448. K. Ota, H. Yoshitake, O. Yamazaki, M. Kuratsuka, K. Yamaki, K. Ando, Y. Iida, N. Kamiya, paper presented at the 4th International Conference on Cold Fusion, Lahaina, HI, December 6-9, 1993.
449. M. R. Swartz, G. Verner, paper presented at the 10th International Conference on Cold Fusion, Cambridge, MA, August 24-29, 2003.
450. I. Dardik, H. Branover, A. El-Boher, D. Gazit, E. Golbreich, E. Greenspan, A. Kapusta, B. Khachatorov, V. Krakov, S. Lesin, B. Michailovitch, G. Shani, T. Zilov, paper presented at the 10th International Conference on Cold Fusion, Cambridge, MA, August 24-29, 2003.
451. I. Dardik, paper presented at the 11th International Conference on Cold Fusion, Marseilles, France, October 31-November 5, 2004.
452. I. Dardik, T. Zilov, H. Branover, A. El-Boher, E. Greenspan, B. Khachatorov, V. Krakov, S. Lesin, M. Tsirlin, paper presented at the 11th International Conference on Cold Fusion, Marseilles, France, October 31-November 5, 2004.
453. I. Dardik, T. Zilov, H. Branover, A. El-Boher, E. Greenspan, B. Khachaturov, V. Krakov, S. Lesin, M. Tsirlin, paper presented at the 12th International Conference on Condensed Matter Nuclear Science, Yokohama, Japan, November 27-December 2, 2005.
454. I. Dardik, T. Zilov, H. Branover, A. El-Boher, E. Greenspan, B. Khachaturov, V. Krakov, S. Lesin, M. Tsirlin, paper presented at the 6th International Workshop on Anomalies in Hydrogen/Deuterium Loaded Metals, Siena, Italy, May 13-15, 2005.
455. I. Dardik, T. Zilov, H. Branover, A. El-Boher, E. Greenspan, B. Khachaturov, V. Krakov, S. Lesin, A. Shapiro, M. Tsirlin, paper presented at the 14th International Conference on Condensed Matter Nuclear Science, Washington, DC, August 10-15, 2008.
456. M. C. McKubre, F. Tanzella, I. Dardik, A. El Boher, T. Zilov, E. Greenspan, C. Sibilia, V. Violante, in *ACS Symposium Series 998,*

Low-Energy Nuclear Reactions Sourcebook, J. Marwan, S. B. Krivit, Eds. (American Chemical Society, Washington, DC, 2008), pp. 219.

457. T. Ohmori, T. Mizuno, Nuclear transmutation reaction caused by light water electrolysis on tungsten cathode under incandescent conditions. *Infinite Energy* **5**, 27, 34 (1999).
458. T. Ohmori, T. Mizuno, paper presented at the 7th International Conference on Cold Fusion, Vancouver, Canada, April 19-24, 1998.
459. T. Mizuno, T. Akimoto, T. Ohmori, paper presented at the 4th Meeting of the Japan Cold Fusion Research Society, Iwate, Japan, October 17-18, 2002.
460. T. Mizuno, T. Ohmori, T. Akimoto, paper presented at the 10th International Conference on Cold Fusion, Cambridge, MA, August 24-29, 2003.
461. J.-F. Fauvarque, P. P. Clauzon, G. Lalleve, G. Le Buzit, paper presented at the 15th International Conference on Condensed Matter Nuclear Science, Rome, Italy, October 5-9, 2009.
462. L. Kowalski, S. R. Little, G. Luce, paper presented at the 12th International Conference on Cold Fusion;, Yokohama, Japan, November 17-December 2, 2005.
463. R. E. Godes, R. George, F. Tanzella, M. C. McKubre, paper presented at the 17th International Conference on Condensed Matter Nuclear Science, Daejeon, Korea, August 12-17, 2012.
464. M. R. Swartz, paper presented at the 14th International Confernce on Condensed Matter Nuclear Science, Washington, DC, August 10-15, 2008.
465. M. R. Swartz, Generality of optimal operating point behavior in low energy nuclear systems. *J. New Energy* **4**, 218-228 (1999).
466. B. Baranowski, S. M. Filipek, M. Szustakowski, J. Farny, W. Woryna, Search for 'cold fusion' in some Me-D systems at high pressures of gaseous deuterium. *J. Less-Common Met.* **158**, 347-357 (1990).
467. A. Stroka, B. Baranowski, S. M. Filipek, Search for ^{3}He and ^{4}He in Pd-D_2 system long term cumulation experiment in high pressure. *Pol. J. Chem.* **67**, 353 (1993).
468. V. Violante, paper presented at the New Advances on the Fleischmann-Pons Effect, European Parliment, Bruxelles, March 6, 2013.
469. E. K. Storms, paper presented at the 10th International Conference on Cold Fusion, Cambridge, MA, August 24-29, 2003.
470. D. Letts, D. Cravens, paper presented at the 5th International Workshop on Anomalies in Hydrogen/Deuterium Loaded Metals, Asti, Italy, March 19-21, 2004.
471. K. Ota, M. Kuratsuka, K. Ando, Y. Iida, H. Yoshitake, N. Kamiya, paper presented at the 3rd International Conference on Cold Fusion, Frontiers of Cold Fusion, Nagoya, Japan, October 21-25, 1992.

472. E. K. Storms, Some characteristics of heat production using the "cold fusion" effect. *Trans. Fusion Technol.* **26**, 96 (1994).
473. E. Storms, paper presented at the 4th International Conference on Cold Fusion, Lahaina, HI, December 6-9, 1993.
474. M. Fleischmann, S. Pons, Calorimetry of the Pd-D_2O system: from simplicity via complications to simplicity. *Phys. Lett. A* **176**, 118 (1993).
475. G. Lonchampt, L. Bonnetain, P. Hieter, paper presented at the 6th International Conference on Cold Fusion, Progress in New Hydrogen Energy, Lake Toya, Hokkaido, Japan, October 13-18, 1996.
476. G. Lonchampt, J.-P. Biberian, L. Bonnetain, J. Delepine, paper presented at the 7th International Conference on Cold Fusion, Vancouver, Canada, April 19-24, 1998.
477. G. Mengoli, M. Bernardini, C. Manduchi, G. Zannoni, Calorimetry close to the boiling temperature of the D_2O/Pd electrolytic system. *J. Electroanal. Chem.* **444**, 155 (1998).
478. M. R. Swartz, G. M. Verner, A. H. Frank, paper presented at the 9th International Conference on Cold Fusion, Condensed Matter Nuclear Science, Tsinghua Univ., Beijing, China, May 19-24, 2002.
479. R. P. Santandrea, R. G. Behrens, A review of the thermodynamics and phase relationships in the palladium- hydrogen, palladium-deuterium and palladium-tritium systems. *High Temperature Materials and Processes* **7**, 149 (1986).
480. T. B. Flanagan, W. A. Oates, The palladium-hydrogen system. *Annu. Rev. Mater. Sci.* **21**, 269 (1991).
481. I. S. Anderson, D. K. Ross, C. J. Carlile, The structure of the γ phase of palladium deuteride. *Phys. Lett. A* **68**, 249 (1978).
482. G. A. Ferguson, A. I. Schindler, T. Tanaka, T. Morita, Neutron diffraction study of temperature-dependent properties of palladium containing absorbed hydrogen. *Phys. Rev.* **137**, 483 (1965).
483. D. H. W. Carstens, W. R. David, "Equilibrium measurements in the beta region of palladium protide and palladium deuteride," (Los Alamos National Laboratory, Los Alamos, 1989).
484. D. M. Nace, J. G. Aston, Palladium hydride. III. The thermodynamic study of Pd_2D between 15 and 303 K. evidence for the Tetragonal PdH_4 structure in palladium hydride. *J. Am. Chem. Soc.* **79**, 3627 (1957).
485. D. M. Nace, J. G. Aston, Palladium hydride. I. The thermodynamic properties of Pd_2H between 273 and 345 K. *J. Am. Chem. Soc.* **79**, 3619 (1957).
486. Y. Fukai, N. Okuma, Evidence of copious vacancy formation in Ni and Pd under a high hydrogen pressure. *Jpn. J. Appl. Phys.* **32**, L1256-1259 (1993).

487. Y. Fukai, N. Okuma, Formation of superabundant vacancies in Pd hydride under high hydrogen pressures. *Phys. Rev. Lett.* **73**, 1640-1643 (1994).
488. Y. Fukai, Y. Shizuku, Y. Kurokawa, Superabundant vacancy formation in Ni-H alloys. *J. Alloys Compds.* **329**, 195-201 (2001).
489. S. Miraglia, D. Fruchart, E. K. Hill, S. S. M. Tavares, D. Dos Santos, Investigation of the vacancy-ordered phases in the Pd-H system. *J. Alloys and Compounds* **317**, 77-82 (2001).
490. E. K. Storms, A study of those properties of palladium that influence excess energy production by the Pons-Fleischmann effect. *Infinite Energy* **2**, 8, 50 (1996).
491. E. Storms, paper presented at the American Physical Society, Atlanta, GA, March 26, 1999.
492. N. F. Mott, H. Jones, *The theory of the properties of metals and alloys.* (Oxford Univ. Press, London, 1936).
493. R. Feenstra, R. Griessen, D. G. de Groot, Hydrogen induced lattice expansion and effective H-H interaction in single phase PdH. *J. Phys. F., Met. Phys.* **16**, 1933 (1986).
494. L. Pauling, Explanations of cold fusion" (section editor's title). *Nature* (London) **339**, 105 (1989).
495. Y. Fukai, Formation of superabundant vacancies in M-H alloys and some of its consequences: a review. *J. Alloys and Compounds* **356-357**, 263-269 (2003).
496. M. C. McKubre, F. Tanzella, paper presented at the 12th International Conference on Condensed Matter Nuclear Science, Yokohama, Japan, November 27-December 2, 2005.
497. B. Baranowski, S. M. Filipek, Diffusion coefficients of deuterium in palladium deuteride during absorption and desorption in high pressures of gaseous deuterium at 298 K. *Polish Journal of Chemistry* **75**, 1051 (2001).
498. B. Baranowski, R. Wisniewski, The electrical resistance of palladium and palladium-gold alloy (50 wt% Au and Pd) in gaseous hydrogen up to 24000 at at 25°C. *Phys. Stat. Sol. A* **35**, 593 (1969).
499. N. Luo, G. H. Miley, paper presented at the 10th International Conference on Cold Fusion, Cambridge, MA, August 24-29, 2003.
500. P. Tripodi, M. C. H. McKubre, F. L. Tanzella, P. A. Honnor, D. Di Gioacchino, F. Celani, V. Violante, Temperature coefficient of resistivity at compositions approaching PdH. *Phys. Lett. A* **A276**, 122 (2000).
501. M. Berrondo, paper presented at the Anomalous Nuclear Effects in Deuterium/Solid Systems, AIP Conference Proceedings 228, Brigham Young Univ., Provo, UT, October 22-23, 1990.

502. L. Bertalot, F. DeMarco, A. DeNinno, R. Felici, A. LaBarbera, F. Scaramuzzi, V. Violante, paper presented at the 4th International Conference on Cold Fusion, Lahaina, HI, December 6-9, 1993.

503. Y. Fukai, paper presented at the High-Pressure Research: Application to Earth and Planetary Sciences, 1992.

504. P. A. Poyser, M. Kemali, D. K. Ross, Deuterium absorption in Pd0.9Y0.1 alloy. *J. Alloys and Compounds* **253-254**, 175 (1997).

505. A. C. Lawson, J. W. Conant, R. Robertson, R. K. Rohwer, V. A. Young, C. L. Talcott, Debye-Waller factors of PdD_x materials by neutron powder diffraction. *J. Alloys and Compounds* **183**, 174 (1992).

506. R. A. Bond, D. K. Ross, The use of Monte Carlo simulations in the study of a real lattice gas and its application to the α Pd-D system. *J. Phys. F: Met. Phys.* **12**, 597 (1982).

507. R. Abbenseth, H. Wipf, Thermal expansion and lattice anharmonicity of Pd-H and Pd-D alloys. *J. Phys. F: Met. Phys.* **10**, 353 (1980).

508. C. L. Talcott, paper presented at the JOWOG-12 Meeting, Atomic Weapons Estab., Aldermaston, September 10-14, 1990.

509. R. Felici, L. Bertalot, A. De Ninno, A. La Barbera, V. Violante, *In situ* measurement of the deuterium (hydrogen) charging of a palladium electrode during electrolysis by energy dispersive x-ray diffraction. *Rev. Sci. Instr.* **66**, 3344 (1995).

510. N. Asami, T. Senjuh, T. Uehara, M. Sumi, H. Kamimura, S. Miyashita, K. Matsui, paper presented at the 9th International Conference on Cold Fusion, Beijing, China, Tsinghua Univ., May 19-25, 2002.

511. J. M. Rowe, J. J. Rush, H. G. Smith, M. Mostoller, H. E. Flotow, Lattice dynamics of a single crystal of $PdD_{0.63}$. *Phys. Rev. Lett.* **33**, 1297-1300 (1974).

512. S. M. Bennington, M. J. Benham, P. R. Stonadge, J. P. A. Fairclough, D. K. Ross, *In-situ* measurements of deuterium uptake into a palladium electrode using time-of-flight neutron diffractometry. *J. Electroanal. Chem.* **281**, 323 (1990).

513. J. E. Worsham Jr., M. K. Wilkinson, C. G. Shull, Neutron-diffraction observations on the palladium-hydrogen and palladium-deuterium systems. *J. Phys. Chem. Solids* **3**, 303 (1957).

514. S. Pyun, C. Lim, K.-B. Kim, An investigation of the electrochemical kinetics of deuterium insertion into a Pd membrane electrode in 0.1M LiOD solution by the a.c. impedance technique. *J. Alloys and Compounds* **203**, 149 (1994).

515. J. K. Baird, Isotope effect in hydrogen atom diffusion in metals. *Phys. Rev. Lett.* **submitted**, (1994).

516. I. I. Astakhov, V. E. Kazarinov, L. A. Reznikova, G. L. Teplitskaya, Diffusion of hydrogen isotopes in palladium hydride and deuteride in the presence of lithium. *Russ. J. Electrochem.* **30**, 1379 (1994).

517. G. L. Powell, R. Lässer, J. R. Kirkpatrick, J. W. Conant, "Surface and Bulk Effects in the Reaction of H and D with Pd," (Oak Ridge National Laboratory, 1991).
518. G. L. Powell, J. R. Kirkpatrick, J. W. Conant, Surface effects in the reaction of H and D with Pd-macroscopic manifestations. *J. Less-Common Met.* **172-174**, 867 (1991).
519. J. Jorné, Unsteady diffusion reaction of electrochemically produced deuterium in palladium rod. *J. Electrochem. Soc.* **137**, 369 (1990).
520. R. G. Leisure, L. A. Nygren, D. K. Hsu, Ultrasonic relaxation rates in palladium hydride and palladium deuteride. *Phys. Rev. B: Mater. Phys.* **33**, 8325 (1986).
521. S. Majorowski, B. Baranowski, Diffusion coefficients of hydrogen and deiterium in highly concentrated palladium hydride and deuteride phases. *J. Phys. Chem. Solid.* **43**, 1119 (1982).
522. M. Tamaki, K. Tasaka, paper presented at the 3rd International Conference on Cold Fusion, Frontiers of Cold Fusion, Nagoya, Japan, October 21-25, 1992.
523. R. C. Brouwer, R. Griessen, Electromigration of hydrogen in alloys: Evidence of unscreened proton behavior. *Phys. Rev. Lett.* **62**, 1760 (1989).
524. A. J. Maeland, T. R. P. Gibb Jr., X-Ray diffraction observations of the Pd-H system through the critical region. *J. Phys. Chem.* **65**, 1270 (1961).
525. H. C. Jamieson, G. C. Weathrely, F. D. Manchester, The β-α phase transformation in palladium-hydrogen alloys. *J. Less-Common Met.* **56**, 85 (1976).
526. J. E. Schirber, B. Morosin, Lattice constants of β-Pd-H_x and β-PdD_x with x near 1.0. *Phys. Rev. B* **12**, 117 (1975).
527. G. Mengoli, M. Fabrizio, C. Manduchi, G. Zannoni, Surface and bulk effects in the extraction of hydrogen from highly loaded Pd sheet electrodes. *J. Electroanal. Chem.* **350**, 57-72 (1993).
528. M. Bertolotti, G. L. Liakhou, R. Li Voti, S. Paoloni, C. Sibilia, V. Violante, paper presented at the 7th International Conference on Cold Fusion, Vancouver, Canada, April 19-24, 1998.
529. A. C. Switendick, Electronic structure and stability of palladium hydrogen (deuterium) systems, PdH(D)n, $1 \leq n \leq 3$. *J. Less-Common Met.* **172-174**, 1363 (1991).
530. P. M. Richards, Molecular-dynamics investigation of deuterium separation in $PdD_{1.1}$. *Phys. Rev. B* **40**, 7966 (1989).
531. S.-H. Wei, A. Zunger, Instability of diatomic deuterium in fcc palladium. *J. Fusion Energy* **9**, 367 (1990).
532. S. H. Wei, A. Zunger, Stability of atomic and diatomic hydrogen in fcc palladium. *Solid State Commun.* **73**, 327 (1990).

533. B. I. Dunlap, D. W. Brenner, R. C. Mowrey, J. W. Mintmire, C. T. White, Linear combination of Gaussian-type orbitals - local-density-functional cluster studies of D-D interactions in titanium and palladium. *Phys. Rev. B* **41**, 9683 (1990).

534. G. V. Fedorovich, Screening of the Coulomb potential in a nondegenerate hydrogen isotope gas. *Fusion Technol.* **25**, 120 (1994).

535. G. Preparata, Cold fusion '93': some theoretical ideas. *Trans. Fusion Technol.* **26**, 397-407 (1994).

536. H. Yuki, T. Satoh, T. Ohtsuki, T. Aoki, H. Yamazaki, J. Kasagi, paper presented at the 6th International Conference on Cold Fusion, Progress in New Hydrogen Energy, Lake Toya, Hokkaido, Japan, October 13-18, 1996.

537. S. E. Jones, E. P. Palmer, J. B. Czirr, D. L. Decker, G. L. Jensen, J. M. Thorne, S. F. Taylor, J. Rafelski, Observation of cold nuclear fusion in condensed matter. *Nature (London)* **338**, 737 (1989).

538. K. P. Sinha, A. Meulenberg, Lochon-mediated low-energy nuclear reactions. *J. Cond. Matter Nucl. Sci.* **6**, 55-63 (2012).

539. E. N. Tsyganov, Cold nuclear fusion. *Phys. Atomic Nuclei* **75**, 153-159 (2011).

540. K. Czerski, paper presented at the 15th International Conference on Condensed Matter Nuclear Science, Rome, Italy, October 5-9, 2009.

541. M. L. Oliphant, P. Harteck, E. Rutherford, Transmutation effects observed with heavy hydrogen. *Nature (London)* **133**, 413-413 (1934).

542. R. J. Beuhler, G. Friedlander, L. Friedman, Cluster-impact fusion [Erratum]. *Phys. Rev. Lett.* **88**, 2108 (1992).

543. R. J. Beuhler, Y. Y. Chu, G. Friedlander, L. Friedman, W. Kunnmann, Deuteron-deuteron fusion by impact of heavy-water clusters on deuterated surfaces. *J. Phys. Chem.* **94**, 7665 (1991).

544. R. J. Beuhler, G. Friedlander, L. Friedman, Cluster-impact fusion. *Phys. Rev. Lett.* **63**, 1292 (1990).

545. R. J. Beuhler, Y. Y. Chu, G. Friedlander, L. Friedman, W. Kunnmann, Deuteron-deuteron fusion by impact of heavy-water clusters on deuterated surfaces. *J. Chem. Phys.* **94**, 7665-7671 (1990).

546. R. J. Beuhler, G. Friedlander, L. Friedman, Cluster-impact fusion. *Phys. Rev. Lett.* **63**, 1292 (1989).

547. M. Rabinowitz, Y. E. Kim, R. A. Rice, G. S. Chulick, paper presented at the Anomalous Nuclear Effects in Deuterium/Solid Systems, AIP Conference Proceedings 228, Brigham Young Univ., Provo, UT, October 22-23, 1990.

548. P. M. Echenique, J. R. Manson, R. H. Ritchie, Cluster-impact fusion. *Phys. Rev. Lett.* **64**, 1413 (1990).

549. K. Czerski, paper presented at the New Advances on the Fleischmann-Pons Effect, European Parliament, Brussels, June 3, 2013.

550. A. S. Roussetski, M. N. Negodaev, A. G. Lipson, paper presented at the 15th International Conference on Condensed Matter Nuclear Science, Rome, Italy, October 5-9, 2009.
551. J. Kasagi, paper presented at the 14th International Conference on Condensed Matter Nuclear Science, Washington, DC, August 10-15, 2008.
552. T. Dairaku, Y. Katayama, T. Hayashi, Y. Isobe, A. Takahashi, paper presented at the 9th International Conference on Cold Fusion, Condensed Matter Nuclear Science, Tsinghua Univ., Beijing, China, May 19-24, 2002.
553. H. Ikegami, Buffer energy nuclear fusion. *Jpn. J. Appl. Phys.* **40**, 6092-6098 (2001).
554. J. Kasagi, H. Yuki, T. Baba, T. Noda, J. Taguchi, W. Galster, Strongly enhanced Li + D reaction in Pd observed in deuteron bombardment on PdLix with energies between 30 and 75 keV. *J. Phys. Soc. Japan* **73**, 608-612 (1998).
555. J. Kasagi, H. Yuki, T. Itoh, N. Kasajima, T. Ohtsuki, A. G. Lipson, paper presented at the 7th International Conference on Cold Fusion, Vancouver, Canada, April 19-24, 1998.
556. J. Kasagi, H. Yuki, T. Baba, T. Noda, J. Taguchi, W. Galster, Energetic protons and alpha particles emitted in 150-keV deuteron bombardment on deuterated Ti. *J. Phys. Soc. Japan* **64**, 777-783 (1995).
557. J. Kasagi, T. Murakami, T. Yajima, S. Kobayashi, M. Ogawa, Measurements of the D + D reaction in Ti metal with incident energies between 4.7 and 18 keV. *J. Phys. Soc. Japan* **64**, 608-612 (1995).
558. T. Iida, M. Fukuhara, H. Miyazaki, Y. Sueyoshi, Sunarno, J. Datemichi, A. Takahashi, paper presented at the 3rd International Conference on Cold Fusion, Frontiers of Cold Fusion, Nagoya, Japan, October 21-25, 1992.
559. T. G. Dignan, M. C. Bruington, R. T. Johnson, R. W. Bland, A search for neutrons from fusion in a highly deuterated cooled palladium thin film. *J. Fusion Energy* **9**, 469 (1990).
560. F. E. Cecil, D. Ferg, T. E. Furtak, C. Mader, J. A. McNeil, D. L. Williamson, Study of energetic charged particles emitted from thin deuterated palladium foils subject to high current densities. *J. Fusion Energy* **9**, 195 (1990).
561. A. T. Budnikov, P. A. Danilov, G. A. Kartamyshev, N. P. Katrich, V. P. Seminozhenko, Study of gases evolving from palladium, nickel and copper, bombarded with D+ ions, from palladium saturated with gases by heavy water electrolysis and by heating in deuterium. *Vopr. At. Nauki Tekh. Ser.: Fiz. Radiats. Povr. Radiats. Materialoved.*, 81 (in Russian) (1990).

562. A. Takahashi, H. Miyadera, K. Ochiai, Y. Katayama, T. Hayashi, T. Dairaku, paper presented at the 10th International Conference on Cold Fusion, Cambridge, MA, August 24-29, 2003.
563. G. Preparata, paper presented at the Anomalous Nuclear Effects in Deuterium/Solid Systems, AIP Conference Proceedings 228, Brigham Young Univ., Provo, UT, October 22-23, 1990.
564. T. Takeda, T. Takizuka, Fractofusion mechanism. *J. Phys. Soc. Japan* **58**, 3073 (1989).
565. K. Yasui, Fractofusion mechanism. *Fusion Technol.* **22**, 400 (1992).
566. B. V. Derjaguin, A. G. Lipson, V. A. Kluev, D. M. Sakov, Y. P. Toporov, Titanium fracture yields neutrons? *Nature* (London) **341**, 492 (issue 6242, 6212. Oct, Scientific Corresp. (1989).
567. A. G. Lipson, B. V. Deryagin, V. A. Klyuev, Y. P. Toporov, M. G. Sirotyuk, O. B. Khavroshkin, D. M. Sakov, Initiation of nuclear fusion by cavitation action on deuterium-containing media. *Zh. Tekh. Fiz.* **62**, 121 (in Russian) (1992).
568. H. O. Menlove, M. M. Fowler, E. Garcia, A. Mayer, M. C. Miller, R. R. Ryan, S. E. Jones, The measurement of neutron emission from Ti plus D_2 gas. *J. Fusion Energy* **9**, 215 (1990).
569. H. O. Menlove, M. M. Fowler, E. Garcia, M. C. Miller, M. A. Paciotti, R. R. Ryan, S. E. Jones, Measurement of neutron emission from Ti and Pd in pressurized D_2 gas and D_2O electrolysis cells. *J. Fusion Energy* **9**, 495 (1990).
570. H. O. Menlove, M. C. Miller, Neutron-burst detectors for cold-fusion experiments. *Nucl. Instr. Methods Phys. Res. A* **299**, 10 (1990).
571. H. O. Menlove, M. A. Paciotti, T. N. Claytor, D. G. Tuggle, paper presented at the 2nd Annual Conference on Cold Fusion, The Science of Cold Fusion, Como, Italy, June 29-July 4, 1991.
572. C. Sánchez, J. Sevilla, B. Escarpizo, F. Fernandez, J. Canizares, paper presented at the Understanding Cold Fusion Phenomena Conference, Varenna, Italy September 15-16, 1989.
573. C. Sánchez, J. Sevilla, B. Escarpizo, F. J. Fernández, J. Canizares, Nuclear products detection during electrolysis of heavy water with titanium and platinum electrodes. *Solid State Commun.* **71**, 1039 (1989).
574. D. Seeliger, M. Bittner, A. Meister, R. Schwierz, T. Streil, paper presented at the 2nd Annual Conference on Cold Fusion, The Science of Cold Fusion, Como, Italy, June 29-July 4, 1991.
575. V. I. Sannikov, V. G. Gorodetskii, E. M. Sulimov, B. G. Polosukhin, V. Y. Kudyakov, Emission of neutrons and gamma-quanta from a titanium electrode polarised by a current in the gas phase over LiD. *Rasplavy*, 86 (in Russian) (1991).
576. V. F. Zelenskii, V. F. Rybalko, Studies of neutron emission by mechanical destruction of Ti and Pd samples, saturated with deuterium.

Vopr. At. Nauki Tekh. Ser.: Fiz. Radiats. Povr. Radiats. Materialoved. **2**, 46 (in Russian) (1991).

577. F. Scaramuzzi, paper presented at the 2nd Annual Conference on Cold Fusion, The Science of Cold Fusion, Como, Italy, June 29-July 4, 1991.
578. F. Cardone, A. Carpinteri, G. Lacidogna, Piezonuclear neutrons from fracturing of inert solids. *Phys. Lett. A* **373**, 4158-4163 (2009).
579. A. Carpinteri, F. Cardone, G. Lacidogna, A. Manuello, O. Borla, paper presented at the 15th International Conference on Condensed Matter Nuclear Science, Rome, Italy, October 5-9, 2009.
580. A. Carpinteri, F. Cardone, G. Lacidogna, Piezonuclear neutrons from brittle fracture: early results of mechanical compression tests. *Strain* **45**, 332-339 (2009).
581. A. Carpinteri, A. Chiodoni, A. Manuello, R. Sandrone, Compositional and microchemical evidence of piezonuclear fission reactions in rock specimens subjected to compression tests. *Strain* (2010).
582. A. Carpinteri, A. J. Manuel, Geomechanical and geochemical evidence of piezonuclear fission reactins in the earth's crust. *Strain* **46**, (2010).
583. A. Carpinteri, A. Chiodoni, A. Manuello, R. Sandrone, Compositional and microchemical evidence of piezonuclear fission reactions in rock specimens subjected to compression tests. *Strain* **47**, 282-292 (2010).
584. B. Y. Moizhes, Formation of a compact D_2 molecule in interstitial sites - a possible explanation for cold nuclear fusion. *Sov. Tech. Phys. Lett.* **17**, 540 (1991).
585. S. E. Weber, F. S. Liu, S. N. Khanna, B. K. Rao, P. Jena, Theory of hydrogen pairing in metals. *J. Less-Common Met.* **172-174**, 485 (1991).
586. P. L. Hagelstein, Constraints on energetic particles in the Fleischmann-Pons experiment. *Naturwissenschaften* **97**, 345 (2010).
587. B. Stritzker, J. Becker, Superconductivity in metastable Pd-alloys produced by ion implantation at low temperatures. *Phys. Lett. A* **51**, 147 (1975).
588. J. E. Schirber, C. J. M. Northrup Jr., Concentration dependence of the superconducting transition temperature In Pd-H and Pd-D. *Phys. Rev. B* **10**, 3818 (1974).
589. V. S. Postnikov, V. V. Postnikov, V. M. Fedorov, Instability and superconductivity in Pd-Ag-D and Pd-H systems. *Phys. Stat. Sol. B* **85**, K115 (1978).
590. F. A. Lewis, *The palladium hydrogen system*. (Academic Press, New York, 1967).
591. G. Shani, C. Cohen, A. Grayevsky, A. Brokman, Evidence for a background neutron enhanced fusion in deuterium absorbed palladium. *Solid State Commun.* **72**, 53 (1989).
592. B. Stella, M. Corradi, F. Ferrarotto, V. Milone, F. Celani, A. Spallone, paper presented at the 3rd International Conference on Cold Fusion, Frontiers of Cold Fusion, Nagoya, Japan, October 21-25, 1992.

593. Y. Fukai, Formation of superabundant vacancies in metal hydrides at high temperatures. *J. Alloys and Compounds* **231**, 35-40 (1995).
594. H. Wipf, V. Erckman, On permeation techniques for electrotransport studies on metal-hydrogen systems. *Scripta Metal.* **10**, 813 (1976).
595. P. L. Hagelstein, Coherent and semicoherent neutron transfer reactions III: phonon frequency shifts. *Fusion Technol.* **23**, 353-361 (1993).
596. K. P. Sinha, P. L. Hagelstein, Electron screening in metal deuterides. *Trans. Am. Nucl. Soc.* **83**, 368 (2000).
597. P. L. Hagelstein, paper presented at the 9th International Conference on Cold Fusion, Condensed Matter Nuclear Science, Tsinghua Univ., Beijing, China, May 19-24, 2002.
598. I. Chaudhary, P. L. Hagelstein, paper presented at the 10th International Conference on Cold Fusion, Cambridge, MA, August 24-29, 2003.
599. P. L. Hagelstein, "Anomalies in metal deuterides," (MIT, 2002).
600. P. L. Hagelstein, paper presented at the 10th International Conference on Cold Fusion, Cambridge, MA, August 24-29, 2003.
601. P. L. Hagelstein, paper presented at the 10th International Conference on Cold Fusion, Cambridge, MA, August 24-29, 2003.
602. I. Chaudhary, P. L. Hagelstein, paper presented at the 10th International Conference on Cold Fusion, Cambridge, MA, August 24-29, 2003.
603. P. L. Hagelstein, M. E. Melich, R. E. Johnson, paper presented at the 14th International Conference on Condensed Matter Nuclear Science, Washington DC, August 10-15, 2008.
604. P. L. Hagelstein, paper presented at the 11th International Conference on Cold Fusion, Marseilles, France, October 31-November 5, 2004.
605. P. L. Hagelstein, paper presented at the 12th International Conference on Condensed Matter Nuclear Science, Yokohama, Japan, November 27-December 2, 2005.
606. I. Chaudhary, P. L. Hagelstein, paper presented at the 12th International Conference on Condensed Matter Nuclear Science, Yokohama, Japan, November 27-December 2, 2005.
607. P. L. Hagelstein, I. Chaudhary, in *ACS Symposium Series 998, Low-Energy Nuclear Reactions Sourcebook,* J. Marwan, S. B. Krivit, Eds. (American Chemical Society, Washington, DC, 2008), pp. 249-267.
608. P. L. Hagelstein, I. Chaudhary, Electron mass shift in nonthermal systems. *J. Phys. B* **41**, 125001 (2008).
609. P. L. Hagelstein, M. C. McKubre, F. Tanzella, paper presented at the 15th International Conference on Condensed Matter Nuclear Science, Rome, Italy, October 5-9, 2009.
610. P. L. Hagelstein, I. Chaudhary, paper presented at the 15th International Conference on Condensed Matter Nuclear Science, Rome, Italy, October 5-9, 2009.
611. P. L. Hagelstein, I. Chaudhary. (2010).

612. P. L. Hagelstein, http://www.youtube.com/watch?v=Al7NMQLvATo (2014).
613. P. L. Hagelstein, Coherent fusion theory. *J. Fusion Energy* **9**, 451-464 (1990).
614. P. L. Hagelstein, S. Kaushik, paper presented at the 4th International Conference on Cold Fusion, Lahaina, HI, December 6-9, 1993.
615. P. L. Hagelstein, Bird's eye view of phonon models for excess heat in the Fleischmann-Pons experiment. *J. Cond. Matter Nucl. Sci.* **6**, 169-180 (2012).
616. P. L. Hagelstein, I. U. Chaudhary, Central and tensor contributions to the phonon-exchange matrix element for the $D_2/^4He$ transition. *J. Cond. Matter Nucl. Sci.* **11**, 15-58 (2013).
617. P. L. Hagelstein, I. U. Chaudhary, Lossy spin-boson model with an unstable upper state and extension to N-level systems. *J. Cond. Matter Nucl. Sci.* **11**, 59-92 (2013).
618. A. B. Karabut, E. A. Karabut, paper presented at the 14th International Conference on Condensed Matter Nuclear Science, Washington DC, August 10-15, 2008.
619. P. L. Hagelstein, I. Chaudhary, Coupling between a deuteron and a lattice. *J. Cond. Matter Nucl. Sci.* **9**, 50-63 (2012).
620. P. L. Hagelstein, I. Chaudhary, Born-Oppenheimer and fixed-point models for second-order phonon exchange in a metal. *J. Cond. Matter Nucl. Sci.* **12**, 69-104 (2013).
621. P. L. Hagelstein, I. U. chaudhary, Phonon-nuclear coupling for anomalies in condensed matter nuclear science. *J. Cond. Matter Nucl. Sci.* **12**, 105-142 (2013).
622. M. R. Swartz, Quasi-one-dimensional model of electrochemical loading of isotopic fuel into a metal. *Fusion Technol.* **22**, 296-300 (1992).
623. M. R. Swartz, Phusons in nuclear reactions in solids. *Fusion Technol.* **31**, 228 (1997).
624. M. R. Swartz, Possible deuterium production from light water excess enthalpy experiments using nickel cathodes. *J. New Energy* **1**, 68 (1996).
625. M. R. Swartz, paper presented at the 7th International Conference on Cold Fusion, Vancouver, Canada, April 19-24, 1998.
626. M. R. Swartz. (1999).
627. M. R. Swartz. (1999).
628. M. R. Swartz, paper presented at the 14th International Conference on Condensed Matter Nuclear Science, Washington DC, August 10-15, 2008.
629. M. R. Swartz, LANR nanostructures and metamaterials driven at their optimal operating point. *J. Cond. Matter Nucl. Sci.* **6**, 149-168 (2012).

630. G. Miley, *Life at the center of the energy crisis: A technologist's seach for a black swan*. (World Scientific, Singapore, 2013).
631. M. Ragheb, G. H. Miley, On the possibility of deuteron disintegration in electrochemically compressed deuterium ion (D+) in a palladium cathode. *Fusion Technol*. **16**, 243. (1989).
632. M. Ragheb, G. H. Miley, Deuteron disintegration in condensed media. *J. Fusion Energy* **9**, 429 (1990).
633. H. Hora, L. Cicchitelli, G. H. Miley, M. Ragheb, A. Scharmann, W. Scheid, Plasma and surface tension model for explaining the surface effect of tritium generation at cold fusion. *Nuovo Cimento Soc. Ital. Fis*. **12D**, 393 (1990).
634. H. Hora, S. Eliezer, F. J. M. Farley, A. K. Ghatak, M. P. Goldsworthy, F. Green, G. W. Kentwell, P. Lalousis, W. Scheid, R. Stening, S. Tapalaga, Eds., *Consequences of high electric fields in laser produced plasmas*, (Plenum Press, New York, 1986), pp. 347.
635. H. Hora, P. Lalousis, R. Postle, Analysis of the inverted double layers produced by nonlinear forces in a laser-produced plasma. *Phys.Rev. Lett*. **53**, 1650 (1984).
636. G. H. Miley, J. U. Patel, J. Javedani, H. Hora, J. C. Kelly, J. Tompkins, paper presented at the 3rd International Conference on Cold Fusion, Frontiers of Cold Fusion, Nagoya, Japan, October 21-25, 1992.
637. G. Miley, E. G. Batyrbekov, H. Hora, J. U. Patel, J. Tompkins, R. K. Zich, paper presented at the International Symposium on Cold Fusion and Advanced Energy Sources, Belarusian State University, Minsk, Belarus, May 24-26, 1994.
638. G. H. Miley, H. Hora, E. G. Batyrbekov, R. L. Zich, Electrolytic cell with multilayer thin-film electrodes. *Trans. Fusion Technol*. **26**, 313 (1994).
639. H. Hora, J. C. Kelly, J. U. Patel, M. A. Prelas, G. H. Miley, J. W. Tompkins, Screening in cold fusion derived from D-D reactions. *Phys. Lett. A* **175**, 138 (1993).
640. H. Hora, G. H. Miley, J. C. Kelly, F. Osman, paper presented at the 9th International Conference on Cold Fusion, Condensed Matter Nuclear Science, Tsinghua Univ., Beijing, China, May 19-24, 2002.
641. G. H. Miley, G. Narne, M. J. Williams, J. A. Patterson, J. Nix, D. Cravens, H. Hora, paper presented at the 6th International Conference on Cold Fusion, Progress in New Hydrogen Energy, Lake Toya, Hokkaido, Japan, October 13-18, 1996.
642. J. A. Patterson. (US Patent # 5,318,675, 1994).
643. J. A. Patterson. (US Patent #5,494,559, 1996).
644. H. Hora, G. H. Miley, New magic numbers from low energy nuclear transmutations predict element (306)X(126) for compound reactions. *Czech. J. Phys*. **48**, 1111 (1998).

645. H. Hora, G. H. Miley, J. C. Kelly, Y. Narne, paper presented at the 7th International Conference on Cold Fusion, Vancouver, Canada, April 19-24, 1998.
646. J. Audouze, S. Vauclair, *An introduction to nuclear astrophysics*. (D. Reidel Publishing Co., Dordrecht, 1980).
647. K. Rutz, M. Bender, T. Bürvenich, T. Schilling, P. G. Reinhard, J. A. Maruhn, W. Greiner, Superheavy nuclei in self-consistent nuclear calculations. *Phys. Rev. C* **56**, 238 (1997).
648. H. Hora, G. H. Miley, Heavy nuclide synthesis by neutrons in astrophysics and by screened protons in host metals. *Czech. J. Phys.* **50**, 433-439 (2000).
649. G. Miley, P. Shrestha, in *ACS Symposium Series 998, Low-Energy Nuclear Reactions Sourcebook,* J. Marwan, S. B. Krivit, Eds. (American Chemical Society, Washington, DC, 2008), pp. 173-218.
650. N. Luo, C. H. Castano, S. O. Kim, A. G. Lipson, G. H. Miley, paper presented at the 9th International Conference on Cold Fusion, Condensed Matter Nuclear Science, Tsinghua Univ., Beijing, China, May 19-24, 2002.
651. C. H. Castano, A. G. Lipson, S. O. Kim, G. H. Miley, paper presented at the 9th International Conference on Cold Fusion, Condensed Matter Nuclear Science, Tsinghua Univ., Beijing, China, May 19-24, 2002.
652. H. Hora, G. Miley, X. Z. Li, J. Kelly, F. Osman, paper presented at the 11th International Conference on Cold Fusion, Marseilles, France, October 31-November 5, 2004.
653. H. Hora, N. Ghahramani, G. H. Miley, M. Ghanaatian, M. Hooshmand, K. Philberth, F. Osman, in *Low-Energy Nuclear Reactions Sourcebook Volume 2,* J. Marwan, S. Krivit, Eds. (Oxford University Press, 2009).
654. G. Miley, X. Yang, K.-J. Kim, H. Hora, paper presented at the 17th International Conference on Condensed Matter Nuclear Science, Daejeon, Korea, August 12-17, 2012.
655. A. Takahashi, T. Iida, F. Maekawa, H. Sugimoto, S. Yoshida, Windows of cold nuclear fusion and pulsed electrolysis experiments. *Fusion Technol.* **19**, 380 (1991).
656. A. Takahashi, Cold fusion research: recent progress. *Kaku Yugo Kenkyu* **68**, 360 (in Japanese) (1992).
657. H. Miyamaru, A. Takahashi, paper presented at the 3rd International Conference on Cold Fusion, Frontiers of Cold Fusion, Nagoya, Japan, October 21-25, 1992.
658. A. Takahashi, T. Iida, T. Takeuchi, A. Mega, Excess heat and nuclear products by D_2O/Pd electrolysis and multibody fusion. *Int. J. Appl. Electromagn. Mater.* **3**, 221-230 (1992).
659. A. Takahashi, Opening possibility of deuteron-catalyzed cascade fusion channel in PdD under D_2O electrolysis. *J. Nucl. Sci. Technol.* **26**, 558. (1989).

660. M. Ohta, A. Takahashi, paper presented at the 8th International Conference on Cold Fusion, Lerici (La Spezia), Italy, May 21-26, 2000.
661. A. Takahashi, N. Yabuuchi, paper presented at the 13th International Conference on Condensed Matter Nuclear Science, Sochi, Russia, June 25-July 1, 2007.
662. A. Takahashi, N. Yabuuchi, On Condensation Force of TSC. *J. Cond. Matter Nucl. Sci.* **1**, 97-105 (2007).
663. A. Takahashi, paper presented at the 10th International Conference on Cold Fusion, Cambridge, MA, August 24-29, 2003.
664. H. Takahashi, paper presented at the 10th International Conference on Cold Fusion, Cambridge, MA, August 24-29, 2003.
665. A. Takahashi, Are Ni + H nuclear reactions possible? *J. Cond. Matter Nucl. Sci.* **9**, 108-127 (2012).
666. A. Takahashi, D. Rocha, paper presented at the 13th Meeting of the Japan Cold Fusion Research Society, Nagoya, Japan, December 8-9, 2012.
667. A. Takahashi, paper presented at the 5th International Workshop on Anomalies in Hydrogen/Deuterium Loaded Metals, Asti, Italy, March 19-21, 2004.
668. A. Takahashi, paper presented at the 17th International Conference on Condensed Matter Nuclear Science, Daejeon, Korea, August 12-17, 2012.
669. A. Takahashi, Production of stable isotopes by selective channel photofission of Pd. *Jpn. J. Appl. Phys. A* **40**, 7031-7046 (2001).
670. G. S. Chulick, Y. E. Kim, R. A. Rice, "Low Energy D-D Fusion Experimental Cross-Sections," (Purdue University, 1989).
671. Y. E. Kim, "Nuclear Theory Hypotheses for Cold Fusion," (Purdue University, 1989).
672. Y. E. Kim, "Neutron-Induced Photonuclear Chain-Reaction Process in Pd Deuteride," (Purdue University report, 1989).
673. Y. E. Kim, "Comment on "Cluster-Impact Fusion", (Purdue University, 1989).
674. Y. E. Kim, "Fission-Induced Inertial Confinement Hot Fusion and Cold Fusion with Electrolysis," (Purdue, 1989).
675. Y. E. Kim, R. A. Rice, G. S. Chulick, "The Electron Screening Effect on Fusion Cross-sections and Rates in Physical Processes," (Purdue University, 1989).
676. D. J. Klepacki, Y. E. Kim, R. A. Brandenburg, "Two-Body Photodisintegration of 3-Helium and 3-Helium Near the Giant Resonance I. Plane-Wave Approximation," (Purdue University, 1989).
677. R. A. Rice, G. S. Chulick, Y. E. Kim, J. Yoon, "The Effect of Velocity Distribution on Cold Deuterium-Deuterium Fusion," (Purdue University, 1989).

678. Y. E. Kim, A. Zubarev, paper presented at the 7th International Conference on Cold Fusion, Vancouver, Canada, April 19-24, 1998.
679. Y. E. Kim, A. Zubarev, Equivalent linear two-body method for Bose-Einstein condensates in time-dependent harmonic traps. *Phys. Rev. A* **66**, 053602 (2002).
680. Y. E. Kim, paper presented at Fusion 03: From a Tunneling Nuclear Microscope to Nuclear Processes in Matter, Matsushita, Japan, November 12-15, 2003.
681. Y. E. Kim, D. Koltick, R. Pringer, J. Myers, R. Koltick, paper presented at the 10th International Conference on Cold Fusion, Cambridge, MA, August 24-29, 2003.
682. Y. E. Kim, D. Koltick, A. Zubarev, paper presented at the 10th International Conference on Cold Fusion, Cambridge, MA, August 24-29, 2003.
683. Y. E. Kim, D. Koltick, R. G. Reifenberger, A. L. Zubarev, paper presented at the 11th International Conference on Cold Fusion, Marseilles, France, October 31-November 5, 2004.
684. Y. E. Kim, T. O. Passell, paper presented at the 11th International Conference on Cold Fusion, Marseilles, France, October 31-November 5, 2004.
685. Y. E. Kim, A. L. Zubarev, paper presented at the 12th International Conference on Condensed Matter Nuclear Science, Yokohama, Japan, November 27-December 2, 2005.
686. H. Y. Kim, paper presented at the 15th International Conference on Condensed Matter Nuclear Science, Rome, Italy, October 5-9, 2009.
687. Y. E. Kim, Bose-Einstein condensate theory of deuteron fusion in metal. *J. Cond. Matter Nucl. Sci.* **5**, 14 (2010).
688. Y. E. Kim, Bose-Einstein condensate theory of deuteron fusion in metal. *J. Cond. Matter Nucl. Sci.* **4**, 188-201 (2011).
689. Y. E. Kim, Generalized theory of Bose-Einstein condensation nuclear fusion for hydrogen-metal system. *J. Nucl. Phys.*, http://www.journal-of-nuclear-physics.com/?p=501 (2011).
690. H. Hora, G. Miley, M. Prelas, K. J. Kim, X. Yang, paper presented at the 17th International Conference on Condensed Matter Nuclear Science, Daejeon, Korea, August 12-17, 2012.
691. Y. E. Kim, T. E. Ward, Bose-Einstein condensation nuclear fusion: role of monopole transition. *J. Cond. Matter Nucl. Sci.* **6**, 101-107 (2012).
692. R. W. Standley, M. Steinback, C. B. Satterthwaite, Superconductivity of PdHx(Dx) from 2K to 4K. *Solid State Commun.* **31**, 801-804 (1979).
693. T. Bressani, E. Del Giudice, G. Preparata, First steps toward an understanding of 'cold' nuclear fusion. *Il Nuovo Cimento* **A101**, 845 (1989).

694. G. Preparata, paper presented at the 1st Annual Conference on Cold Fusion, University of Utah Research Park, Salt Lake City, UT, March 28-31, 1990.
695. G. Preparata, paper presented at the 2nd Annual Conference on Cold Fusion, The Science of Cold Fusion, Como, Italy, June 29-July 4, 1991.
696. G. Preparata, Some theories of 'cold' nuclear fusion: a review. *Fusion Technol.* **20**, 82-92 (1991).
697. G. Preparata, A new look at solid-state fractures, particle emission and 'cold' nuclear fusion. *Nuovo Cimento A* **104**, 1259 (1991).
698. T. Bressani, G. Preparata, What makes a crystal stiff enough for the Mossbauer effect? *Il Nuovo Cimento, Note Brevi* **14D**, 345-349 (1992).
699. G. Preparata, paper presented at the 3rd International Conference on Cold Fusion, Frontiers of Cold Fusion, Nagoya, Japan, October 21-25, 1992.
700. G. Preparata, paper presented at the 4th International Conference on Cold Fusion, Lahaina, HI, December 6-9, 1993.
701. M. Fleischmann, S. Pons, G. Preparata, Possible theories of cold fusion. *Nuovo Cimento* **107A**, 143-156 (1994).
702. M. Fleischmann, paper presented at the 1st Annual Conference on Cold Fusion, University of Utah Research Park, Salt Lake City, UT, March 28-31, 1990.
703. M. Fleischmann, paper presented at the 9th International Conference on Cold Fusion, Condensed Matter Nuclear Science, Tsinghua Univ., Beijing, China, May 19-24, 2002.
704. A. J. Leggett, G. Baym, Exact upper bounds on barrier penetration probabilities in many-body systems: Application to 'cold fusion'. *Phys. Rev. Lett.* **63**, 191 (1989).
705. A. J. Leggett, G. Baym, Can solid-state effects enhance the cold-fusion rate? *Nature (London)* **340**, 45 . (1989).
706. J. Schwinger, Cold fusion: a hypothesis. *Z. Naturforsch.* **45A**, 756 (1990).
707. M. C. H. McKubre, R. C. Rocha-Filho, S. I. Smedley, F. L. Tanzella, S. Crouch-Baker, T. O. Passell, J. Santucci, paper presented at the 2nd Annual Conference on Cold Fusion, The Science of Cold Fusion, Como, Italy, June 29-July 4, 1991.
708. E. K. Storms, C. L. Talcott, paper presented at the 1st Annual Conference on Cold Fusion, University of Utah Research Park, Salt Lake City, UT, March 28-31, 1990.
709. D. Lewis, K. Sköld, A phenomenological study of the Fleischmann-Pons effect. *J. Electroanal. Chem.* **294**, 275 (1990).
710. J. Schwinger, paper presented at the 1st Annual Conference on Cold Fusion, University of Utah Research Park, Salt Lake City, UT, March 28-31, 1990.

711. P. L. Hagelstein, paper presented at the 1st Annual Conference on Cold Fusion, University of Utah Research Park, Salt Lake City, UT, March 28-31, 1990.
712. M. Rabinowitz, paper presented at the 4th International Conference on Cold Fusion, Lahaina, HI, December 6-9, 1993.
713. J. M. Blatt, V. F. Weisskopf, *Theoretical Nuclear Physics*. (Wiley, New York, 1952).
714. J. Schwinger, Nuclear energy in an atomic lattice. 1. *Z. Phys. D: At., Mol. Clusters* **15**, 221 (1990).
715. S. R. Chubb, T. A. Chubb, "Nuclear Fusion in a Solid via a Bose Bloch Concentrate," (Naval Research Laboratory, Memorandum Report 6617, Washington, 1990).
716. T. A. Chubb, S. R. Chubb, Cold fusion as an interaction between ion band states. *Fusion Technol.* **20**, 93-99 (1991).
717. S. R. Chubb, T. A. Chubb, paper presented at the 3rd International Conference on Cold Fusion, Frontiers of Cold Fusion, Nagoya, Japan, October 21-25, 1992.
718. S. R. Chubb, T. A. Chubb, Ion band state fusion: reactions, power density, and the quantum reality question. *Fusion Technol.* **24**, 403 (1993).
719. S. R. Chubb, T. A. Chubb, The role of hydrogen ion band states in cold fusion. *Trans. Fusion Technol.* **26**, 414 (1994).
720. T. A. Chubb, S. R. Chubb, paper presented at the 5th International Conference on Cold Fusion, Monte-Carlo, Monaco, April 9-13, 1995.
721. S. R. Chubb, T. A. Chubb, paper presented at the 6th International Conference on Cold Fusion, Progress in New Hydrogen Energy, Lake Toya, Hokkaido, Japan, October 13-18, 1996.
722. T. A. Chubb, S. R. Chubb, paper presented at the 6th International Conference on Cold Fusion, Progress in New Hydrogen Energy, Lake Toya, Hokkaido, Japan, October 13-18, 1996.
723. T. A. Chubb, S. R. Chubb, paper presented at the 7th International Conference on Cold Fusion, Vancouver, Canada, April 19-24, 1998.
724. S. R. Chubb, T. A. Chubb, paper presented at the 8th International Conference on Cold Fusion, Lerici (La Spezia), Italy, May 21-26, 2000.
725. T. A. Chubb, S. R. Chubb, paper presented at the 8th International Conference on Cold Fusion, Lerici (La Spezia), Italy, May 21-26, 2000.
726. T. A. Chubb, S. R. Chubb, paper presented at the 9th International Conference on Cold Fusion, Beijing, China, May 19-25, 2002.
727. S. R. Chubb, T. A. Chubb, paper presented at the 9th International Conference on Cold Fusion, Condensed Matter Nuclear Science, Tsinghua Univ., Beijing, China, May 19-24, 2002.

728. T. A. Chubb, paper presented at the 9th International Conference on Cold Fusion, Condensed Matter Nuclear Science, Tsinghua Univ., Beijing, China, May 19-24, 2002.
729. T. A. Chubb, paper presented at the 10th International Conference on Cold Fusion, Cambridge, MA, August 24-29, 2003.
730. T. A. Chubb, paper presented at the 10th International Conference on Cold Fusion, Cambridge, MA, August 24-29, 2003.
731. S. R. Chubb, paper presented at the 10th International Conference on Cold Fusion, Cambridge, MA, August 24-29, 2003.
732. S. R. Chubb, paper presented at the 10th International Conference on Cold Fusion, Cambridge, MA, August 24-29, 2003.
733. T. A. Chubb, paper presented at the 11th International Conference on Cold Fusion, Marseilles, France, October 31-November 5, 2004.
734. T. A. Chubb, paper presented at the 11th International Conference on Cold Fusion, Marseilles, France, October 31-November 5, 2004.
735. T. A. Chubb, paper presented at the 11th International Conference on Cold Fusion, Marseilles, France, October 31-November 5, 2004.
736. T. A. Chubb, M. Daehler, Lattice-Assisted Nuclear Fusion. *Infinite Energy*, **17**, 101, 22-28 (2012).
737. R. T. Bush, paper presented at the Winter Annual Meeting of Am. Soc. Mechan. Eng., San Francisco, December 10-15, 1989.
738. R. T. Bush, R. D. Eagleton, 'Cold nuclear fusion': A hypothetical model to probe an elusive phenomenon. *J. Fusion Energy* **9**, 397-408 (1990).
739. R. T. Bush, A unifying model for cold fusion. *Trans. Fusion Technol.* **26**, 431-441 (1994).
740. K. Kunimatsu, N. Hasegawa, A. Kubota, N. Imai, M. Ishikawa, H. Akita, Y. Tsuchida, paper presented at the 3rd International Conference on Cold Fusion, Frontiers of Cold Fusion, Nagoya, Japan, October 21-25, 1992.
741. R. T. Bush, Consequences of lattice occupational symmetry. *Infinite Energy* **2**, 12, 34 (1997).
742. L. Turner, Thoughts unbottled by cold fusion. *Phys. Today* Sept., 140 (1989).
743. L. Turner, Peregrinations on cold fusion. *J. Fusion Energy* **9**, 447 (1990).
744. L. Turner, Peregrinations on cold fusion. *J. Fusion Energy* **9**, 447 (1994).
745. R. H. Parmenter, W. E. Lamb Jr., Cold fusion in palladium: a more realistic calculation. *Proc. Natl. Acad. Sci. USA* **87**, 8652-8654 (1990).
746. R. H. Parmenter, W. E. Lamb Jr., More cold fusion in metals: corrected calculations and other considerations. *Proc. Natl. Acad. Sci. U.S.A.* **87**, 3177 (1990).
747. R. H. Parmenter, A possible scenario for the onset of cold fusion in deuterated metals. *Infinite Energy* **4**, 21, 41 (1998).

748. R. H. Parmenter, Enhancement of cold fusion processes in palladium by catalytic agents. *Infinite Energy* **8**, 43, 66 (2002).
749. R. W. Bass, paper presented at the 1st Annual Conference on Cold Fusion, University of Utah Research Park, Salt Lake City, UT, March 28-31, 1990.
750. R. W. Bass, M. C. McKubre, paper presented at the 10th International Conference on Cold Fusion, Cambridge, MA, August 24-29, 2003.
751. R. W. Bass, M. Swartz, paper presented at the 14th International Conference on Condensed Matter Nuclear Science, Washington, DC, August 10-15, 2008.
752. R. W. Bass, paper presented at the ACS, March 21, 2010.
753. S. Martellucci, A. Rosati, F. Scaramuzzi, V. Violante, *Cold Fusion, The History of Research in Italy*. (ENEA Italian National Agency for New Technologies, 2008).
754. A. De Ninno, V. Violante, paper presented at the 3rd International Conference on Cold Fusion, "Frontiers of Cold Fusion", Nagoya Japan, October 21-25, 1992.
755. V. Violante, A. De Ninno, paper presented at the 6th International Conference on Cold Fusion, Progress in New Hydrogen Energy, Lake Toya, Hokkaido, Japan, October 13-18, 1996.
756. V. Violante, A. De Ninno, paper presented at the 5th International Conference on Cold Fusion, Monte-Carlo, Monaco, April 9-13, 1995.
757. V. Violante, Lattice ion-trap confinement for deuterons and protons: Possible interaction in condensed matter. *Fusion Technol.* **35**, 361 (1999).
758. V. Violante, A. Torre, G. Dattoli, Lattice ion trap: classical and quantum description of a possible collision mechanism for deuterons in metal lattices. *Fusion Technol.* **34**, 156 (1998).
759. V. Violante, paper presented at the 7th International Conference on Cold Fusion, Vancouver, Canada, April 19-24, 1998.
760. V. Violante, A. De Ninno, Lattice ion trap: a possible mechanism inducing a strong approach between two deuterons in condensed matter. *Fusion Technol.* **31**, 219 (1997).
761. V. Violante, paper presented at the Asti Workshop on Anomalies in Hydrogen/Deuterium Loaded Metals, Rocca d'Arazzo, Italy, November 27-30, 1997.
762. V. Violante, A. Torre, G. Selvaggi, G. H. Miley, Three-dimensional analysis of the lattice confinement effect on ion dynamics in condensed matter and lattice effect on the d-d nuclear reaction channel. *Fusion Technol.* **39**, 266 (2001).
763. A. K. Vijh, Electrode potentials and interface plasmons in the metal/gaseous electrolyte (*i.e.*, plasma) interphasic region. *Mater. Chem. Phys.* **14**, 47 (1986).

764. E. Castagna, C. Sibilia, S. Paoloni, V. Violante, F. Sarto, paper presented at the 12th International Conference on Condensed Matter Nuclear Science, Yokohama, Japan, November 27-December 2, 2005.
765. A. De Ninno, V. Violante, A. La Barbera, Consequences of lattice expansive strain gradients on hydrogen loading in palladium. *Phys. Rev. B* **56**, 2417-2420 (1997).
766. J. A. S. Green, F. A. Lewis, Overvoltage component at palladized cathodes of palladium and palladium alloys prior to and during bubble evolution. *Trans. Faraday Soc.* **60**, 2234 (1964).
767. A. Widom, L. Larsen, Absorption of nuclear gamma radiation by heavy electrons on metallic hydride surfaces. *arXiv:cond-mat/0509269v1*, (2005).
768. A. Widom, L. Larsen, Ultra low momentum neutron catalyzed nuclear reactions on metallic hydride surfaces. *Eur. Phys. J.* **C46**, 107 (2006).
769. D. Bushnell, paper presented at the International Low Energy Nuclear Reactions Symposium (ILENRS-12), William & Mary, Williamsburg, VA, July 1-3, 2012.
770. G. C. Fralick, J. D. Wrbanek, S. Y. Wrbanek, J. M. Niedra, M. G. Millis, D. J. Spry, R. Meredith, J. Mazo, "Investigation of Anomalous Heat Observed in Bulk Palladium," (NASA, www.nasa.gov, 2009).
771. S. Ciuchi, L. Maiani, A. D. Polosa, V. Riquer, R. Ruocco, M. Vignati, Low energy neutron production by inverse beta decay in metallic hydride surfaces. *arXiv:1209.6501v1 [nucl-th]*, (2012).
772. E. Tennfors, On the idea of low-energy nuclear reactions in metallic lattices by producing neutrons from protons capturing “heavy” electrons. *Eur. Phys. J. Plus* **128**, 1 (2013).
773. P. L. Hagelstein, Electron mass enhancement and the Widom-Larsen model. *J. Cond. Matter Nucl. Sci.* **12**, 18-40 (2013).
774. A. Widom, Y. N. Srivastava, L. Larsen, Erroneous wave functions of Ciuchi et al. for collective modes in neutron production on metallic hydride cathodes. *arXiv:1210.5212v1 [nucl-th]*, (2012).
775. J. M. Zawodny, S. Krivit, Eds., *Widom-Larsen theory: Possible explanation of LENRs* (Wiley & Sons, NJ, Singapore, 2011), pp. 595.
776. A. Widom, L. Larsen, Nuclear abundances in metallic hydride electrodes of electrolytic chemical cells. *arXiv:cond-mat/062472 v1*, (2006).
777. L. Larsen, A. Widom. (Lattice Energy, USA, 2011).
778. Y.-F. Chang, Potential exploration of cold fusion and its quantitative theory of physical-chemical-nuclear multistage chain reaction mechanism. *Intern. J. Modern Chem.* **5**, 29-43 (2013).
779. R. E. Godes, The quantum fusion hypothesis. *Infinite Energy* **14**, 82, 15-23 (2008).
780. J. C. Fisher, Polyneutrons as agents for cold nuclear reactions. *Fusion Technol.* **22**, 511 (1992).

781. J. C. Fisher, Liquid-drop model for extremely neutron rich nuclei. *Fusion Technol.* **34**, 66 (1998).
782. R. A. Oriani, J. C. Fisher, Anomalous power generation produced by stirring water solutions. *Trans. Am. Nucl. Soc.* **83**, 368 (2000).
783. J. C. Fisher, paper presented at the 10th International Conference on Cold Fusion, Cambridge, MA, August 24-29, 2003.
784. R. A. Oriani, J. C. Fisher, paper presented at the 11th International Conference on Cold Fusion, Marseilles, France, October 31-November 5, 2004.
785. J. C. Fisher, paper presented at the 6th International Workshop on Anomalies in Hydrogen/Deuterium Loaded Metals, Siena, Italy, May 13-15, 2005.
786. J. C. Fisher, paper presented at the 12th International Conference on Condensed Matter Nuclear Science, Yokohama, Japan, November 27-December 2, 2005.
787. J. C. Fisher, paper presented at the 8th International Workshop on Anomalies in Hydrogen/Deuterium Loaded Metals, Catania, Italy, October 13-18, 2007.
788. J. C. Fisher, Palladium fusion triggered by polyneutrons. *J. Cond. Matter Nucl. Sci.* **1**, 1 (2007).
789. H. Kozima, Neutron Mossbauer effect and the cold fusion in inhomogeneous materials. *Il Nuovo Cimento* **107 A**, 1781 (1994).
790. H. Kozima, How the cold fusion occurs? . **28**, 31 (1994).
791. H. Kozima, S. Watanabe, paper presented at the 5th International Conference on Cold Fusion, Monte-Carlo, Monaco, April 9-13, 1995.
792. H. Kozima, Neutron drop: condensation of neutrons in metal hydrides and deuterides. *Fusion Technol.* **37**, 253-258 (2000).
793. H. Kozima, *The science of the cold fusion phenomenon*. (Elsevier Science, 2006), pp. 208.
794. J. A. Maly, J. Va'vra, Electron transitions on deep Dirac levels I. *Fusion Technol.* **24**, 307 (1993).
795. J. A. Maly, J. Va'vra, Electron transitions on deep Dirac levels II. *Fusion Technol.* **27**, 59 (1995).
796. R. L. Mills, W. R. Good, Fractional quantum energy levels of hydrogen. *Fusion Technol.* **28**, 1697 (1995).
797. R. A. Rice, Y. E. Kim, M. Rabinowitz, A. L. Zubarev, paper presented at the 4th International Conference on Cold Fusion, Lahaina, HI, December 6-9, 1993.
798. R. A. Rice, Y. E. Kim, Comments on 'Electron transitions on deep Dirac levels I'. *Fusion Technol.* **26**, 111 (1994).
799. R. A. Rice, Y. E. Kim, M. Rabinowitz, Reply to 'Response to "Comments on "Electron Transitions on Deep Dirac Levels I. *Fusion Technol.* **26**, 348-349 (1995).

800. J. A. Maly, J. Va'vra, Response to 'Comments on 'Electron transitions on deep Dirac levels I''. *Fusion Technol.* **26**, 112 (1994).
801. J. A. Maly, J. Va'vra, Reply to 'Letter to the Editor' Fusion Technol. 27, 348 [1995]''. *Fusion Technol.* **30**, 386 (1996).
802. A. Meulenberg Jr., From the naught orbit to the ^{4}He excited state. *J. Cond. Matter Nucl. Sci.* **10**, 15-29 (2013).
803. K. P. Sinha, A theoretical model for low-energy nuclear reactions. *Infinite Energy* **5**, 29, 54-57 (2000).
804. K. P. Sinha, A. Meulenberg Jr., Lochon catalyzed D-D fusion in deuterated palladium in the solid state. *National Acad. Sci. Lett.* **30**, 243 (2007).
805. K. P. Sinha, A. Meulenberg, Laser stimulation of low-energy nuclear reactions in deuterated palladium. *Current Sci.* **91**, 907-912 (2006).
806. A. Meulenberg, Tunneling beneath the ^{4}He* fragmentation energy. *J. Cond. Matter Nucl. Sci.* **4**, 241-255 (2011).
807. A. Meulenberg Jr., K. P. Sinha, Lochon and extended-lochon models for low-energy nuclear reactions in a lattice. *Infinite Energy* **19**, 112, 29-32 (2013).
808. J. P. Vigier, paper presented at the 3rd International Conference on Cold Fusion, Frontiers of Cold Fusion, Nagoya, Japan, October 21-25, 1992.
809. J. P. Vigier, On cathodically polarized Pd/D systems. *Phys. Lett. A* **221**, 138 (1996).
810. A. Dragic, Z. Maric, J. P. Vigier, New quantum mechanical tight bound states and 'cold fusion'. *Phys. Lett. A* **265**, 163-167 (2000).
811. J. Dufour, An introduction to the Pico-chemistry working hypothesis. *J. Cond. Matter Nucl. Sci.* **10**, 40-45 (2013).
812. J. C. Dufour, Very sizeable increase of gravitation at Picometer distance: a novel working hypothesis to explain anomalous heat effects and apparent transmutations in certain metal/hydrogen systems. *J. Cond. Matter Nucl. Sci.* **1**, 47 (2007).
813. J. Dufour, J. H. Foos, X. J. C. Dufour, paper presented at the 7th International Conference on Cold Fusion, Vancouver, Canada, April 19-24, 1998.
814. J. Dufour, J. Foos, J. P. Millot, X. Dufour, paper presented at the 6th International Conference on Cold Fusion, Progress in New Hydrogen Energy, Lake Toya, Hokkaido, Japan, October 13-18, 1996.
815. J. J. Dufour, J. Foos, J. C. Dufour, paper presented at the Asti Workshop on Anomalies in Hydrogen/Deuterium Loaded Metals, Rocca d'Arazzo, Italy, November 27-30, 1997.
816. J. Dufour, D. Murat, X. Dufour, J. Foos, paper presented at the 8th International Conference on Cold Fusion, Lerici (La Spezia), Italy, May 21-26, 2000.

817. J. Dufour, Response to 'Comments on 'Interaction of palladium/hydrogen and palladium/ deuterium to measure the excess energy per atom for each isotope'. *Fusion Technol.* **33**, 385 (1998).
818. R. Mills, Comments on "Interaction of palladium/hydrogen and palladium/deuterium to measure excess energy per atom for each isiotope". *Fusion Technol.* **33**, 384 (1998).
819. R. L. Mills, W. R. Good, R. M. Shaubach, Dihydrino molecule identification. *Fusion Technol.* **25**, 103-119 (1994).
820. R. Mills, Novel hydrogen compounds from a potassium carbonate electrolytic cell. *Fusion Technol.* **37**, 157 (2000).
821. R. Mills, W. R. Good, J. J. Farrell, *Unification of Spacetime, the Forces, Matter and Energy*. (Science Press, Ephrata, PA, 1992), pp. 220.
822. R. Mills, *The grand unified theory of classical quantum mechanics*. (Cadmus Professional Communications, Ephrata, PA, 2006), pp. 1450.
823. R. L. Mills, P. Ray, Spectral emission of fractional quantum energy levels of atomic hydrogen from a helium-hydrogen plasma and the implications for dark matter. *J. Hydrogen Eng.* **27**, 301 (2002).
824. R. L. Mills, P. Ray, Vibrational spectral emission of fractional-principal-quantum-energy-level hydrogen molecule ion. *J. Hydrogen Eng.* **27**, 533 (2002).
825. R. L. Mills, P. Ray, The grand unified theory of classical quantum mechanics. *J. Hydrogen Eng.* **27**, 565 (2002).
826. R. L. Mills, P. C. Ray, B. Dhandapani, R. M. Mayo, J. He, comparison of excessive Balmer alpha line broadening of glow discharge and microwave hydrogen plasmas with certain catalysts. *J. Appl. Phys.* **92**, 7008 (2002).
827. R. L. Mills, B. Dhandapani, M. Nansteel, H. J., A. Voigt, Identification of compounds containing novel hydride ions by nuclear magnetic resonance spectroscopy. *J. Hyd. Energy* **26**, 965 (2001).
828. R. Mills, W. R. Good, J. Phillips, A. I. Popov. (USA, 2000).
829. J. Phillips, C.-K. Chen, K. Ahktar, B. Dhandapani, R. Mills, Evidence of the production of hot atomic hydrogen in RF generated hydrogen/argon. *Int. J. Hydrogen Energy* **32**, 3010-3025 (2007).
830. R. Mills, Y. Lu, Editorial about "Time-resolved hydrino continuum transitions with cutoffs at 22.8 nm and 10.1 nm". *Eur. Phys. J.* **D64**, 63 (2011).
831. R. Mills. (US 6,024,935, Application 20110114075, 2011).
832. A. Rathke, A critical analysis of the hydrino model. *New J. Phys* **7**, 127 (2005).
833. R. L. Mills, P. Kneizys, Excess heat production by the electrolysis of an aqueous potassium carbonate electrolyte and the implications for cold fusion. *Fusion Technol.* **20**, 65 (1991).

834. F. Piantelli, W. J. M. F. Collis, paper presented at the 10th International Workshop on Anomalies in Hydrogen Loaded Metals, Siena, Italy, April 10-14, 2012.
835. A. Rossi, Rossi Cold Fusion & E-cat News. 2012.
836. A. Rossi, Journal of Nuclear Physics. 2012.
837. M. Lewan, *An Impossible Invention*. (Vulkan.se, 2014), pp. 399.
838. R. Long, *Cold Fusion for Everything, How Andrea Rossi's E-Cat could change the world*. (self, 2012).
839. Y. E. Kim, J. Hadjichristos, paper presented at the 18th International Conference on Cold Fusion, Columbia, MO, July 2013.
840. X. Z. Li, D. Z. Jin, L. Chang, paper presented at the 3rd International Conference on Cold Fusion, Frontiers of Cold Fusion, Nagoya, Japan, October 21-25, 1992.
841. X. Z. Li, The 3-dimensional resonance tunneling in chemically assisted nuclear fission and fusion reactions. *Trans. Fusion Technol.* **26**, 480-485 (1994).
842. X. Z. Li, A new approach towards fusion energy with no strong nuclear radiation. *J. New Energy* **1**, 44 (1996).
843. X. Z. Li, Overcoming of the Gamow tunneling insufficiencies by maximizing the damp-matching resonant tunneling. *Czech. J. Phys.* **49**, 985 (1999).
844. X. Z. Li, C. X. Li, H. F. Huang, Maximum value of the resonant tunneling current through the Coulomb barrier. *Fusion Technol.* **36**, 324 (1999).
845. X. Z. Li, J. Tian, M. Y. Mei, C. X. Li, Sub-barrier fusion and selective resonant tunneling. *Phys Rev. C: Nucl. Phys.* **61**, 4610 (2000).
846. X. Z. Li, B. Liu, X. Z. Ren, J. Tian, D. X. Cao, S. Chen, G. H. Pan, D. l. Ho, Y. Deng, paper presented at the 9th International Conference on Cold Fusion, Condensed Matter Nuclear Science, Tsinghua Univ., Beijing, China, May 19-24, 2002.
847. X. Z. Li, A Chinese view on the summary of the condensed matter nuclear science. *J. Fusion Energy* **23**, 217 (2004).
848. J. Louthan, M. R., J. Caskey, G. R., J. A. Donovan, J. Rawl, D. E., Hydrogen embrittlement of metals. *Mater. Sci. and Eng.* **10**, 357 (1972).
849. X. Xu, M. Wen, S. Fukuyama, K. Yokogawa, Simulation of hydrogen embrittlement at crack tip in nickel single crystal by embedded atom method. *Mater. Trans.* **42**, 2283-2289 (2001).
850. S. P. Lynch, Hydrogen embrittlement and liquid-metal embrittlement in nickel single crystals. *Scr. Metall.* **13**, 1051-1056 (1979).
851. S. P. Lynch, A fractographic study of hydrogen-assisted cracking and liquid-metal embrittlement in nickel. *J. Mater. Sci.* **21**, 692-704 (1986).
852. H. Vehoff, W. Rothe, Gaseous hydrogen embrittlement in FeSi- and Ni-single crystals. *Acta Metall.* **31**, 1781-1793 (1983).

853. H. Vehoff, H. K. Klameth, Hydrogen embrittlement and trapping at crack tips in Ni-single crystals. *Acta Metall.* **33**, 955-962 (1985).
854. F. Frisone, Study of the probability of interaction between the plasmons of metal and deuterons. *Nuovo Cimento Soc. Ital. Fis.* **18D**, 1279 (1996).
855. F. Frisone, paper presented at the Asti Workshop on Anomalies in Hydrogen/Deuterium Loaded Metals, Rocca d'Arazzo, Italy, November 27-30, 1997.
856. F. Frisone, paper presented at the 7th International Conference on Cold Fusion, Vancouver, Canada, April 19-24, 1998.
857. F. Frisone, Can variations in temperature influence deuteron interaction within crystalline lattices? *Nuovo Cimento Soc. Ital. Fis.* **20 D**, 1567 (1998).
858. F. Frisone, paper presented at the 8th International Conference on Cold Fusion, Lerici (La Spezia), Italy, May 21-26, 2000.
859. F. Frisone, Calculation of deuteron interactions within microcracks of a D_2 loaded crystalline lattice at room temperature. *J. Cond. Matter Nucl. Sci.* **1**, 41 (2007).
860. F. Frisone, paper presented at the 15th International Conference on Condensed Matter Nuclear Science, Rome, Italy, October 5-9, 2009.
861. E. K. Storms, Measurements of excess heat from a Pons-Fleischmann-type electrolytic cell using palladium sheet. *Fusion Technol.* **23**, 230 (1993).
862. E. Storms, paper presented at the 3rd International Conference on Cold Fusion, Frontiers of Cold Fusion, Nagoya, Japan, October 21-25, 1992.
863. E. K. Storms, paper presented at the 14th International Conference on Condensed Matter Nuclear Science, Washington, DC, August 10-15, 2008.
864. E. K. Storms, B. Scanlan, paper presented at the American Physical Society Conference, New Orleans, March 10, 2008.
865. E. K. Storms, A Theory of LENR Based on Crack Formation. *Infinite Energy* **19**, 112, 24-27 (2013).
866. E. K. Storms, paper presented at the 18th International Conference on Cold Fusion, Columbia, MO, July 2013.
867. E. K. Storms, paper presented at the International Low Energy Nuclear Reactions Symposium (ILENRS-12), William & Mary, Williamburg, VA, July 1-3, 2012.
868. E. K. Storms, The role of voids as the location of LENR. *J. Cond. Matter Nucl. Sci.* **11**, 123-141 (2013).
869. E. K. Storms, Cold fusion from a chemist's point of view. *Infinite Energy* **18**, 108, 13-18 (2013).
870. E. K. Storms, An explanation of low-energy nuclear reactions (cold fusion). *J. Cond. Matter Nucl. Sci.* **9**, 85-107 (2012).

871. E. K. Storms, B. Scanlan, What is real about cold fusion and what explanations are plausible? *J. Cond. Matter Nucl. Sci.* **4**, 17-31 (2011).
872. R. McIntyre, paper presented at the 10th International Conference on Cold Fusion, Cambridge, MA, August 24-29, 2003.
873. P. I. Golubnichii, V. A. Kurakin, A. D. Filonenko, V. A. Tsarev, A. A. Tsarik, A possible mechanism for cold nuclear fusion. *J Kratk. Soobshch. Fiz.*, 56 (in Russian) (1989).
874. V. Godbole, paper presented at the 17th International Conference on Condensed Matter Nuclear Science, Daejeon, Korea, August 12-17, 2012.
875. F. Frisone, paper presented at the 11th International Conference on Cold Fusion, Marseilles, France, October 31-November 5, 2004.
876. F. Frisone, paper presented at the 10th International Conference on Cold Fusion, Cambridge, MA, August 24-29, 2003.
877. F. Frisone, paper presented at the 9th International Conference on Cold Fusion, Condensed Matter Nuclear Science, Tsinghua Univ., Beijing, China, May 19-24, 2002.
878. F. Frisone, Theoretical model of the probability of fusion between deuterons within deformed crystalline lattices with microcracks at room temperature. *Fusion Sci. & Technol.* **40**, 139 (2001).
879. F. Frisone, Deuteron interaction within a microcrack in a lattice at room temperature. *Fusion Technol.* **39**, 260 (2001).
880. F. Frisone, paper presented at the AIP Conf. Proc. 513 (Nuclear and Condensed Matter Physics), 2000.
881. J. O. M. Bockris, The complex conditions needed to obtain nuclear heat from D-Pd systems. *J. New Energy* **1**, 210 (1996).
882. L. H. Bagnulo, paper presented at the 2nd Annual Conference on Cold Fusion, The Science of Cold Fusion, Como, Italy, June 29-July 4, 1991.
883. M. H. Miles, K. B. Johnson, M. A. Imam, paper presented at the 6th International Conference on Cold Fusion, Progress in New Hydrogen Energy, Lake Toya, Hokkaido, Japan, October 13-18, 1996.
884. M. H. Miles, M. A. Imam, M. Fleischmann, Calorimetric analysis of a heavy water electrolysis experiment using a Pd-B alloy cathode. *Proc. Electrochem. Soc.* **2001**, 194 (2001).
885. A. B. Karabut, E. A. Karabut, paper presented at the 14th International Conference on Condensed Matter Nuclear Science, Washington DC, August 10-15, 2008.
886. R. T. Bush, paper presented at the International Symposium on Cold Fusion and Advanced Energy Sources, Belarusian State University, Minsk, Belarus, May 24-26, 1994.
887. T. Ohmori, M. Enyo, Iron formation in gold and palladium cathodes. *J. New Energy* **1**, 15 (1996).

888. R. T. Bush, A light water excess heat reaction suggests that 'cold fusion' may be 'alkali-hydrogen fusion'. *Fusion Technol.* **22**, 301-322 (1992).

889. E. Wigner, H. B. Huntington, On the possibility of a metallic modification of hydrogen. *J. Chem. Phys.* **3**, 764 (1935).

890. W. J. Nellis, A. L. Ruoff, I. F. Silvera, Has metallic hydrogen been made in a diamond anvil cell? *arXiv:1201.0407 [cond-mat.other]*, (2012).

891. N. W. Ashcroft, Metallic hydrogen: a high-temperature superconductor? *Phys. Rev. Lett.* **2126**, 1748-1749 (1968).

892. R. L. Liboff, Fusion via metallic deuterium. *Phys. Lett.* **71A**, 361 (1979).

893. C. J. Horowitz, Cold nuclear fusion in dense metallic hydrogen. *Astrophys. J.* **367**, 288 (1991).

894. P. Kalman, T. Keszthelyi, Nuclear processes in solids: basic 2nd-order processes. *arXiv:1303.1078v1 [nucl-th]*, (2013).

895. D. Letts, P. L. Hagelstein, paper presented at the 14th International Conference on Condensed Matter Nuclear Science, Washington DC, August 10-15, 2008.

896. E. Brillas, J. Esteve, G. Sardin, J. Casado, X. Domenech, J. A. Sanchez-Cabeza, Product analysis from D_2O electrolysis with Pd and Ti cathodes. *Electrochim. Acta* **37**, 215 (1992).

897. D. R. Coupland, M. L. Doyle, J. W. Jenkins, J. H. F. Notton, R. J. Potter, D. T. Thompson, paper presented at the 3rd International Conference on Cold Fusion, Frontiers of Cold Fusion, Nagoya, Japan, October 21-25, 1992.

898. M. Okamoto, H. Ogawa, Y. Yoshinaga, T. Kusunoki, O. Odawara, paper presented at the 4th International Conference on Cold Fusion, Lahaina, HI, December 6-9, 1993.

899. J. H. N. Van Vucht, K. H. J. Buschow, Note on the occurrence of intermetallic compounds in the lithium-palladium system. *J. Less-Common Met.* **48**, 345 (1976).

900. J. Loebich, D., C. J. Raub, Das Zustandsdiagramm Lithium-Palladium und die Magnetischen Eigenschaften der Li-Pd Legierungen. *J. Less-Common Met.* **55**, 67 (1977).

901. R. A. Howald, Calculation on the palladium-lithium system for cold fusion. *CALPHAD* **14**, 1 (1990).

902. S. Szpak, P. A. Mosier-Boss, F. Gordon, J. Dea, M. Miles, J. Khim, L. Forsley, paper presented at the 14th International Conference on Condensed Matter Nuclear Science, Washington DC, August 10-15, 2008.

903. M. H. Miles, paper presented at the 8th International Conference on Cold Fusion, Lerici (La Spezia), Italy, May 21-26, 2000.

904. A. G. Lipson, B. F. Lyakhov, D. M. Sakov, V. A. Kuznetsov, T. S. Ivanova, paper presented at the 6th International Conference on Cold Fusion, Progress in New Hydrogen Energy, Lake Toya, Hokkaido, Japan, October 13-18, 1996.

INDEX

A

Adamenko, S., 92, 95
adiabatic calorimeter, 108
Afonichev, D., 13, 44, 88
^{107}Ag, 52
^{109}Ag, 27, 231
^{27}Al, 52
Al_2O_3, 78, 242
Albagli, D., 29
Alessandrello, A., 29
alpha decay, 23, 238
alpha emission, 29, 73, 78, 79, 80-82, 94, 95, 153, 234, 235, 238
alpha energy, 81, 93, 153
alpha phase, 122, 124, 125, 127, 129
alpha radiation, 79, 82, 153
aluminum, 57, 67, 95
annihilation radiation, 74, 207
Aoki, T., 33
Apicella, M., 39, 96
Aqua Regia, 249
Arata, Y., 15, 27, 35, 40, 99
arc formation, 64, 65
arsenic, 57
Ashcroft, N., 240
Associazione Euratom ENEA, 194
Au/Pd/PdO, 78
Auger electron spectrometry, 46
autoradiograph, 13, 63
average positive charge, 128

B

^{10}B, 21, 25, 232, 239
^{11}B, 232, 239
$^{10}B/^{11}B$, 232
^{137}Ba, 59
Badiei, S., 98
ball lightning, 92
Baranowski, B., 118
barium, 59
Barmina, E., 97
Bass, R., 104, 189, 192-194
Baym, G., 185
Bazhutov, Y., 91
^{8}Be, 73, 180, 181, 197
^{9}Be, 73
beat frequency, 101-104
Beltyukov, I., 96
Bernardini, M., 87
Berrondo, M., 134
Bertolotti, M., 132
beta decay, 22, 199, 203, 227
beta phase, 70, 114, 122, 124, 125-129, 132, 136, 150
Beuhler, R.., 138
BF_3, 75
biological effects, 92
Blacklight Power Inc., 203
Bockris, J., 15, 16, 31
Bohr level, 203
Bohr orbit, 163, 199, 202
Bose-Bloch Condensate, 189
Bose-Einstein Condensate (BEC), 151, 156, 157, 170, 179, 183, 184, 190
bosons, 183, 190, 201
Botta, E., 33
β-PdH, 122, 124
β-PdD, 61, 97, 126, 137
^{80}Br, 52
Bremsstrahlung, 55, 91, 93, 94
Bressani, T., 185
Brillouin Energy Corp., 197
brittle compound, 247
Brudanin, V., 29
Bush, B., 27, 30, 35, 37
Bush, R., 64, 84, 86, 87, 190, 191, 239

C

^{12}C, 25, 73
cadmium, 72, 231, 239
calcium, 61, 64
California State Polytechnic University, 190

calorimeter, 34, 35, 38, 105-109, 115-117, 120, 135
CaO, 54, 232
Case, L., 37
Castellano, M., 68
cathode surface, 16, 22, 29, 64, 90, 117, 246
^{142}Ce, 59
centrifugal potential barrier, 206
cesium, 57, 72
cesium iodide, 75
Chang, Y., 197
charge density, 134
chemical activity, 125
chemical lattice, 141, 144, 149, 150, 157, 159, 160, 167, 171, 179, 181, 183, 195, 207, 210, 242
chemical system, 134, 135, 146, 149, 156, 210
Chernov, I., 95
chlorine, 113, 249
Chubb, S., 189
Chubb, T., 189
Cirillo, D., 66
Ciuchi, S., 196
Clarke, B., 15
Claytor, T., 13
cloud chamber, 83
cluster, 133, 135, 136, 141, 145, 149, 150, 151, 153, 157, 170-172, 178, 183, 204
^{60}Co, 96
CO_2, 121, 249
co-deposition, 80, 110, 112, 113, 246
coherent oscillation, 185, 186
coherent radiation, 92
coherent resonance, 222
cold fusion, 1, 2, 4-6, 21, 23, 76, 138-140, 144, 145, 149, 156-158, 175, 179, 185-188, 191, 197, 200, 212, 252
conceptual space, 143
conduction band, 125, 126, 163
conservation of energy, 144
copper, 15, 16, 54, 65, 95, 204, 205, 239, 249
Coulomb barrier, xi, 3, 5, 6, 43, 46, 53, 58, 72, 78, 136, 137, 140, 141, 143, 147, 154, 156-158, 163, 175, 176, 179, 180, 182-185, 187, 189, 195, 200-202, 204, 206, 208-210, 221, 228, 240, 250
CR-39, 66, 73, 74, 76-81, 92-95, 199, 239
crack, 29, 68, 69, 72, 76, 90, 96, 97, 114, 124, 129, 137, 140, 141, 144, 150, 161, 175, 195, 200, 207, 218, 219, 226, 232, 243, 244, 246, 247
cracking, 86
craters, 47
Cravens, D., 99-101
crystal structure, 133, 136, 147, 150, 185
^{63}Cu, 70, 167
^{65}Cu, 70
Cu/Pd, 83

D

D/H ratio, 14, 20, 23
D^+ clusters, 138
D+D+D fusion, 182
d+p fusion, 171
d+p=^3He, 188
D_2O, 10, 14-17, 23, 29-31, 33, 39, 56, 65, 96, 98, 109, 111-112, 117, 120, 121, 174, 198, 232
D_2O+Li, 29, 30, 31, 33, 37, 38, 39
D_2O+Li+200 ppm Al, 35
$D_2O+LiCl+PdCl_2$, 76
$D_2O+Na_2CO_3$, 29
$D_2O-K_2CO_3$, 87
D_2O-Li, 31, 55, 79, 81, 82, 84, 88, 99, 174, 179
D_2O-Na, 78, 98
$D_2O-PdSO_4$, 113
Daehler, M., 189
Dardik et al., 117
dark matter, 204
deep Dirac level, 200-202
deep trapping potential, 181
Defkalion Green Technology Corp., 205

Del Giudice, E., 185
DeNinno, A., 39, 40
density, 124
density-of-states, 103
deutex, 203
diffusion, 3, 4, 111, 119, 128, 130-132, 145, 150, 178, 195, 213-215, 237
dineutron, 27, 57, 197, 198
Dipartimento di Fisica dell'Universita, 185
Dirac, 200-201
dislocations, 44, 129, 137, 144, 150, 161, 220, 249
double occupancy, 150
dual laser, 99, 101, 103
Dufour, J., 202-203
dummy cathode, 249
Dunlap, B., 134

E

Eagleton, R., 84, 86, 190
Echenique, P., 138
electrolysis, 10, 13, 15, 19, 25, 31, 33, 37, 44, 54, 62, 64, 79, 81-84, 98, 99, 109, 110, 124, 131, 204, 219, 242, 246
electromagnetic spectrum, 103
electromigration, 33, 111, 163
electromigration and diffusion, 111
electron bombardment, 80, 95
electron bonding, 133
electron capture, 224
electron dispersive X-rays, 44
electron flux, 95
electron screening, 157, 186
electroplating, 110
end-of-life, 248
energetic alpha emission, 29, 73, 79, 153, 234, 238
energetic radiations, 6
enthalpy, 135
entropy, 135
environmental tritium, 17, 18
Enyo, M., 64, 239
Erzion, 91
EVO, 92
excess volume, 124-126, 130

F

Farrell, J., 203
fat neutron, 200
fcc lattice, 128, 161, 185
^{26}Fe, 65
^{57}Fe, 55
^{58}Fe, 55, 65
Fe_2O_3+K, 98
Fermi's golden rule, 181
Fisher, J., 81, 92, 198
flashes of light, 112
flaws in a theory, 169
Fleischmann, M., ix, xi, 29, 108, 120, 137, 185, 219, 252, 253
flow calorimeter, 108
flow calorimetry, 33
flux of D, 114, 206
Focardi, S., 205
fractional quantum states, 203
fractionation, 172
fractofusion, 15, 90, 140
fragmentation, 5, 43, 47, 72, 151, 180, 182, 185, 200, 201, 209, 230, 232-234, 239
fragments, 2, 47, 144, 233, 236
frequency of resonance, 224, 244
Frisone, F., 208
Fukai, Y., 124, 127, 160, 161
fused salt, 31
fusion-fission, 47, 49, 55, 57, 59, 61-67, 69, 70, 74, 77, 94, 236, 237

G

Galileo Project, 80
Galkin, E., 13
gamma-ray shield, 197
Gamow factor, 184
gaps, 137, 144
gas (glow) discharge, 13, 23, 44, 54, 78, 79, 90, 93, 110, 208
gas loading, 13, 37, 54, 110, 207, 208
Ge detector, 83
Geiger-Müller Detector, 75

germanium, 57, 66, 75
Gibbs energy, 125, 135, 136, 150, 151, 157, 160, 163, 178, 222
glow discharge, 13, 38, 62, 78, 80, 82, 92, 110, 219
GM detector, 85, 92
Godes, R., 53, 117, 197
gold, 15, 32, 78, 99, 105
Gozzi, D., 33
grain boundary, 24, 201, 219, 242
graphite, 65
Green function, 179

H

^{4}H, 197, 198, 201-203, 225, 226, 227, 230, 238, 250
^{6}H, 198
H^- ions, 204
H/D ratio, 20, 242
H_2 pressure, 124
H_2O, 23, 30, 49, 64-66, 109, 120, 121, 239, 249
H_2O+K_2CO_3, 64, 204
H_2O+Rb_2CO_3, 64
H_2O-Li, 55, 79-81, 83, 84, 85
H_2O-Na_2CO_3, 87
H_2O-Rb_2CO_3, 64, 87
Hagelstein, P., 76, 101, 103, 153, 170-173, 180, 184, 186, 191, 196
Hagelstein energy limit, 180
Hamiltonian, 147, 155, 172, 188
harmonic oscillation, 194
^{3}He, 11, 15, 23, 29, 31, 32, 38, 74, 75, 171, 181, 188, 192, 193, 197, 235, 238, 239
^{4}He, 9, 15, 23, 25, 31, 32, 33, 38, 62, 73, 74, 121, 153, 166, 171, 184, 188, 193, 197, 203, 206, 213, 225, 226, 227, 230, 238, 239, 250
heat capacity, 108, 116, 148
heavy electrons, 179, 196, 197
Heisenberg confinement energy, 198
helium, 2, 10, 23-43, 79, 94, 110, 118, 121, 141, 151, 154, 166, 171, 172, 188, 194, 197, 199, 209, 224-226, 229, 238
He-Ne laser, 70
^{201}Hg, 172
Hioki, T., 44
Hora, H., 174, 175, 177
Horowitz, C., 241
hot fusion, xi, 2-6, 20, 62, 76-81, 138-140, 144, 199, 200, 241, 251
hydrex, 203
hydrino, 200, 203, 204, 240
Hydroton, 208, 213, 218, 221-227, 229, 230, 232-235, 238-241, 244, 250
Hydroton structure, 218, 240, 250

I

Indian Institute of Science, 200
indium, 66, 231
induced fluorescence, 98
inductively coupled plasma, 45
Innoventek Inc., 192
internal nuclear transitions, 172
ion bombardment, 6, 137, 138, 140, 157, 183, 214, 248
ion counting, 11
ion traps, 183
IR, 74
IR camera, 178
IR radiation calorimeter, 109
Irion, C., 28
iron, 55, 64, 65, 72, 239
Isobe, Y., 38
isoperibolic calorimeter, 107, 108
ITER reactor, x
Iwamura, Y., 15, 54, 57, 59, 72, 82, 229, 232, 239
Iyengar, P., 88

J

Jiang, X., 65
Johnson-Matthey, xi
Jones, S., 137
Jupiter, 241

K

K_2CO_3, 66, 239

Kalma, P., 241
Kalman-Leverrier, 193
Kamada, K., 95
Karabut, A., 38, 90, 172, 222
Karabut, E., 38, 90, 172, 222
Kaushik, S., 171
k-capture, 207
KCl-LiCl-LiD, 31
Keeney, F., 90
Keszthelyi, T., 241
Kim, Y., 183, 184
Kozima, H., 46, 50, 199
^{80}Kr, 52
Krivit, S., 26, 27, 81, 196
krypton, 52, 57
Kunimatsu, K., 191

L

Lagowski, J., 27, 35, 37
Lamb, W., 192
large energy barrier, 133
Larsen, L., 26, 196, 198
laser, 39, 67-72, 73, 90, 96-105, 172, 202, 217, 243, 244, 245, 248
laser radiation, 67, 68, 70, 72, 79, 96, 98, 99, 101, 121, 202, 243, 244
lattice-assisted nuclear reactions, 174, 189
Lattice Energy, LLC, 196
lattice parameter, 124, 129, 130, 160, 193, 247, 248
lattice structure, 128, 141, 144, 160, 187, 190, 207, 249
lattice vibrations, 172
Lawrence Livermore National Laboratory, x
Laws of Thermodynamics, 134, 141, 148, 153, 160, 161, 167, 178, 182, 183
lead, 65, 82, 85
Leggett, A., 185
Letts, D., 99-101, 103
Lewis, E., 92
^{4}Li, 181
^{6}Li, 21, 73, 232
^{7}Li, 21, 73, 197, 232
Li, X., 206
Li_2SO_4-H_2O, 78
Liaw, B., 31
Liboff, R., 240
life-after-death, 206, 207
$LiNi_5$, 32
Lipson, A., 77, 81
liquid scintillation, 11
lithium, 23, 26, 27, 28, 37, 43, 73, 109, 117, 191, 193, 197, 199, 232, 246
local melting, 48, 95, 178, 189, 195, 227
Lochak, G., 92
lochon, 201, 202
Los Alamos National Laboratory, 16, 208
lossy-spin-boson model, 171
Luo, N., 127, 176
Luch, 62

M

magic numbers, 175
magnesium, 57, 67
magnetic field, 76, 95, 99, 101, 105, 110, 243, 244
Maly, J., 200
mass spectrometer, 11-13, 15, 24, 30, 45
Matsuda, J., 29
Matsumoto, T., 65, 91
Matsunaka, M., 95
McKubre, M., 15, 24, 37, 99, 113, 114, 191
melting, 95
Mengoli, G., 131
metal atom vacancies, 124, 129-131, 136, 150, 160
metallic hydrogen, 98, 222, 226, 240, 241
Meulenberg, A., 200-202, 240
^{27}Mg, 52
MgO, 54
micro/nanometric, 204
microcracks, 208
Miles, M., 28, 31, 33, 38, 240

Miley, G., 47, 49, 51, 127, 170, 174-178
Mills, R., 200, 202-204
Miraglia, S., 124
MIT, 29, 170, 173
Mizuno, T., 65, 66, 117
Mn, 65
MnO_x, 32
molybdenum, 58
momentum shocks, 190
Monte Carlo calculation, 195
Morrey, J., 29
Mosier-Boss, P., 76
muon, 5, 6, 156, 200
mystery of LENR, 212, 221, 250

N

$(NH_4)_2MoO_4$, 65
n/T ratio, 21
NaI detector, 84
Nakamura, K., 65
nanogaps, 8, 73, 208, 242, 244-246, 249
Nassisi, V., 67, 69, 70
National Ignition Facility, x
natural distribution, 175
naught electrons, 201
naught orbit, 200
Naval Research Laboratory, 189
^{146}Nd, 59
NE-213 detector, 78
neutrino, 10, 11, 151, 158, 167, 201, 202, 224, 225
neutron activation, 45, 49, 65
neutron bursts, 82, 88, 140
neutron creation, 171
neutron detection, 75
neutron diffraction, 128
neutron drop, 199
neutron emission, 65, 80, 153, 159, 179
neutron exchange, 191
neutron formation, 196
neutron production, 139
neutron transfer, 171, 174
neutrons, 6, 10, 13, 15, 21-23, 26, 27, 45, 52, 55, 66, 72-77, 88, 137, 141, 152, 153, 158, 159, 162, 171, 178, 186, 196-201, 209, 228, 233
^{58}Ni, 237
Ni, 49, 65, 83, 182
nickel, 53, 54, 82, 121, 173, 194, 204, 205, 235, 247, 254
nickel cathode, 87, 121
Ni-Cu, 89, 121
NiH, 70, 142, 145
Nishioka, T., 32
Northeastern University, 196
Notoya, R., 64, 87
novel particles, 164
nuclear active environment (NAE), 2-4, 8, 15, 19, 20, 37, 43, 55, 67, 72, 73, 95, 96, 105, 109, 111, 114, 116-119, 141, 142, 144, 145, 158, 161, 162, 189, 192, 195, 207, 211, 212-215, 217-222, 227, 230, 235, 237, 238, 242, 244, 245, 247, 249, 250, 253, 254
nuclear active sites, 212, 213, 218, 249

O

octahedral, 130, 136, 150, 160
Ohmori, T., 239
Ohnishi, T., 87
Ohta, M., 179
Oppenheimer-Phillips, 190, 192
optical phonons, 172
optical potential, 202
optimal operating point (OOP), 118, 173, 174
orbital mixing, 177
ordered phase, 122
ordered structure, 161
ordinary hydrogen, 120, 238
organic dye, 95
Oriani, R., 81, 92, 114
oscillating electron cloud, 194
Oshawa, G., 64

P

P-10, 75
palladium, 13, 15, 19, 27-29, 31, 35, 39, 42, 47, 49, 52, 54, 55, 58, 62, 64, 76, 82, 109, 110, 114, 118, 120-123, 126, 132, 159, 163, 174, 220, 225, 231, 233, 242, 246-249, 254
palladium-black, 15, 27, 35
Paneth, F., 28
Parmenter, R., 192
particle size, 1, 2, 49, 121, 243
Passell, T., 25, 27, 49, 232, 239
Patterson, J., 175
$^{110}Pd/^{108}Pd$, 27
Pd/PdO, 77, 79, 80
Pd+Ag alloy, 84
Pd_3H_4, 124, 125, 127
Pd-Ag, 121
Pd-Ag alloy, 14, 31, 118, 247
Pd-B, 121
Pd-B alloy, 38, 220, 247
Pd-black, 35
Pd-CaO, 57, 59
Pd-Ce, 28
$PdCl_2$, 246, 249
PdD, 4, 24, 63, 67, 68, 79, 80, 93-95, 99, 101, 118, 120, 121, 125, 134-137, 142, 144, 145, 146, 149, 151, 155-160, 163, 170, 172, 175, 181, 185, 188, 190-195, 206, 207
PdD films, 68, 69, 73
Pd-Li, 246
p-e-p fusion, 15, 232
Peters, K., 28
phase change calorimeter, 109
phonon, 155, 165, 171, 172, 173, 184, 188, 194, 197, 198, 201, 203
photographic film, 13, 202
photon, 8, 74, 75, 78, 80, 90, 151, 154, 164, 165, 171, 182, 197, 203, 208, 222-227
photon emission, 27, 65, 82, 163, 165, 222, 225, 230
photon pairs, 226
photon radiation, 9, 84, 90, 93, 95, 152, 182, 239, 244, 250
Phusor model, 173
physical randomness, 135
Piantelli, F., 82, 204, 205
Pico-chemistry, 203
piezonuclear, 141
plasma, 45, 46, 54, 64, 65, 110, 117, 136, 138, 139, 146, 155, 175, 185
plasma electrolysis, 64, 110
plasmarons, 243
plasmons, 70, 195, 208, 243
platinum, 49, 55, 66, 99, 109, 111, 114, 117, 234, 246, 249
polaritons, 195, 243
polyethylene, 83
polyneutrons, 81, 92, 198
Pons, S., ix, xi, 29, 108, 120, 137, 185, 252, 253
pores, 161
positron, 151, 165, 171, 207
positron annihilation, 171
powder, 248
powdered nickel, 205
power production, 213
praseodymium, 57, 72
pre-cleaning, 249
Preparata, G.,185-187
pressure, 118
Profusion Energy Inc., 197
Pryakhin, E., 92
pulse width, 97
pulsed glow-discharge, 78
pulsed laser, 67, 98
Purdue University, 183

Q

Qiao, G., 35
quantum electrodynamics, 187
quantum mechanics, 134, 146, 186, 209, 250
quasi-one-dimensional model, 173

R

Rabinowitz, M., 138, 186
radiation, 83, 85, 88, 91, 92, 95, 96, 151, 153, 162, 231

radioactive, 49, 52, 62, 87, 88, 162, 171, 235
radioactivity, 61, 152, 159
Rafelski, J., 91
rapid variations, 116
Reifenschweiler, O., 18
resistance, 75, 126, 127
resistance ratio, 126, 127
resonance, 82, 155, 156, 182, 184, 191, 192, 195, 201, 206, 207, 222, 224, 241, 244
resonance cycle, 224
resonates, 241
RF, 74
Richards, P., 133
Romodanov, V., 14
Rossi, A., 53, 54, 205, 216
Roussetski, A., 81
Rutz, K., 175
Rydberg matter, 98
Rydberg molecules, 98

S

$^{33}S/^{32}S$, 57
$^{34}S/^{32}S$, 57
Sakaguchi, H., 32
samarium, 59
Savvatimova, I., 91
scandium, 55
Scanlan, B., 78, 82, 83, 85, 92
Schwinger, J., 186-188
Schwinger ratio, 193, 194
screen, 137
screening, 137, 140, 157, 179
screening energy, 139
Second Law of Thermodynamics, 6, 134, 159, 195
secondary ion mass spectrometry, 45
secondary radiation, 91, 153, 165, 171, 184, 199
Seebeck calorimeter, 35, 108
selenium, 57
semiconductor detectors, 75
Shell Research, 202
Shirai, O., 112
Shoulders, K., 92
Shrestha, P., 49, 176, 177
Si film, 69
SiC, 242
silicon, 57, 68, 242
silver, 72, 177, 231, 247
SIMS, 33
Singh, M., 64
Sinha, K., 200-202, 240
^{146}Sm, 59
sodium iodide, 75
sonic, 111
SPAWAR, 76
special orientation, 97
spot welding, 249
^{86}Sr, 239
^{89}Sr, 239
stabilized neutrons, 158, 159, 171
Storms, E., 13, 16, 24, 26, 78, 82, 83, 85, 92, 99, 106, 111, 113, 114, 124, 144, 189, 208, 218
"strange" radiation, 6, 74, 91, 92, 96
stress, 44, 48, 195, 226, 242
stress relief, 8, 207, 220, 246, 249
Stringham, R., 39
strontium, 57, 58, 64
sub-nano-holes, 181
sulfur, 44, 57
Sundaresan, R., 64
superabundant vacancies, 160
super-heavy nucleus, 47, 176
super-wave, 117, 220
surface, 48, 114, 242
surface analysis, 102
surface plasmons, 70
Swartz, M., 49, 91, 99, 112, 118, 121, 172-174, 189, 193
swimming electron layer, 175
swimming electrons, 175, 176, 178
Switendick, A., 133
Szpak, S., 112

T

Takahashi, A., 37, 170, 178, 179, 181, 182
Talcott, C., 16
Tanzella, F., 92, 113

Taubes, G., 18
Technova Inc., xi, 178
temperature, 119, 131, 172, 214
temperature coefficient of resistance, 127
temperature gradient, 107, 108
tetrahedral sites, 126, 130, 131, 134, 136, 150, 194
tetrahedral symmetric condensate, 179
thermal conductivity, 132, 215
thermoelectric converters, 108
thin coatings, 114, 121, 246
thin deposits, 114
thin films, 70, 83, 114
thin layers, 114
three-body cascade, 179
three-body process, 179
Ti, 33, 79, 81, 98
Tian, J., 71, 97
TiD, 79, 80, 90, 93-95, 138, 182
TiH, 79
tin, 231
titanium, 18, 55, 77, 87, 88, 141
titanium alloy, 88
TNCF model, 199
transition to cold fusion, 138
transmutation, 6-10, 12, 26-28, 43, 44, 46, 47, 49, 50, 52-55, 62, 64-70, 72-77, 83, 92, 95, 96, 105, 110, 120, 137, 141, 151, 154, 162, 163, 165, 166, 175, 177, 182, 197, 199, 201, 204, 205, 209, 210, 214, 221, 227-230, 232, 233, 235-240, 245, 250
transmutation reactions, 28, 46, 47, 62, 77, 92, 154, 162, 238
tritium, 9-24, 26, 29, 31-33, 39, 43, 83, 90, 96, 98, 110, 137, 141, 151, 152, 154, 171, 186, 190, 191, 193, 202, 204, 206, 208, 213, 225-227, 245, 250
tritium contamination, 20
TSC structure, 181
Tsinghua University, 206
tungsten, 28, 61, 66, 117, 249
tungsten filament, 35
tunneling, 6, 140, 157, 181, 185, 187, 205, 206, 208, 209
Turner, L., 192
two lasers, 101-103

U

Urutskoev, L., 92
Universita di Siena, 204
Universiti Sains Malaysia, 200
University of Bologna, 205
University of Catania, 208
University of Illinois, 174
University of Lecce, 67
University of New South Wales, 175
upper phase boundary, 126, 127

V

vacancies, 124, 129-131, 136, 150, 160, 161, 171, 172, 201, 220, 249
vacant sites, 144
vanadium, 68
Va'vra, J., 200
Verner, G., 91, 117
Vigier, J., 202
Violante, V., 70, 83, 96, 194, 195
virtual neutrons, 177, 178
volcano-like, 47
Vysotskii, V., 92, 95

W

Ward, T., 184
wave behavior, 190
Wei, S., 133
Wendt, G., 28
Widom, A., 26, 196, 198

X

$^{306}Xe_{126}$,, 175
XeCl laser, 67
XPS, 46, 57
X-radiation, 82, 90
X-ray, 15, 44, 74, 82, 91, 94, 96, 158
X-ray emission, 55, 78, 80
X-ray film, 30, 34, 91

X-ray fluorescence, 45
X-ray photoelectron spectroscopy, 46

Y

Y_2O_3, 54
YAG laser, 71
Yamaguchi, E., 32
Young's modulus, 195

Z

Zhang, Y., 35
Zhang, Q., 33
zinc, 68, 70
zinc sulphide, 75
Zn, 65, 69, 70
Zr+O, 233
Zr-Ni-Pd, 40
Zywocinski, A., 31